Quorum Sensing

Quorum Sensing

Molecular Mechanism and Biotechnological Application

Edited by

Giuseppina Tommonaro

Researcher, Dr. Institute of Biomolecular Chemistry, CNR,
Pozzuoli, Napoli, ITA

ACADEMIC PRESS

An imprint of Elsevier

Academic Press is an imprint of Elsevier
125 London Wall, London EC2Y 5AS, United Kingdom
525 B Street, Suite 1650, San Diego, CA 92101, United States
50 Hampshire Street, 5th Floor, Cambridge, MA 02139, United States
The Boulevard, Langford Lane, Kidlington, Oxford OX5 1GB, United Kingdom

Notices
Knowledge and best practice in this field are constantly changing. As new research and experience broaden our
understanding, changes in research methods, professional practices, or medical treatment may become necessary.

Practitioners and researchers must always rely on their own experience and knowledge in evaluating and using any
information, methods, compounds, or experiments described herein. In using such information or methods they
should be mindful of their own safety and the safety of others, including parties for whom they have a
professional responsibility.

To the fullest extent of the law, neither the Publisher nor the authors, contributors, or editors, assume any liability
for any injury and/or damage to persons or property as a matter of products liability, negligence or otherwise,
or from any use or operation of any methods, products, instructions, or ideas contained in the material herein.

Library of Congress Cataloging-in-Publication Data
A catalog record for this book is available from the Library of Congress

British Library Cataloguing-in-Publication Data
A catalogue record for this book is available from the British Library

ISBN 978-0-12-814905-8

For information on all Academic Press publications
visit our website at https://www.elsevier.com/books-and-journals

Working together
to grow libraries in
developing countries

www.elsevier.com • www.bookaid.org

Publisher: Andre G. Wolff
Acquisition Editor: Glyn Jones
Editorial Project Manager: Carlos Rodriguez
Production Project Manager: Maria Bernard
Cover Designer: Matthew Limbert

Typeset by SPi Global, India

Dedication

To Michela, Serena, and Sara: the best of my Life.

Contents

Chapter 3: Quorum Sensing in Marine Biofilms and Environments.........................55
Raphaël Lami

Chapter 4: Quorum Sensing in Extremophiles..97

Gennaro Roberto Abbamondi, Margarita Kambourova, Annarita Poli, Ilaria Finore,
Barbara Nicolaus

PART 2 Inter-Kingdom Communication......................................125

Chapter 5: Quorum Sensing in Phytopathogenesis.........................127

Onur Kırtel, Maxime Versluys, Wim Van den Ende, Ebru Toksoy Öner

Chapter 8: Effect of Polyphenols on Microbial Cell-Cell Communications............195

Filomena Nazzaro, Florinda Fratianni, Antonio d'Acierno, Vincenzo De Feo,
Fernando Jesus Ayala-Zavala, Adriano Gomes-Cruz, Daniel Granato, Raffaele Coppola

PART 4 Applications...225

Chapter 9: Pseudomonas aeruginosa *Quorum Sensing and Biofilm Inhibition*.....227

Barış Gökalsın, Didem Berber, Nüzhet Cenk Sesal

Chapter 10: Cyclic Peptides in Neurological Disorders: The Case of Cyclo(His-Pro)...257

Ilaria Bellezza, Matthew J. Peirce, Alba Minelli

Contributors

Gennaro Roberto Abbamondi   National Research Council of Italy—Institute of Biomolecular Chemistry, Pozzuoli, Italy

Fernando Jesus Ayala-Zavala   Center for Research in Nutrition and Development, A. C (CIAD AC), Hermosillo, Mexico

Ilaria Bellezza   Department of Experimental Medicine, University of Perugia, Perugia, Italy

Didem Berber   Marmara University, Department of Biology, Faculty of Arts and Sciences, Istanbul, Turkey

Raffaele Coppola   DiAAA, Department of Agricultural, Environmental and Food Sciences, University of Molise, Campobasso, Italy

Valeria Costantino   TheBlueChemistryLab, Department of Pharmacy, University of Naples Federico II, Napoli, Italy

Adele Cutignano   National Research Council of Italy—Institute of Biomolecular Chemistry, Pozzuoli, Italy

Antonio d'Acierno   Institute of Food Science, ISA-CNR, Avellino, Italy

Vincenzo De Feo   Department of Pharmacy, University of Salerno, Salerno, Italy

Gerardo Della Sala   Laboratory of Pre-Clinical and Translational Research, IRCCS-CROB, Referral Cancer Center of Basilicata, Rionero in Vulture, Italy

Germana Esposito   TheBlueChemistryLab, Department of Pharmacy, University of Naples Federico II, Napoli, Italy

Ilaria Finore   National Research Council of Italy—Institute of Biomolecular Chemistry, Pozzuoli, Italy

Florinda Fratianni   Institute of Food Science, ISA-CNR, Avellino, Italy

Barış Gökalsın   Marmara University, Department of Biology, Institute of Pure and Applied Sciences, Istanbul, Turkey

Adriano Gomes-Cruz   Federal Institute of Education, Science and Technology of Rio de Janeiro (IFRJ), Department of Food, Rio de Janeiro, Brazil

Daniel Granato   State University of Ponta Grossa (UEPG), Department of Food Engineering, Ponta Grossa, Brazil

Angel G. Jimenez   Department of Microbiology, University of Texas Southwestern Medical Center, Dallas, TX, United States

Margarita Kambourova   Institute of Microbiology—Bulgarian Academy of Sciences, Sofia, Bulgaria

Onur Kırtel   Industrial Biotechnology and Systems Biology Research Group, Marmara University, Bioengineering Department, Istanbul, Turkey

Raphaël Lami   Sorbonne Université, CNRS, Laboratoire de Biodiversité et Biotechnologies Microbiennes, LBBM, Banyuls-sur-Mer, France

Alba Minelli   Department of Experimental Medicine, University of Perugia, Perugia, Italy

Filomena Nazzaro   Institute of Food Science, ISA-CNR, Avellino, Italy

Barbara Nicolaus   National Research Council of Italy—Institute of Biomolecular Chemistry, Pozzuoli, Italy

Ebru Toksoy Öner   Industrial Biotechnology and Systems Biology Research Group, Marmara University, Bioengineering Department, Istanbul, Turkey

Matthew J. Peirce   Department of Experimental Medicine, University of Perugia, Perugia, Italy

Annarita Poli   National Research Council of Italy—Institute of Biomolecular Chemistry, Pozzuoli, Italy

Wim J. Quax   Groningen Research Institute of Pharmacy, Department of Chemical and Pharmaceutical Biology, University of Groningen, Groningen, The Netherlands

Nüzhet Cenk Sesal   Marmara University, Department of Biology, Faculty of Arts and Sciences, Istanbul, Turkey

Vanessa Sperandio   Department of Biochemistry, University of Texas Southwestern Medical Center, Dallas, TX, United States

Roberta Teta   TheBlueChemistryLab, Department of Pharmacy, University of Naples Federico II, Napoli, Italy

Wim Van den Ende   Laboratory of Molecular Plant Biology, KU Leuven, Leuven, Belgium

Maxime Versluys   Laboratory of Molecular Plant Biology, KU Leuven, Leuven, Belgium

Jan Vogel   Groningen Research Institute of Pharmacy, Department of Chemical and Pharmaceutical Biology, University of Groningen, Groningen, The Netherlands

Preface

Quorum sensing is a process of bacterial cooperative behavior that has an effect on gene regulation. This cell-cell communication system involves the production of signaling molecules according to cell density and growth stage. Virulence, the ability to infest a habitat and cause disease, is also governed by such communication signals. This volume considers quorum sensing as a way to discover new antibiotics, and gain improved knowledge of the microbial world.

This book also provides different approaches on the study of the quorum sensing mechanism and its biotechnological application in several fields (biomaterial design, drug delivery, marine biofilm, etc.). It contributes to the understanding of the molecular basis that regulates this mechanism and describes new findings in fields of application. It is a handbook useful for researchers working in the fields of microbiology, environmental and food sciences, marine biology, and chemistry who already investigate quorum sensing, or beginners would like to learn more about this exciting field of study.

The book is organized in four sections (1. General: Chemistry and Microbiology; 2. Interkingdom Communication; 3. Quorum Sensing Inhibition; 4. Applications) covering the chemistry of signal molecules, the most recent outcomes from *extremophiles*, interkingdom communication in plants and humans, different molecules as quorum sensing inhibitors, and some examples of biotechnological studies.

I would like to thank all the authors for accepting the invitation to collaborate on this project with their excellent contributions and for helping me make this book a high-quality and comprehensive text. I also thank the Elsevier staff for their professionality and kind support. At last but non least, I would like to thank Dr. Salvatore De Rosa (Chemist, 1946–2014), my mentor, and Mr. Carmine Iodice, my indispensable laboratory assistant, for their trust in my research work.

Thank you!

Giuseppina Tommonaro, Editor

PART 1

General: Chemistry and Microbiology

CHAPTER 1

The Chemical Language of Gram-Negative Bacteria

Gerardo Della Sala*, Roberta Teta[†], Germana Esposito[†], Valeria Costantino[†]

*Laboratory of Pre-Clinical and Translational Research, IRCCS-CROB, Referral Cancer Center of Basilicata, Rionero in Vulture, Italy, [†]TheBlueChemistryLab, Department of Pharmacy, University of Naples Federico II, Napoli, Italy

1 Introduction

Roughly 40 years ago, two papers were published reporting the same breaking news: bacteria are social organisms that communicate each other using a chemical language to coordinate group activities. In 1965 *Nature* published an article by Tomasz (1965) reporting the first example of a regulatory mechanism in *Pneumococcus* that uses a chemical factor, named *competence factor*, while only a few years later, Nealson et al. (1970) reported "The discovery of autoinducer activity in *Vibrio fischeri*" in the *Journal of Bacteriology*.

Since then, it took almost 30 years to accept the idea that bacteria are social organisms able to communicate with each other using a cell-signaling system. Finally, Fuqua et al. (1994) introduced the term *quorum sensing* to describe a population-density-responsive gene regulation system able to sense population density and to react only when a *quorum* of cells is reached. In which way does this system act? Signaling molecules, called *autoinducers*, provide the means of communication. They are produced and accumulated in the environments. When the *quorum* (i.e., the right concentration of chemical signals) is reached, these molecules bind the receptor protein and activate changes in the gene expression (Hense and Schuster, 2015) that result in the activation of the cascade responsible for the production of a number of factors, including virulence and biofilm formation (Fig. 1).

Since then, a number of publications have reported on this signaling system in different bacteria species; it is now generally described as *quorum sensing* (QS) (Fuqua and Greenberg, 2002). Nowadays, QS is considered a general feature of bacteria, either Gram negative either Gram positive, differing in the architecture of their peculiar chemical signals, that allows bacteria to coordinate their collective behaviors in a way that often mimics that of multicellular organisms. This concept of cell-to-cell communication as social activity among

Quorum Sensing. https://doi.org/10.1016/B978-0-12-814905-8.00001-0

3

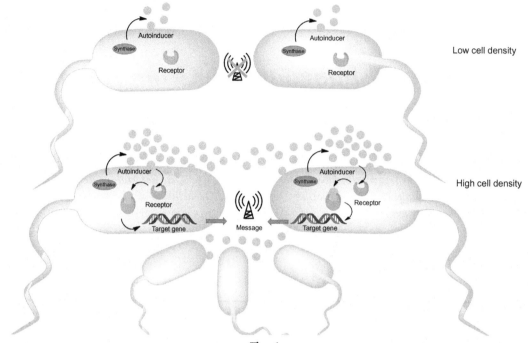

Fig. 1

Quorum sensing system in Gram-negative bacteria. QS relies on the synthesis of autoinducing signals which are produced in a population density dependent manner; when a threshold concentration (quorum) is reached, these molecules interact with a transcriptional regulator, allowing the expression of specific genes at a high cell density.

bacteria has been termed *social microbiology*. Parsek and Greenberg (2005) introduced this term in 2005 to describe this social phenomenon that includes biofilm formation and control of virulence factors.

Many Gram-negative bacteria synthesize *N*-acyl homoserine lactones (AHLs) as chemical signals and then use LuxR family protein to sense the concentration of AHLs in the environment and activate LuxI synthase, the cognate protein. The first QS system of Gram-negative bacteria described in detail is that of *Vibrio fischeri*, which colonizes the light organ of the bobtail squid, *Euprymna scolopes* (Ruby and Lee, 1998). Recent studies proposed that some bacterial species display LuxR homologues without a cognate autoinducer synthase LuxI. They are called LuxR "solos" or "orphans" (Brameyer et al., 2014). They exert roles in bacterial interspecies and interkingdom communication.

Beside AHLs, also known as autoinducer-1, the last 15 years of research has revealed the existence of other classes of signaling molecules and, consequently, novel QS pathways, which for several bacterial species coexist to weave a complex network governing bacterial virulence and physiology.

2 Quorum Sensing System Architectures in Gram-Negative Bacteria

Gram-negative bacteria utilize two main classes of autoinducers: AHLs, also known as autoinducer-1, and autoinducer-2 (AI-2, which is also used by Gram-positive bacteria). As the study of QS has been expanded, novel QS signals have been discovered, such as autoinducer-3 (AI-3), quinolone signal (PQS), and diffusible signal factor (DSF).

Herein, we describe briefly QS systems based upon different autoinducers for a comprehensive understanding of QS-dependent virulence mechanisms used by bacterial pathogens.

2.1 AHL-Based Quorum Sensing

The most common signaling molecules for QS systems are AHLs, which mediate communication exclusively in Gram-negative bacteria and between Gram-negative bacteria and their host (Britstein et al., 2018; Parsek and Greenberg, 2000). AHLs are small molecules featuring a homoserine lactone ring linked to a fatty acyl chain that can vary in length, oxidation state at β-position, and saturation degree. Typically, AHLs are biosynthesized (with few exceptions) by LuxI-type AHL synthases, catalyzing condensation between *S*-adenosylmethionine (SAM) and an acylated acyl carrier protein (acyl-ACP) (La Sarre and Federle, 2013).

At high cell densities, when the autoinducer AHL reaches a critical concentration (*quorum*), it binds a LuxR-type receptor in the cytoplasm to form a transcriptional regulator complex. The LuxR-AHL complex can then bind to specific promoter sequences (*lux*-boxes) found upstream of QS-regulated genes affecting their expression.

Chromobacterium violaceum harbors a typical LuxI/LuxR QS system, which involves CviI (LuxI-type AHL synthase), CviR (LuxR-type AHL receptor), and the AHL *N*-Hexanoyl-L-homoserine lactone (C6-HSL) for the regulation of violacein production.

In alternative to cytosolic LuxR receptors, membrane-integrated sensor kinases can interact with the AHL ligand: in *Vibrio harvey*, a LuxN sensor kinase can recognize *N*-(3-hydroxybutanoyl)-L-homoserine lactone (3-OH-C4HSL) at high cell density and catalyze the dephosphorilation of LuxU, which is a phosphorelay protein modulating the transcriptional response regulator LuxO.

Notably, each AHL receptor displays a certain degree of selectivity in AHL recognition and recruiting, strikingly related to the length and the oxidation and saturation rates of the fatty acyl chain of the ligand. It has become a sort of paradigm that a bacterial species expresses a cognate synthase/receptor pair, which is able to produce and respond to a specific AHL signal; however, some exceptions to this commonly accepted rule arose from several findings. Indeed, along with QS profiling in *Pseudomonas aeruginosa*, it could be deduced that there are species using multiple synthase/receptor pairs, responsible for simultaneous biosynthesis and transduction of different chemical signals. In addition, "solo" LuxR-type receptors orphans of their cognate LuxI synthases also occur in bacteria, such as the QscR of *P. aeruginosa* and the SidA of *Escherichia coli*.

The human pathogen *P. aeruginosa* uses a complex *QS* network that has been characterized in depth over the years, mainly because of its impact on human health. QS in *P. aeruginosa* (Fig. 2) displays at least three distinct signal/receptor pairs, coordinated in a hierarchical way. This pathogen expresses two LuxI/LuxR-type QS systems, namely LasI/LasR and RhlI/RhlR, producing and sensing small molecules: *N*-(3-Oxo-dodecanoyl)-L-homoserine lactone (3-O-C12-HSL) and *N*-*bu*tanoyl-L-homoserine lactone (C4-HSL), respectively. The Las system regulates these two QS systems in *P. aeruginosa*: upon recognition of the proper AHL, the LasR receptor drives simultaneous expression of LasI synthase and RhlR receptor. Then, the RhlR/C4-HSL complex autoinduces transcription of *rhlI* gene. Both Las and Rhl systems are involved in modulation of the third QS circuit based upon PQS (the **p**seudomonas **q**uinolone **s**ignal; see Section 2.3). Particularly, LasR and RhlR regulate in a positive and negative manner, respectively, the expression of genes involved in the PQS QS system. PQS autoinduces its own synthesis and activates RhlR expression, thus self-limiting its production by a negative-feedback mechanism. Additionally, each QS system has direct regulatory activity of various virulence factor-encoding genes. In this QS network, the "solo" LuxR-type protein, QscR, plays a relevant role: QscR recognizes 3-O-C12-HSL (the LasR ligand) and negatively modulates both the Las and Rhl QS systems, closing this complex self-regulating circuit.

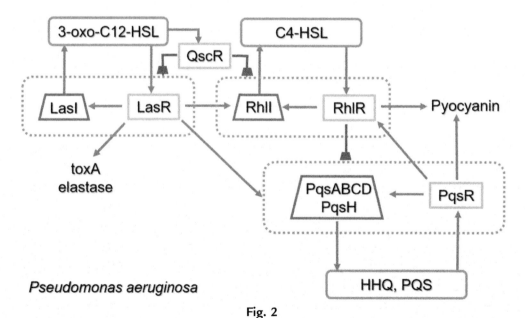

Fig. 2

Quorum sensing network in *Pseudomonas aeruginosa*. QS in *P. aeruginosa* displays three distinct signal/ receptor pairs, LasI/LasR, RhlI/RhlR, and PQS/PqsR, and the "solo" LuxR-type protein, QscR.

2.2 AI-2 Based Quorum Sensing

Autoinducer-2 is the name used for a family of QS signals represented by cyclic furanone compounds (Guo et al., 2013). The chemical structure of AI-2 in *V. harveyi* has been elucidated as the boron ester of (2*R*,4*S*)-2-methyl-2,3,3,4-tetrahydroxytetrahydro-furan, whereas in *Salmonella enterica* serovar Typhimurium, AI-2 was found to lack the borate.

Generally speaking, AI-2-type molecules are biosynthesized in two main steps: first, the cleavage of *S*-adenosyl-*L*-homocysteine by the nucleosidase MTAN (also known as Pfs) produces *S*-ribosyl-L-homocysteine (SRH) by removal of adenine; second, SRH is subsequently converted to homocysteine and the AI-2 precursor 4,5-dihydroxy-2,3-pentanedione (DPD) by the metalloenzyme LuxS. DPD is unstable and undergoes spontaneous cyclization and rearrangements, thus resulting in different cyclic furanone compounds representing the family of AI-2-type QS signals (Fig. 5).

In *V. harvey* and *V. cholerae*, when AI-2 reaches high concentration, it can diffuse across the cell envelope and interact with the LuxP/LuxQ receptor/sensor kinase complex, which is integrated in the bacterial membrane (Fig. 3). The AI-2/LuxPQ complex dephosphorylates LuxU, which in turn also dephosphorylates and inactivates LuxO, thus inhibiting expression of regulatory small RNAs and, therefore, switching on expression of virulence determinants.

In *E. coli* and *S. typhimurium*, the AI-2-mediated QS circuit reveals a different pathway. When the *quorum* is reached, AI-2 is internalized into the cell via a transporter protein, LsrB, and

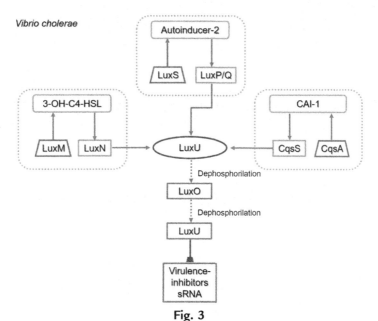

Fig. 3

Quorum sensing network in *Vibrio cholerae*. *V. cholerae* harbors three different QS circuits: AHL-, AI-2-, and CAI-1-based QS systems.

then phosphorylated by a kinase, LsrK; the phospho-AI-2 can bind the repressor LsrR to de-repress the *lsr* operon and activate target genes, some of which responsible for virulence (such as those involved in biofilm formation in *E. coli*).

2.3 QS Systems Using Other Autoinducers

4-Hydroxy-2-alkylquinolines (HAQs) are an additional class of QS signals, being reported for several species of *Pseudomonas* and *Burkholderia* (Kim et al., 2010). This class includes derivatives of 4-hydroxy-2-heptylquinoline (HHQ) and the corresponding dihydroxylated derivatives, such as 2-heptyl-3,4-dihydroxyquinoline (also known as PQS, **p**seudomonas **q**uinolone **s**ignal). In *P. aeruginosa*, HHQ is biosynthesized by PqsABCD from the precursor anthranilic acid and, consequently, PQS is obtained by hydroxylation of HHQ by the monooxygenase PqsH. Both HHQ and PQS activate the multiple virulence factor regulator PqsR (also called MvfR), driving production of QS molecules as well as toxins (pyocyanin) and biofilm formation (Allegretta et al., 2017).

The diffusible signaling factor (**DSF**) family include *cis*-2-unsaturated fatty acids with different chain length and branching. DSF signals are emerging as widely distributed messengers of cell-cell communication mechanisms among Gram-negative bacteria (Zhou et al., 2017).

Typically, biosynthesis of such molecules is in charge of the DSF synthase, showing a dual enzymatic activity, as it acts as both dehydratase and thioesterase of 3-hydroxyacyl-ACP substrates. The biosynthetic pathway of the DSF family of signals likely branches from the canonical fatty acid synthetic pathway.

In *Xanthomonas campestris*, upon accumulation in the cellular environment, DSF is recognized and binds the sensor kinase RpfC, triggering a phosphorelay-cascade mechanism to activate RpfG. RpfG is the response regulator and degrades cyclic di-GMP (c-di-GMP), an inhibitory ligand of the global transcription factor Clp. As a matter of fact, derepressed Clp activates the transcription of several genes, including those encoding virulence factor production. This type of QS system has been functionally characterized for *Xanthomonas* sp., *Xylella fastidiosa*, *Lysobacter enzymogenes*, and *Stenotrophomonas maltophilia* (Zhou et al., 2017). Actually, the opportunistic pathogens *Burkholderia cenocepacia* and *Cronobacter turicensis* harbor a DSF-dependent QS circuit, with the main difference being represented by the employment of a novel sensor RpfR, regulating intracellular levels of c-di-GMP (Zhou et al., 2017).

In addition, it has been recently reported that *P. aeruginosa* uses also a DSF-type QS system, with a characteristic gene cluster for signal production, perception and transduction (Zhou et al., 2017).

Autoinducer 3 (AI-3) is produced by human intestinal microflora (Parker et al., 2017). Particularly, AI-3-type signals are sensed through the histidine sensor kinase QseC by the enterohemorrhagic *E. coli*. QseC modulates activity of three response regulators (RRs), namely as QseB, QseF, and KdpE, controlling expression of virulence determinants. Indeed, phosphorylation by QseC of these RRs triggers a signaling cascade resulting in transcription of target genes, such as those responsible for flagellar motility, the type III secretion system encoded by the locus of enterocyte effacement (LEE), and Shiga toxin. Interestingly, the same QseC-dependent mechanism is also induced upon bacterial detection of epinephrine and norepinephrine released by the host.

Several species of *Vibrio*, such as *V. cholerae* and *V. harvey*, contain a QS system based upon the signaling molecule CAI-1, also known as cholera autoinducer-1 (Fig. 3). Biosynthesis of CAI-1 requires the CqsA synthase and the substrates (*S*)-2-aminobutyrate and decanoyl coenzyme A. CqsA produces amino-CAI-1, thus converting the amino-CAI-1 to CAI-1 in a following CqsA-independent step. CAI-1 interacts with the sensor kinase CqsS, which drives virulence factor expression via a phosphorelay cascade process targeting LuxU and LuxO. In *Vibrio* species, there is the coexistence and integration of the AHL-, AI-2-, and CAI-1-dependent QS circuits, all of them sharing a common LuxU-dependent downstream cascade pathway for transcriptional regulation of virulence encoding genes.

3 Chemistry of Autoinducers

The term *autoinducers* refers to diffusible chemical signals produced by bacteria in response to changes in the population density and employed to communicate within and between different species.

Bacteria release a wide variety of small products in the extracellular environment; to be classified as a QS signal molecule, four criteria have to be respected as reported by Winzer et al. (2002):

(i) the production of the autoinducer is triggered during specific growth phases, in certain physiological conditions, or in consequence of environmental changes;

(ii) the autoinducer diffuses and accumulates in the extracellular environment and binds a specific bacterial receptor;

(iii) only when the autoinducer concentration has reached a threshold level (*quorum*), a collective action follows;

(iv) the cellular response consists of a series of events beyond to metabolize or detoxify the molecule.

Among Gram-negative bacteria, QS is primarily mediated by AHLs (also known as autoinducer-1); different classes of other autoinducers have been described to date. Here examples of each of the major types of autoinducers are presented.

3.1 N-Acyl Homoserine Lactones

AHL signaling molecules (also known as autoinducer-1) are made of a homoserine lactone ring (HSL) linked through an amide bond to an acyl side chain. AHLs differ in length (Britstein et al., 2016), substitution at C-3 of the acyl chain, and the degree of unsaturation (Fig. 4A) (Saurav et al., 2016). These differences confer signal specificity to LuxR transcriptional regulators. Acyl side chains, putatively arising from fatty acid biosynthesis, are made of 4 to 18 carbons, usually by increments of two carbon units (C4, C6, C8, etc.). Most of the acyl side chains are unbranched, saturated, or monounsaturated, and even-numbered, corresponding to fatty acids readily available in microbial cells. The acyl chain of a specific length is indicated herein by Cn for the number of carbons in the chain (i.e., octananoyl is C8). The substitution type and position are designated as 3-O (3-oxo) or 3-OH (3-hydroxy), and HSL refers to D/L-homoserine lactone (e.g., 3-O-C6HSL) (Fig. 4B). AHL refers to *N*-acyl-HSL, with any specified chain length or degree of substitution.

Recently, AHLs bearing a di-unsaturated acyl chain have been also described in several alphaproteobacteria, such as *Methylobacterium extorquens*, *Dinoroseobacter shibae* (Wagner-Döbler et al., 2005). C7HSL from *Rhizobium leguminosarum* (Schripsema et al., 1996) was the first AHL reported with an odd-numbered acyl moiety; generally odd-numbered AHLs occur in trace amounts in comparison with the even-numbered, except in *Sulfitobacter*

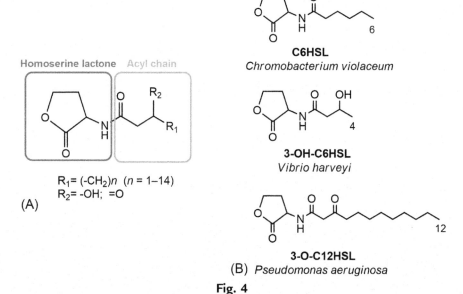

R$_1$= (-CH$_2$)n (n = 1–14)
R$_2$= -OH; =O

(A)

C6HSL
Chromobacterium violaceum

3-OH-C6HSL
Vibrio harveyi

3-O-C12HSL
(B) *Pseudomonas aeruginosa*

Fig. 4

N-Acyl Homoserine Lactone (AHL) signaling molecules: (A) general structure of acyl homoserine lactones; (B) chemical structures of AHLs from *Chromobacterium violaceum*, *Vibrio harveyi*, and *Pseudomonas aeruginosa*.

sp. D13, where the 9-C17:1HSL is synthesized in large quantities (Ziesche et al., 2015). The presence of methyl branches has been described for isoC9HSL and 3-OH-isoC9HSL from *Aeromonas culicicola* (Thiel et al., 2009). The substitution on C3 belongs to three types: (a) simple acyl, (b) 3-hydroxyacyl, and (c) 3-oxoacyl; this substitution into the acyl chain introduces a new stereogenic center, in addition to the one of the L-configured homoserine lactone. So far, the absolute configuration at this center has only been identified as (3*R*,7*Z*)-*N*-(3-hydroxy-7-tetradecenoyl) homoserine lactone, from *Rhizobium leguminosarum*. Short-chain AHLs, such as C4HSL, diffuse freely through the cell membrane while a carrier and efflux pump exports 3-O-C12HSL (Smith and Iglewski, 2003).

3.2 Cyclic Furanone Compounds

Unlike other autoinducers, which are specific to a particular species of bacteria, autoinducer 2 (AI-2) is found in many (~70) species of both Gram-negative and Gram-positive bacteria (Guo et al., 2013); it is an interspecies autoinducer and goes under the appellative "universal autoinducer." It is the boron ester of (2*R*,4*S*)-2-methyl-2,3,3,4-tetrahydroxytetrahydro-furan, therefore composed by two fused five-membered rings, stabilized within the LuxP binding site by numerous polar interactions. AI-2 derives from the spontaneous cyclization of the linear (*S*)-4,5-dihydroxypentanedione (DPD) (*S*) into the two isomeric forms of tetrahydroxytetrahydrofuran (*R*-THMF and *S*-THMF) that coexist in a dynamic equilibrium. In presence of environmental boron, (Carrano et al., 2009) borate complexes such as THMF-borate are formed (Fig. 5). A peculiarity of AI-2 signaling is that diverse bacteria have different AI-2 receptors, which recognize distinct forms of AI-2. So far, two AI-2 receptors have been identified: LuxP binds THMF-borate, while LsrB binds *R*-THMF, which lacks the borate (Rui et al., 2012).

3.3 Others

Quinolones are molecules structurally derived from the heterobicyclic aromatic compound quinoline. 2-hydroxyquinoline and 4-hydroxyquinoline, which predominantly exist as 2(1H)-quinolone and 4(1H)-quinolone, respectively, form the core structure of many alkaloids that were isolated from plant sources. Several different animal and bacterial species also produce compounds of the quinolone class. These differ not only in the varied substitutions in the carbocyclic and heteroaromatic rings but also have other rings fused to the quinolone nucleus (Heeb et al., 2011).

2-Heptyl-3-hydroxy-4(1H)-quinolone (PQS) and its immediate precursor, 2-heptyl-4-hydroxyquinoline (HHQ), are the primary HAQs (4-hydroxy-2-alkylquinolines) involved in QS, although other active AQ (alkylquinoline) analogues, such as the C9 congeners, 2-nonyl-3-hydroxy-4(1H)-quinolone (C9-PQS), and 2-nonyl-4-hydroxyquinoline (NHQ), are produced by *P. aeruginosa* in similar concentrations (Ilangovan et al., 2013).

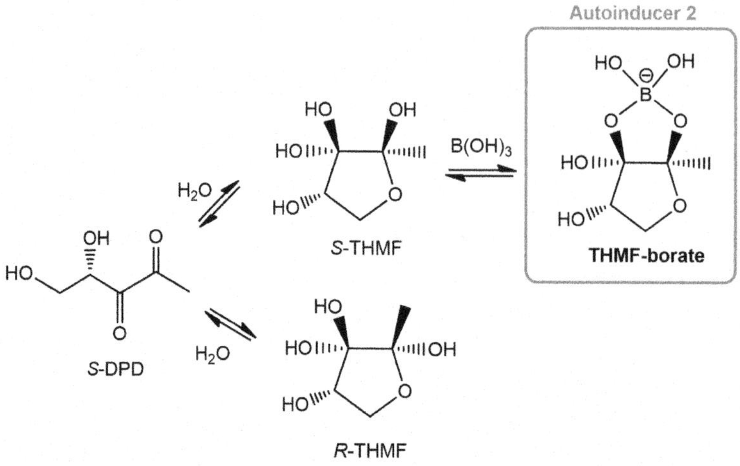

Fig. 5

Formation and structure of AI-2. AI-2 is formed through spontaneous cyclisation of the linear (S)-4,5-dihydroxypentanedione (DPD) into the two isomeric forms of 2-methyl-2,3,3,4-tetrahydroxytetrahydrofuran (R-THMF and S-THMF). These species coexist in a dynamic equilibrium and can form borate complexes in presence of environmental boron.

Substitution at C3 and C6 positions has impact on the activity profile (Fig. 6) (Kamal et al., 2017).

The diffusible signal factor (DSF) family of signals comprises unsaturated fatty acids of different chain lengths, branching pattern and double bond configuration. The α,β-double bond and *cis*-configuration of fatty acid are critical for QS signaling activity, but in a species-specific manner. The first family member described was the *cis*-11- methyl-dodecenoic acid in *Xanthomonas campestris*. Other family members include *cis*-2-dodecenoic acid (BDSF)

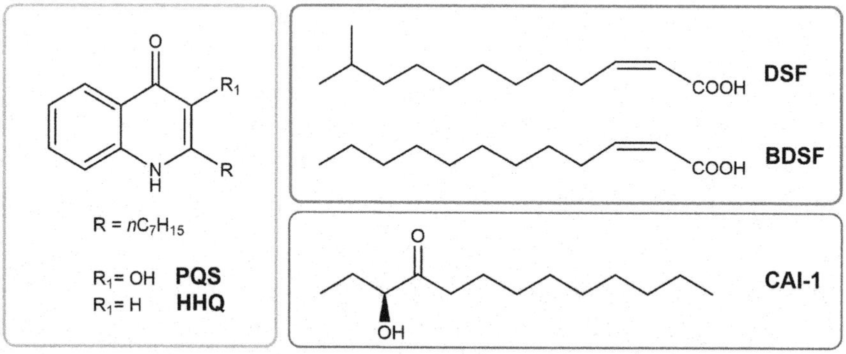

Fig. 6

Other QS signals. Representatives of 4-hydroxy-2-alkylquinoline-, diffusible signaling factor- and α-hydroxyketone- families of signaling molecules.

from *Burkholderia cenocepacia*, *cis*-2-decenoic acid from *Pseudomonas aeruginosa*, and *cis*-2-hexadecenoic acid (XfDSF2) from *Xylella fastidiosa* (Fig. 6) (Dow, 2017).

AI-3 is an interkingdom signaling between prokaryotic and eukaryotic cells (Hughes and Sperandio, 2008); it helps bacterial species (*Escherichia coli* and *Salmonella serovar Typhimurium*) to "crosstalk" with mammalian epinephrine/norepinepherine hormones during infection (Sperandio et al., 2003). AI-3 is an aromatic aminated compound, but its full structure has been not determined yet. The major QS signal in *V. cholerae* is CAI-1; its structure has been identified as (*S*)-3-hydroxytridecan-4-one, and it represents the first molecule of a further class of QS signals, namely AHK (α-hydroxyketones) (Kong et al., 2011). The biological activity of CAI-1 is sensitive to side chain length, as the 13-carbon molecule has 8-fold greater activity than the 12-carbon molecule. The configuration of the hydroxyl group has a relatively small effect on activity (2-fold differences) (Fig. 6).

3.4 Quorum Sensing in Marine Environments

Although QS was first observed in a marine bacterium nearly 4 decades ago, only in the past decade has there been an interest in the role that QS plays in the ocean. QS is involved in marine carbon and organic phosphorus cycles, in the health of coral reef ecosystems, in trophic interactions between a range of eukaryotes and their bacterial associates, and in the large-scale bioluminescent event coincident with algal blooms (Hmelo, 2017). The most well studied QS systems in the ocean occur in surface attached (biofilm) communities and are mediated by AHLs.

It is generally believed that AHL-QS systems would be used only by bacteria growing in enclosed niches (such as in sponges), where the signal molecules do not diffuse away and can concentrate to threshold levels. In addition, at the high pH of seawater the AHL-signal molecules produced by a free-living bacterium undergo lactonolysis, the opening of the lactone ring, preventing the efficient interaction of AHL-signal molecules. On the contrary, the pH inside host tissue allows the function of AHLs as signaling molecules. Thus the achievement of *quorum* is influenced not only by AHL accumulation, but also by the rate of AHL-turnover, which depends on environmental conditions (e.g., pH, temperature) and *N*-acyl chain length. Moreover, different length-sized-AHLs have diverse stability to pH, with longer side-chain AHLs and AHLs lacking 3-oxo substituents being less susceptible to inactivation by hydrolysis of the lactone ring at higher pH.

Bioassays of crude samples and extracts of several marine invertebrate and macroalgal samples suggested that associated bacteria produce AHLs in situ (La Sarre and Federle, 2013; Taylor et al., 2004). These initial observations have subsequently been corroborated by sensitive and advanced mass spectrometric techniques.

Identification of AHLs from the bacterium *Paracoccus* sp. Ss63, associated with the marine sponge *Sarcotragus*, has been easily achieved through high performance liquid chromatography-high resolution tandem mass spectrometry (LC-HRMS/MS), using surface induced dissociation (SID), taking advantage from the characteristic homoserine lactone product ion at *m/z* 102.05 (Saurav et al., 2016).

To date, AHLs have been structurally identified in extracts of marine snow (Guo et al., 2013), colonies of the cyanobacterium *Trichodesmium* (Van Mooy et al., 2012), stromatolites (Decho et al., 2009), sponges (Britstein et al., 2016), and sea anemone (Ransome et al., 2014).

In addition to AHLs, aryl-HSLs (e.g., *p*-coumaroyl-HSL) have been observed in marine bacterial cultures, and AI-2 is only known to function as a signal in *Vibrio* species (Schaefer et al., 2008).

4 Quorum Quenching Molecules

Inhibition of the QS system could represent a challenging task in order to design novel antivirulence drugs. The idea is to design drugs able to reduce virulence in pathogen bacteria, rather than to inhibit their growth or kill them. Inhibition of the QS system can be accomplished by two kinds of molecules that can be grouped as following, based on their structure and mechanism of action:

(A)   Molecules that structurally mimic the AHLs or AIPs. Those molecules interfere with the QS communication by binding to the receptor because of their structure similarity, or accelerating the LuxR turnover, or covalently modifying and inactivating LuxS.
(B)   Enzyme inhibitors: small molecules that are able to inhibit enzymes essentials in the AHL biosynthesis, such as triclosan that inhibits the enoyl-ACP reductase.

In addition, enzymatic degradation is an alternative way to inhibit QS system. Bacterial AHLases that catalyze the hydrolysis of AHLs have been isolated from various strain of Gram negative bacteria.

Here, we report the most interesting examples of quorum quenching (QQ) molecules reported in the literature.

4.1 A Case Study of Inhibitors of AHL Receptors: LasR Antagonists

QS inhibitors targeting LuxR receptors have raised a great interest in order to design pharmaceutical drugs turning off virulence mechanisms in bacteria. The strategy of this QQ approach is to develop analogues of native signals and known QQ molecules maintaining signal-receptor interactions but disrupting downstream biochemical events through the formation of inactive ligand-LuxR complexes. Studies over the past few years have shown that QS inhibitory mechanisms are widely distributed in many prokaryotic organisms to gain advantage over competitors (Delago et al., 2016). These naturally occurring QQ systems rely on small molecules, which can be used as lead compounds for designing a new generation of antivirulence drugs. In this contest, several studies have emerged from the literature depicting the microbial fermenters marine sponges as a fruitful environment for QQ processes involved in microbe-microbe interactions and host-microbe interactions. Plakofuranolactone (Costantino et al., 2017) is the first QQ molecule isolated from a sponge holobiome; it acts as a LuxR antagonist in in vitro bioassays, thus revealing it is worthwhile to exploit sponges as natural factories of QS inhibitory compounds in the future.

Even though QQ is still a potential strategy for antivirulence therapy, promising steps have been made toward treating some of the most important human pathogens (*V. cholerae*, *P. aeruginosa*) with QQ approaches, using both in vitro and in vivo animal models (Duan and March, 2010; Saeidi et al., 2011; Skindersoe et al., 2008).

There is an urgent need to inhibit AHL-mediated QS in *P. aeruginosa*, which is currently plaguing human health through opportunistic induction of hospital-acquired infections and chronic lung infections associated with cystic fibrosis. A number of studies have focused on QQ molecules targeting the LasR receptor, which is the central regulator of different co-occurring QS circuits in *P. aeruginosa* (Fig. 2).

Ligand recognition by LasR allows for the binding of molecules with surprising structural diversity, making it difficult to predict chemical requirements for agonistic and antagonist activity. In general, SAR (structure-activity relationship) and molecular modeling studies have highlighted that both agonistic and antagonistic profiles need sufficient ability to form hydrogen bonds and van der Waals interactions to allow ligand accommodation in LasR and balance any bulky effect exerted by nonnative ligand substituents.

Overall, some insights have emerged from past studies, which we try to summarize herein.

Molecules mimicking AHL structure require proper acyl chain length to exert agonistic/ antagonistic bioactivity on LasR (Fig. 7A) (Passador et al., 1996). AHL mimics displaying an acyl chain (more than six carbons) similar to the endogenous LasR ligand, 3-oxo-C12-HSL, can interact with LasR. As an example, the non-3-oxo-C10-HSL is the most potent LasR inhibitor among AHLs with alkyl chain (Geske et al., 2008a). The carbonyl group at position 3 of the acyl chain in the native AHL is important but not essential for activity, and modification at this site results in an antagonistic AHL-inspired ligand in most cases (Passador et al., 1996).

Kline et al. (1999) demonstrated that an extended and flexible chain geometry is strikingly required for LasR activation, because constrained enolic AHL-analogues, blocked into irreversible conformers, were not able to interact with the binding pocket.

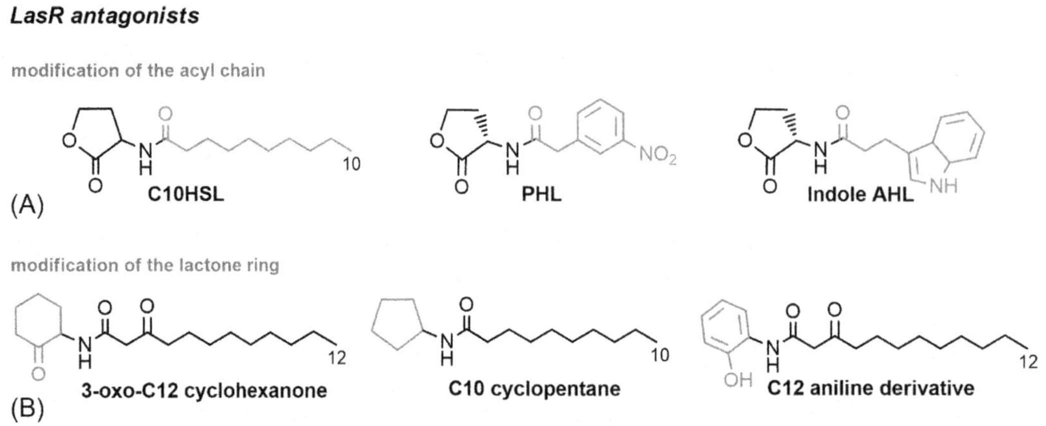

LasR antagonists

modification of the acyl chain

(A)   C10HSL   PHL   Indole AHL

modification of the lactone ring

(B)   3-oxo-C12 cyclohexanone   C10 cyclopentane   C12 aniline derivative

Fig. 7

LasR antagonists mimicking endogenous AHLs. Chemical structures of LasR antagonists derived from modification of the acyl chain (A) and the lactone ring (B) of prototypical AHL-type autoinducers.

Incorporation of aromatic functionalities to design AHL analogues with nonnative acyl groups led Geske et al. (2008a) to identify novel LasR inhibitors, represented by phenylacetanoil homoserine lactones (PHLs) and indole AHLs. Interestingly, slight modifications in the phenyl ring of PHLs, as well as moving a substituent from *para* to *meta* position, resulted in increased inhibitory activity against LasR (Fig. 7A) (Geske et al., 2008a).

Several studies have been performed aiming to elucidate the effects of modifications of the lactone ring in AHL mimics (Fig. 7B). Lactone ring replacement with bioisostere groups is generally well tolerated, with few exceptions (i.e., the γ-lactam ring produces loss of any kind of activity) (Passador et al., 1996). Introduction of γ-thiolactone or cyclopentanone has been reported to activate the LasR receptor, whereas cyclohexanone and cyclopentane, with the latter completely lacking H-bond acceptors, were useful substitutions to design LasR antagonists, keeping the proper size of the acyl chain (Ishida et al., 2007; Passador et al., 1996; Suga and Smith, 2003). More interestingly, synthetic analogues where the lactone ring was replaced by aniline (Smith et al., 2003) exhibited inhibitory effect on LasR receptor. These findings led to the hypothesis that aromatic ring and, in general unsaturation, in place of the lactone moiety produces LasR antagonists, as in the case of halogenated furanones.

Furthermore, the stereochemistry of the lactone ring deserves some comments. It is known that the L-stereoisomers of AHLs are the natural ligands of LuxR receptors. However, not all D-AHLs are inactive (Geske et al., 2008a), and even a nonnatural AHL derivative with D-stereochemistry has been found to act as a LasR agonist (Geske et al., 2008b).

Through the years, efforts have been made to delineate key interactions of ligands with LasR binding domain and, consequently, highlight conformational changes of the receptors upon ligand recognition. Here, we report molecular insights into the interaction of native ligand and some QS inhibitors with LasR, according to the model from Bottomley et al. (2007). This model has been outlined through a combination of X-ray data and molecular modeling studies.

.The LasR ligand-binding domain displays a symmetrical dimeric structure, where each monomer has a α-β-α fold (6 α-helices and 5 β-sheets), roughly similar to those ones reported for other LuxR homologues, such as TraR (from *Agrobacterium tumefaciens*) and SdiA (from *E. coli*). The LasR native ligand, the 3-oxo-C12-HSL, is placed parallel to the β-sheet, in a binding pocket formed between the β-sheet and helices α3, α4, and α5. 3-oxo-C12-HSL can form six hydrogen bonds in the binding pocket, involving Tyr-56, Trp-60, Arg-61 (hydrogen bond water-mediated), Asp-73, Thr-75, and Ser-129. It is worthy to note that Tyr-56, Trp-60, Asp-73, and Ser-129 are highly conserved in all LuxR proteins, as these residues are essential for the interaction of the shared homoserine-lactone (HSL) core with the binding site and, consequently, for receptor activation. Despite these similarities, LasR binding domain includes several additional residues (e.g., Leu-40, Tyr-47, Cys-79, and Thr-80), which are absent in TraR and SdiA (selective for shorter AHL ligands) and are responsible for selective accommodation of the long acyl chain of 3-oxo-C12-HSL.

Even if LuxR homologues share key amino-acid residues in the binding pocket for accommodation of the HSL, this moiety establishes different H-bonding interactions according to the LuxR type. In the LasR receptor, the autoinducer forms H-bonds simultaneously with Tyr-56 and Ser-129 through its 1-oxo group, and with the Arg-61 side chain through its 3-oxo group. On the other hand, in the TraR receptor, the HSL 1-oxo group forms an H-bonding exclusively with Tyr-53 (corresponding to Tyr-56 in LasR), whereas the 3-oxo group interacts with both Thr-129 side chain and the backbone carbonyl of Ala-38. Because of these different H-bonding networks, the acyl chains of the relevant autoinducers extend in opposite directions in LasR and TraR binding sites, starting from the 3-oxo group of the AHL. Although sharing high homology, LasR and TraR harbor their ligands in a different way. In addition, LasR and TraR have binding pockets modeled exactly on the size of their cognate ligands, 3-oxo-C12-HSL and 3-oxo-C8-HSL, respectively; indeed, LasR has a larger binding pocket ($670 \, \text{Å}^3$) compared with TraR ($440 \, \text{Å}^3$).

Furanones (Costantino et al., 2017; Wu et al., 2004), patulin and penicillic acid (Rasmussen et al., 2005) (Fig. 8) are well known QS inhibitors targeting the LuxR receptors. It is

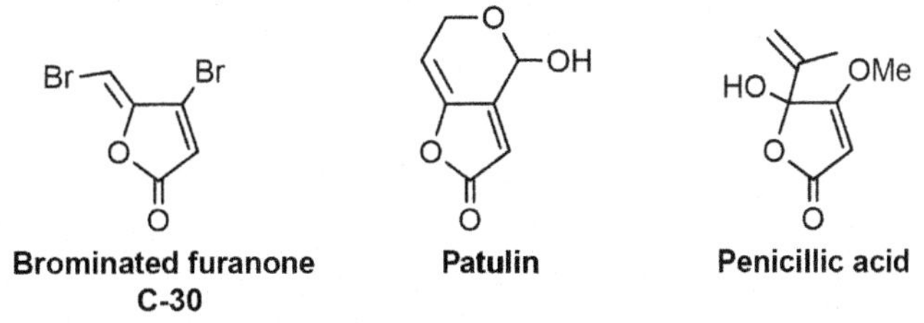

Brominated furanone C-30   **Patulin**   **Penicillic acid**

Fig. 8

Chemical structures of QS inhibitors targeting the LuxR receptors.

current opinion that these inhibitors act neither by displacing the native AHL ligand from the binding site nor by interaction with a putative allosteric site (which is actually absent) of the LuxR homologue. The AHL is so deeply hidden in the binding pocket that it cannot be removed by an antagonist molecule. It is more likely that QSIs can successfully compete with the native ligand in binding with the nascent LuxR protein, interfering with the AHL-mediated folding and packing of the functional receptor. Recently, using an in vitro assay, Suneby et al. (2017) demonstrated several antagonists to function by binding to and constraining LasR in a conformation that it is not able to bind DNA.

Brominated furanones are reported to inhibit virulence factor production (e.g., pyocyanin, protease), and biofilm formation, as well as to increase antibiotic sensitivity in *P. aeruginosa*. Docking studies by Bottomley et al. (2007) clearly suggest that halogenated furanones interact with LasR similarly to 3-oxo-C12-HSL, through H-bonding with Trp-60 and hydrophobic interactions with aromatic side chains of several amino acid residues. Nevertheless, due to different chemical features from AHLs, furanones cannot engage other essential canonical H-bonds, and lack of a long acyl chain for the correct formation of LasR hydrophobic core. As a result, it is conceivable that the formation of an aberrant ligand-receptor complex may occur when furanones are bounded. Particularly, if the bromine atoms give lipophilicity to allow accommodation of furanones in LasR binding domain, steric hindrance of such halogens disrupts a water-mediated internal H-bond between Tyr-93 and Leu-110 in LasR protein, affecting dramatically protein stability. Accordingly, it has been observed that modifications at C-4 of the lactone ring result in the generation of LasR antagonists, as breaking key interaction for proper folding induction.

Patulin is another QQ molecule mimicking AHL structure. The in silico model (Bottomley et al., 2007) of the LasR-patulin complex revealed patulin to establish a canonical H-bond with Trp-60 and two additional H-bonds through the oxygen of its dihydropyranic ring and its hydroxyl group with Tyr-93 and Asp-73 (or Thr-75), respectively. However, patulin and furanones share a common feature, as they do not display the same long acyl chain as 3-oxo-C12-HSL, thus hampering correct conformation of the lipophilic pocket.

An alternative emerging trend in QSI discovery is the design of QQ molecules having novel chemical scaffolds, not mimicking AHL structure, such as triphenyl scaffold-based compounds (O'Reilly and Blackwell, 2016). Mutational studies of LasR protein highlighted Trp60, Tyr56, and Ser129 as H-bonding sites crucial in determining activation or inhibition of the receptor also in the case of non-lactone LasR modulators (Gerdt et al., 2014).

It is commonly known that the triphenyl scaffold-based compound TP-1 can function as a LasR agonist, being able to (a) form three H-bonds with Tyr-56, Trp-60, and Ser-129, and (b) occupy the entire binding pocket in LasR receptor thanks to its triphenyl moiety (Bottomley et al., 2007). Although sharing a common core with TP-1, TP-5 acts as a LasR antagonist; indeed, the absence of a methylene group in the triphenyl system of TP-5 leads to loss of rotational freedom and increased steric interference compared to TP-1 (Fig. 9). Consequently,

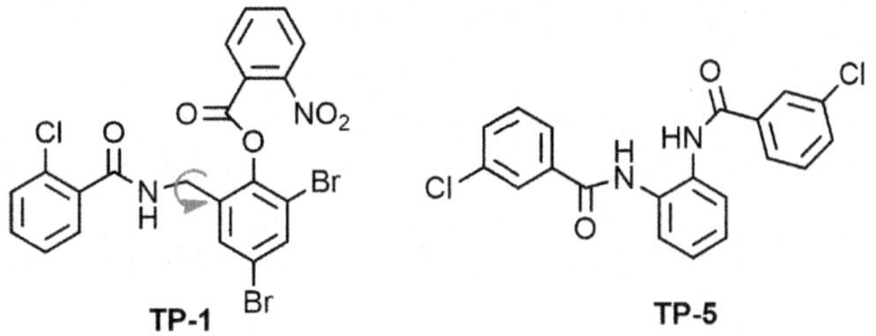

Fig. 9

Chemical structures of LasR modulators having nonlactone chemical scaffolds. TP-1 (LasR agonist) and TP-5 (LasR antagonist) are triphenyl scaffold-based compounds.

TP-5 cannot establish all the fold-promoting interactions with the binding domain. Zou and Nair (2009) hypothesized that TP-5 acts as an inhibitor because it (a) disrupts the native conformation of LasR through the steric clash between the chlorine atom and Leu-125 and (b) displays poor alignment of the chlorine atom with the Trp60 side chain NH for a hydrogen bond. Afterward, confirming the previous hypothesis, O'Reilly and Blackwell (2016) reported that mutation of Trp60 to Phe does not affect the antagonistic activity of TP-5 against LasR, thus revealing TP-5 does not form an H-bond with Trp60 to interact with the binding site.

4.2 Enzymatic Inhibition of QS

Enzymatic inhibition of QS can be realized through two possible strategies:

- Enzymatic degradation and inactivation of AHLs or other QS signaling molecules
- Inhibition of the biosynthesis of QS signaling molecules

4.2.1 Enzymatic degradation and inactivation of AHLs

AHL-lactonases, AHL-acylases, and AHL-oxidoreductases (La Sarre and Federle, 2013) are enzymes that work in this direction. AHL-lactonases are basically metalloproteins responsible for the hydrolysis of the ester bond of homoserine lactone ring to get acyl-homoserine compounds. The alkaline pH can promote this hydrolysis naturally, and this reaction is reversible if the pH becomes acid and the lactone ring is restored. Because of the presence of homoserine lactone in all AHL molecules, the lactonases exhibit a wide range of action. The first AHL lactonase identified was AiiA (autoinducer inactivation) from *Bacillus* sp. strain 240B1 (Dong et al., 2000). The *aiiA* homogenous genes are widespread in *Bacillus* species. Heterologous expression of *aiiA* in numerous pathogenic bacteria, including *P. aeruginosa*, *E. carotovora*, *B. thailandensis*, resulted in reduced release of AHL signals and decrease of their virulence expression. Moreover, transgenic plants expressing AiiA have been shown to be significantly less susceptible to infection by *E. carotovora*, indicating the potential use of AiiA as a strategy for antivirulence therapy (Dong et al., 2000).

AHL-acylases hydrolyze, in a nonreversible way, the acyl-amide bond existing between the acyl-chain and the lactone ring, yielding a fatty acid chain and the corresponding homoserine lactone. Differently from lactonase, they are highly substrate-specific, depending on the acyl chain length and on the substitution eventually present on the third position of acyl chains. The first AHL acylase identified was the acylase AiiD identified in 2003 from *Ralstonia* sp. XJ12B. (Lin et al., 2003) AiiD acylase shares similarity with several cephalosporin and penicillin acylases. Actually, it has been shown that AiiD does not degrade penicillin G or ampicillin, indicating that AHLs are its unique substrates (Lin et al., 2003). Like AHL lactonase, it was hypothesized that AHL acylase can interfere with the QS system of bacterial pathogens. Expression of AiiD in *P. aeruginosa* PAO1 decreased the accumulation of 3OC12HSL and C4HSL, consequently altering virulence factor production and swarming motility. Furthermore, expression of AiiD in *P. aeruginosa* weakened its ability to produce elastase and phycocyanin, and to paralyze nematodes.

AHL-oxidoreductases inactivate the AHLs modifying, by oxidation or reduction, acyl chains (Fig. 10).

This class is the less abundant and less studied among these of enzymes targeting AHLs. This type of AHL inactivation was first discovered in *Rhodococcus erythropolis* in which AHLs with 3-oxo substituents were rapidly degraded by reduction of the keto group at the β position, yielding the corresponding 3-hydroxy derivative AHLs (Uroz et al., 2005).

The gene responsible for this activity has not yet been identified. Moreover, the NADH-dependent enzyme BpiB09 was discovered, and it was identified by metagenomic analysis as able to inactivate 3OC12HSL. Expression of *bpiB09* in *P. aeruginosa* decreased its swimming motilities, phycocyanin production, and biofilm formation and thereby the pathogenicity to *C. elegans,* (Bijtenhoorn et al., 2011), definitely confirming the evidence that oxidoreductases can work as inhibitors of QS systems.

4.2.2 Enzymatic degradation and inactivation of non-AHLs quorum sensing signals

- To date, a number of studies focused on inactivation of AHLs, while only several reports described enzymatic inactivation of other QS autoinducers. A recent report showed the

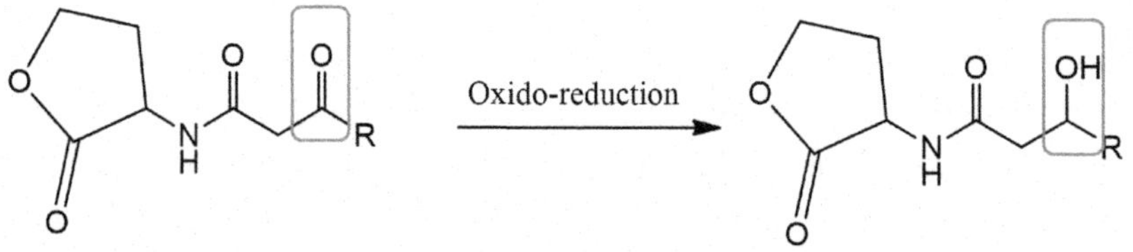

Fig. 10

Enzymatic degradation of AHLs is a QQ strategy adopted in Gram-negative bacteria. AHL-oxidoreductases inactivate the AHLs modifying acyl chains by oxido-reduction.

possibility to interfere with the AI-2-mediated QS. In particular, the ability of LsrK to exploit AI-2 signaling was studied. LsrK is the cytoplasmic enzyme responsible for phosphorylation of AI-2.

- AI-2 is transported into the cell and subsequently is phosphorylated by LsrK (Xavier et al., 2007). Roy et al. (2010) found that by providing LsrK exogenously to *E. coli* or *S. Typhimurium* cultures, they could inhibit QS activation by blocking AI-2 import (the negative charge of phospho-AI-2 prevents its transport via the Lsr transporter).

Members of the DSF family of molecules are used in QS by both *Xylella fastidiosa* and several *Xanthomonas* species. It is important to underline that DSF signals affect the virulence of these two pathogens (Deng et al., 2011). Newman et al. (2008) identified *carA* and *carB* as the genes responsible for the inhibition activity of DSF signals. CarA and CarB are the subunits of the heterodimeric complex responsible for the synthesis of carbamoylphosphate, a required precursor for arginine and pyrimidine biosynthesis (Llamas et al., 2003). CarAB function to inactivate DSF in multiple bacterial species (Newman et al., 2008).

4.2.3 Biosynthesis inhibition of AHLs molecules

AHLs molecules are synthesized by AHL synthases belonging to the LuxI and AinS family (Gilson et al., 1995; Parsek et al., 1999).

S-adenosyl methionine (SAM) and an acyl carrier protein (ACP) are the substrates responsible for the synthesis of AHLs. SAM is the amino donor for generation of the homoserine lactone ring moiety, and the acyl-ACP is the precursor for the acyl side chain of the AHL signal. Therefore, the AHLs biosynthesis inhibition depends from

- SAM biosynthesis inhibition
- Interference with acyl-ACP production
- Various analogs of SAM, such as S-adenosylhomocysteine, S-adenosyl cysteine, and sinefungin, have been demonstrated to be potent inhibitors of AHL synthesis catalyzed by the *P. aeruginosa* RhlI protein.
- FabI (NADH-dependent enoyl-ACP reductases) is the bacterial enzyme responsible for the catalysis of the final step of the acyl-ACP biosynthesis. In *P. aeruginosa*, FabI is involved in the biosynthesis of butyryl-ACP for the production of C4HSL. Triclosan is a FabI inhibitor able to inhibit the C4HSL production (Hoang and Schweizer, 1999). *P. aeruginosa* is resistant to Triclosan (Schweizer, 2003); it can be considered as a lead compound in search of other derivatives.

4.2.4 Biosynthesis inhibition of AI-2

S-adenosylhomocysteine (SAH) is the substrate responsible for the biosynthesis of AI-2 that is catalyzed by two enzymes:

- The 5′-methylthioadenosine/S-adenosylhomocysteine nucleosidase, MTAN (also Pfs), that catalyzes the removal of adenine from SAH to yield S-ribosyl-L-homocysteine (SRH)
- The metalloenzyme LuxS that converts SRH in homocysteine and DPD (Schauder et al., 2001), which is unstable in aqueous solution and undergoes spontaneous rearrangement into multiple different isomers of cyclic furanone compounds that are in equilibrium with each other. These isomers of cyclic furanone compounds as a group are termed AI-2 (Guo et al., 2013).

LuxS inhibition occurs with two SRH analogues: S-anhydroribosyl-L-homocysteine and S-homoribosyl-L-cysteine. They are both competitive inhibitors that block the first and the last steps of catalytic mechanism (Alfaro et al., 2004).

The research of new inhibitors led to the conclusion that the amino acid moiety and the link to the metallic center play a crucial role in the LuxS inhibition (Shen et al., 2006).

The MTAN enzyme is involved in the biosynthesis of both AHL and AI-2 autoinducers. In the AHL biosynthesis, MTAN is responsible for depurinating methylthioadenosine (MTA), a by-product of AHL synthesis, while in the AI-2 biosynthesis, MTAN catalyzes the removal of adenine from SAH to yield SRH. MTA analogs could inhibit MTAN (Ferro et al., 1976). Particularly, several papers demonstrated that the transition state analogs of MTA hydrolysis strongly inhibited MTAN from several bacteria, including *S. pneumoniae*, *E. coli*, and *V. cholera* (Gutierrez et al., 2009; Singh et al., 2006). Immucillin-A (ImmA) analogs aim to mimic an early dissociative transition state where ribosyl and adenine bond is partially broken while DADMe-ImmA analogs mimic a late transition state whereby adenine is fully dissociated. MTA is also a substrate for the human MTA phosphorylase; hence it is possible that some MTA analogs could inhibit the human enzyme to cause toxicity. There are, however, structural differences between the bacterial MTA nucleosidase and the human MTA phosphorylase allowing selective targeting of the bacterial enzyme (Lee et al., 2004; Longshaw et al., 2010).

5 Conclusion

Bacterial QS, a cell-cell communication process, controls, among other functions, many virulence factors (i.e., toxins, proteases, biofilm formation) and antibiotic tolerance. In the QS system, secreted signaling molecules (such as AHLs) coordinate the behavior of an entire bacterial population to express virulence determinants in a synchronized manner. Inhibition of QS signals may make pathogens harmless, because they lose the ability to produce coordinated virulence responses in their host. This appears to be a promising disease control strategy. Studies over the past few years have shown that QS inhibitory mechanisms, also known as quorum quenching (QQ) mechanisms, are widely distributed in many prokaryotic organisms. These naturally occurring QQ mechanisms are essential to gain advantage over competitors, and rely on small molecules, which can be used as lead compounds for designing a new generation of antimicrobial drugs. QQ compounds are able to interfere with virulence regulatory mechanisms of pathogens rather than killing them. Therefore, in the context of the incessant outbreak of antibiotic-resistant infections, this emerging therapeutic approach significantly reduces selective pressures for the occurrence of antibiotic resistance in pathogens.

Glossary

Aromatic aminated compound chemical compound displaying an aromatic ring and an amine functional group

Dihydropyranic ring monounsaturated six-membered ring, having five carbon atoms and one oxygen atom

Enolic compounds chemical compounds in which one carbon of a double-bonded pair is attached to a hydroxyl group

Furanone five-membered α,β-unsaturated lactone

Heteroaromatic ring aromatic ring containing at least one heteroatom (noncarbon atom)

Isomeric forms chemical compounds sharing the same molecular formula but having different structural arrangement of the atoms in space and, therefore, different properties

Lactone cyclic esters of hydroxy carboxylic acids

Metagenomic analysis analysis of DNA isolated from an assemblage of microorganisms

Molecular modeling studies in silico exploration of molecular structures and properties through computational chemistry and graphical visualization techniques, aiming to figure out a plausible 3D representation under given settings

Stereogenic center an atom with three or more different ligands; interchanging of two of these ligands leads to another stereoisomer

Abbreviations

ACP	acyl carrier protein
AHK	α-hydroxyketones
AHL	*N*-acyl homoserine lactone
AI-2	autoinducer-2
AI-3	autoinducer-3
AiiA	autoinducer inactivation
CAI-1	cholera autonducer-1
DPD	4,5-dihydroxy-2,3-pentanedione
DSF	diffusible signal factor
HAQ	4-hydroxy-2-alkylquinolines
HHQ	4-hydroxy-2-heptylquinoline
HSL	homoserine lactones

LC-HRMS/MS	liquid chromatography-high resolution tandem mass spectrometry
NHQ	2-nonyl-4-hydroxyquinoline
PHL	phenylacetanoil homoserine lactone
PQS	pseudomonas quinolone signal
QQ	quorum quenching
QS	quorum sensing
SAM	*S*-adenosylmethionine
SID	surface induced dissociation
SRH	*S*-ribosyl-L-homocysteine
THMF	tetrahydroxymethyltetrahydrofuran

References

Alfaro, J.F., Zhang, T., Wynn, D.P., Karschner, E.L., Zhaohui, S.Z., 2004. Synthesis of LuxS inhibitors targeting bacterial cell-cell communication. Org. Lett. 6 (18), 3043–3046.

Allegretta, G., Maurer, C.K., Eberhard, J., Maura, D., Hartmann, R.W., Rahme, L., Empting, M., 2017. In-depth profiling of MvfR-regulated small molecules in *Pseudomonas aeruginosa* after quorum sensing inhibitor treatment. Front. Microbiol. 8, 1–12.

Bijtenhoorn, P., Mayerhofer, H., Müller-Dieckmann, J., Utpatel, C., Schipper, C., Hornung, C., Szesny, M., Grond, S., Thürmer, A., Brzuszkiewicz, E., Daniel, R., Dierking, K., Schulenburg, H., Streit, W.R., 2011. A novel metagenomic short-chain dehydrogenase/reductase attenuates *Pseudomonas Aeruginosa* biofilm formation and virulence on *Caenorhabditis elegans*. PLoS One. 6(10), e26278.

Bottomley, M.J., Muraglia, E., Bazzo, R., Carfi, A., 2007. Molecular insights into quorum sensing in the human pathogen *Pseudomonas aeruginosa* from the structure of the virulence regulator LasR bound to its autoinducer. J. Biol. Chem. 282 (18), 13592–13600.

Brameyer, S., Kresovic, D., Bode, H.B., Heermann, R., 2014. LuxR solos in *Photorhabdus* species. Front. Cell. Infect. Microbiol. 4, 1–11.

Britstein, M., Devescovi, G., Handley, K.M., Malik, A., Haber, M., Saurav, K., Teta, R., Costantino, V., Burgsdorf, I., Gilbert, J.A., Sher, N., Venturi, V., Steindler, L., 2016. A new N-acyl homoserine lactone synthase in an uncultured symbiont of the red sea sponge *Theonella swinhoei*. Appl. Environ. Microbiol. 82, 1274–1285.

Britstein, M., Saurav, K., Teta, R., Della Sala, G., Bar-Shalom, R., Stoppelli, N., Zoccarato, L., Costantino, V., Steindler, L., 2018. Identification and chemical characterization of N-acyl-homoserine lactone quorum sensing signals across sponge species and time. FEMS Microbiol. Ecol. 94 (2), 1–7.

Carrano, C.J., Schellenberg, S., Amin, S.A., Green, D.H., Küpper, F.C., 2009. Boron and marine life: a new look at an enigmatic bioelement. Mar. Biotechnol. 11, 431–440.

Costantino, V., Sala, G.D., Saurav, K., Teta, R., Bar-Shalom, R., Mangoni, A., Steindler, L., 2017. Plakofuranolactone as a quorum quenching agent from the Indonesian sponge *Plakortis* cf. lita. Mar. Drugs 15, 1–12.

Decho, A.W., Visscher, P.T., Ferry, J., Kawaguchi, T., He, L., Przekop, K.M., Norman, R.S., Reid, R.P., 2009. Autoinducers extracted from microbial mats reveal a surprising diversity of N-acylhomoserine lactones (AHLs) and abundance changes that may relate to diel pH. Environ. Microbiol. 11, 409–420.

Delago, A., Mandabi, A., Meijler, M.M., 2016. Natural quorum sensing inhibitors—small molecules, big messages. Israel J. Chem. 56, 310–320.

Deng, Y., Wu, J., Tao, F., Zhang, L.H., 2011. Listening to a new language: DSF-based quorum sensing in gram-negative bacteria. Chem. Rev. 111, 160–179.

Dong, Y.H., Xu, J.L., Li, X.Z., Zhang, L.H., 2000. AiiA, an enzyme that inactivates the acylhomoserine lactone quorum-sensing signal and attenuates the virulence of *Erwinia carotovora*. Proc. Natl. Acad. Sci. U. S. A. 97, 3526–3531.

Dow, J.M., 2017. Diffusible signal factor-dependent quorum sensing in pathogenic bacteria and its exploitation for disease control. J. Appl. Microbiol. 122, 2–11.

Duan, F., March, J.C., 2010. Engineered bacterial communication prevents *Vibrio cholerae* virulence in an infant mouse model. Proc. Natl. Acad. Sci. U. S. A. 107 (25), 11260–11264.

Ferro, A.J., Barrett, A., Shapiro, S.K., 1976. Kinetic properties and the effect of substrate analogues on 5'-methylthioadenosine nucleosidase from *Escherichia coli*. Biochim. Biophys. Acta 438, 487–494.

Fuqua, C., Greenberg, E.P., 2002. Listening in on bacteria: Acyl-homoserine lactone signaling. Nat. Rev. Mol. Cell Biol. 3, 685–695.

Fuqua, W.C., Winans, S.C., Greenberg, E.P., 1994. Quorum sensing in bacteria: the LuxR–LuxI family of cell density-responsive transcriptional regulators. J. Bacteriol. 176, 269–275.

Gerdt, J.P., McInnis, C.E., Schell, T.L., Rossi, F.M., Blackwell, H.E., 2014. Mutational analysis of the quorum-sensing receptor LasR reveals interactions that govern activation and inhibition by non-lactone ligands. Chem. Biol. 21 (10), 1361–1369.

Geske, G.D., O'Neill, J.C., Blackwell, H.E., 2008a. Expanding dialogues: from natural autoinducers to non-natural analogues that modulate quorum sensing in gram-negative bacteria. Chem. Soc. Rev. 37 (7), 1432–1447.

Geske, G.D., O'Neill, J.C., Miller, D.M., Wezeman, R.J., Mattmann, M.E., Lin, Q., Blackwell, H.E., 2008b. Comparative analyses of *N*-Acylated Homoserine lactones reveal unique structural features that dictate their ability to activate or inhibit quorum sensing. Chembiochem 9 (3), 389–400.

Gilson, L., Kuo, A., Dunlap, P.V., 1995. AinS and a new family of autoinducer synthesis proteins. J. Bacteriol. 177 (23), 6946–6951.

Guo, M., Gamby, S., Zheng, Y., Sintim, H.O., 2013. Small molecule inhibitors of AI-2 signaling in bacteria: State-of-the-art and future perspectives for anti-quorum sensing agents. Int. J. Mol. Sci. 14, 17694–17728.

Gutierrez, J.A., Crowder, T., Rinaldo-Matthis, A., Ho, M.C., Almo, S.C., Schramm, V.L., 2009. Transition state analogs of 5'-methylthioadenosine nucleosidase disrupt quorum sensing. Nat. Chem. Biol. 5, 251–257.

Heeb, S., Fletcher, M.P., Chhabra, S.R., Diggle, S.P., Williams, P., Cámara, M., 2011. Quinolones: from antibiotics to autoinducers. FEMS Microbiol. Rev. 35, 247–274.

Hense, B.A., Schuster, M., 2015. Core principles of bacterial autoinducer systems. Microbiol. Mol. Biol. Rev. 79, 153–169.

Hmelo, L.R., 2017. Quorum sensing in marine microbial environments. Annu. Rev. Mar. Sci. 9, 257–281.

Hoang, T.T., Schweizer, H.P., 1999. Characterization of *Pseudomonas aeruginosa* enoyl-acyl carrier protein reductase (FabI): A target for the antimicrobial triclosan and its role in acylated homoserine lactone synthesis. J. Bacteriol. 181, 5489–5497.

Hughes, D.T., Sperandio, V., 2008. Inter-kingdom signaling: communication between bacteria and their hosts. Nat. Rev. Microbiol. 6, 111–120.

Ilangovan, A., Fletcher, M., Rampioni, G., Pustelny, C., Rumbaugh, K., Heeb, S., Cámara, M., Truman, A., Chhabra, S.R., Emsley, J., Williams, P., 2013. Structural basis for native agonist and synthetic inhibitor recognition by the *Pseudomonas aeruginosa* quorum sensing regulator PqsR (MvfR). PLoS Pathog. 9(7). e1003508.

Ishida, T., Ikeda, T., Takiguchi, N., Kuroda, A., Ohtake, H., Kato, J., 2007. Inhibition of quorum sensing in Pseudomonas aeruginosa by N-acyl cyclopentylamides. Appl. Environ. Microbiol. 73 (10), 3183–3188.

Kamal, A.A.M., Petrera, L., Eberhard, J., Hartmann, R.W., 2017. Structure-functionality relationship and pharmacological profiles of *Pseudomonas aeruginosa* alkylquinolone quorum sensing modulators. Org. Biomol. Chem. 15, 4620–4630.

Kim, K., Kim, Y.U., Koh, B.H., Hwang, S.S., Kim, S.H., Lépine, F., Cho, Y.H., Lee, G.R., 2010. HHQ and PQS, two *Pseudomonas aeruginosa* quorum-sensing molecules, down-regulate the innate immune responses through the nuclear factor-κB pathway. Immunology 129, 578–588.

Kline, T., Bowman, J., Iglewski, B.H., de Kievit, T., Kakai, Y., Passador, L., 1999. Novel synthetic analogs of the *Pseudomonas* autoinducer. Bioorg. Med. Chem. Lett. 9 (24), 3447–3452.

Kong, S., Application, F., Data, P., 2011. (12) United States Patent 2, 12–15.

La Sarre, B., Federle, M.J., 2013. Exploiting quorum sensing to confuse bacterial pathogens. Microbiol. Mol. Biol. Rev. 77, 73–111.

Lee, J.E., Settembre, E.C., Cornell, K.A., Riscoe, M.K., Sufrin, J.R., Ealick, S.E., Lynne Howell, P., 2004. Structural comparison of MTA Phosphorylase and MTA/AdoHcy Nucleosidase explains substrate preferences and identifies regions exploitable for inhibitor design. Biochemistry 43 (18), 5159–5169.

Lin, Y.H., Xu, J.L., Hu, J., Wang, L.H., Leong Ong, S., Renton Leadbetter, J., Zhang, L.H., 2003. Acyl-homoserine lactone acylase from *Ralstonia* strain XJ12B represents a novel and potent class of quorum-quenching enzymes. Mol. Microbiol. 47, 849–860.

Llamas, I., Suarez, A., Quesada, E., Bejar, V., del Moral, A., 2003. Identification and characterization of the carAB genes responsible for encoding carbamoylphosphate synthetase in *Halomonas eurihalina*. Extremophiles 7, 205–211.

Longshaw, A.I., Adanitsch, F., Gutierrez, J.A., Evans, G.B., Tyler, P.C., Schramm, V.L., 2010. Design and synthesis of potent "sulfur-free" transition state analogue inhibitors of 5′-methylthioadenosine nucleosidase and 5′-methylthioadenosine phosphorylase. J. Med. Chem. 53, 6730–6746.

Nealson, K.H., Platt, T., Hastings, J.W., 1970. Cellular control of the synthesis and activity of the bacterial luminescent system. J. Bacteriol. 104, 313–322.

Newman, K.L., Chatterjee, S., Ho, K.a., Lindow, S.E., 2008. Virulence of plant pathogenic bacteria attenuated by degradation of fatty acid cell-to-cell signaling factors. Mol. Plant-Microbe Interact. 21, 326–334.

O'Reilly, M.C., Blackwell, H.E., 2016. Structure-based design and biological evaluation of Triphenyl scaffold-based hybrid compounds as hydrolytically stable modulators of a LuxR-type quorum sensing receptor. ACS Infect. Dis. 2 (1), 32–38.

Parker, C.T., Russell, R., Njoroge, J.W., Jimenez, A.G., Taussig, R., Sperandio, V., 2017. Genetic and mechanistic analyses of the Periplasmic domain of the Enterohemorrhagic *Escherichia coli*. J. Bacteriol. 199, 1–15.

Parsek, M.R., Greenberg, E.P., 2000. Acyl-homoserine lactone quorum sensing in gram-negative bacteria: a signaling mechanism involved in associations with higher organisms. Proc. Natl. Acad. Sci. U. S. A. 97 (16), 8789–8793.

Parsek, M.R., Greenberg, E.P., 2005. Sociomicrobiology: the connections between quorum sensing and biofilms. Trends Microbiol. 13 (1), 27–33.

Parsek, M.R., Val, D.L., Hanzelka, B.L., Cronan, J.E., Greenberg, E.P., 1999. Acyl homoserine-lactone quorum-sensing signal generation. Proc. Natl. Acad. Sci. 96, 4360–4365.

Passador, L., Tucker, K.D., Guertin, K.R., Journet, M.P., Kende, A.S., Iglewski, B.H., 1996. Functional analysis of the *Pseudomonas aeruginosa* autoinducer PAI. J. Bacteriol. 178 (20), 5995–6000.

Ransome, E., Munn, C.B., Halliday, N., Cámara, M., Tait, K., 2014. Diverse profiles of N-acyl-homoserine lactone molecules found in cnidarians. FEMS Microbiol. Ecol. 87, 315–329.

Rasmussen, T.B., Skindersoe, M.E., Bjarnsholt, T., Phipps, R.K., Christensen, K.B., Jensen, P.O., Andersen, J.B., Koch, B., Larsen, T.O., Hentzer, M., Eberl, L., Hoiby, N., Givskov, M., 2005. Identity and effects of quorum-sensing inhibitors produced by *Penicillium* species. Microbiology 151 (5), 1325–1340.

Roy, V., Fernandes, R., Tsao, C.Y., Bentley, W.E., 2010. Cross species quorum quenching using a native AI-2 processing enzyme. ACS Chem. Biol. 5, 223–232.

Ruby, E.G., Lee, K.H., 1998. The *Vibrio fischeri-Euprymna scolopes* light organ association: current ecological paradigms. Appl. Environ. Microbiol. 64, 805–812.

Rui, F., Marques, J.C., Miller, S.T., Maycock, C.D., Xavier, K.B., Ventura, M.R., 2012. Stereochemical diversity of AI-2 analogs modulates quorum sensing in *Vibrio harveyi* and *Escherichia coli*. Bioorg. Med. Chem. 20, 249–256.

Saeidi, N., Wong, C.K., Lo, T.-M., Nguyen, H.X., Ling, H., Leong, S.S.J., Poh, C.L., Chang, M.W., 2011. Engineering microbes to sense and eradicate *Pseudomonas aeruginosa*, a human pathogen. Mol. Syst. Biol. 7, 521.

Saurav, K., Burgsdorf, I., Teta, R., Esposito, G., Bar-Shalom, R., Costantino, V., Steindler, L., 2016. Isolation of marine *Paracoccus* sp. Ss63 from the sponge *Sarcotragus* sp. and characterization of its quorum-sensing chemical-signaling molecules by LC-MS/MS analysis. Israel J. Chem. 56, 330–340.

Schaefer, A.L., Greenberg, E.P., Oliver, C.M., Oda, Y., Huang, J.J., Bittan-Banin, G., Peres, C.M., Schmidt, S., Juhaszova, K., Sufrin, J.R., Harwood, C.S., 2008. A new class of homoserine lactone quorum-sensing signals. Nature 454, 595–599.

Schauder, S., Shokat, K., Surette, M.G., Bassler, B.L., 2001. The LuxS family of bacterial autoinducers: Biosynthesis of a novel quorum-sensing signal molecule. Mol. Microbiol. 41 (2), 463–476.

Schripsema, J., De Rudder, K.E.E., Van Vliet, T.B., Lankhorst, P.P., De Vroom, E., Kijne, J.W., Van Brussel, A.A.N., 1996. Bacteriocin small of *Rhizobium leguminosarum* belongs to the class of N-acyl-L-homoserine lactone molecules, known as autoinducers and as quorum sensing co-transcription factors. J. Bacteriol. 178, 366–371.

Schweizer, H.P., 2003. Efflux as a mechanism of resistance to antimicrobials in *Pseudomonas aeruginosa* and related bacteria: unanswered questions. Genet. Mol. Res. 2 (1), 48–62.

Shen, G., Rajan, R., Zhu, J., Bell, C.E., Pei, D., 2006. Design and synthesis of substrate and intermediate analogue inhibitors of S-ribosylhomocysteinase. J. Med. Chem. 49 (10), 3003–3011.

Singh, V., Shi, W., Almo, S.C., Evans, G.B., Furneaux, R.H., Tyler, P.C., Painter, G.F., Lenz, D.H., Mee, S., Zheng, R., Schramm, V.L., 2006. Structure and inhibition of a quorum sensing target from *Streptococcus pneumoniae*. Biochemistry 45 (43), 12929–12941.

Skindersoe, M.E., Alhede, M., Phipps, R., Yang, L., Jensen, P.O., Rasmussen, T.B., Bjarnsholt, T., Tolker-Nielsen, T., Høiby, N., Givskov, M., 2008. Effects of antibiotics on quorum sensing in *Pseudomonas aeruginosa*. Antimicrob. Agents Chemother. 52 (10), 3648–3663.

Smith, R.S., Iglewski, B.H., 2003. *Pseudomonas aeruginosa* quorum sensing as a potential antimicrobial target. J. Clin. Investig. 112, 1460–1465.

Smith, K.M., Bu, Y., Suga, H., 2003. Induction and inhibition of *Pseudomonas aeruginosa* quorum sensing by synthetic autoinducer analogs. Chem. Biol. 10 (1), 81–89.

Sperandio, V., Torres, A.G., Jarvis, B., Nataro, J.P., Kaper, J.B., 2003. Bacteria-host communication: the language of hormones. Proc. Natl. Acad. Sci. 100, 8951–8956.

Suga, H., Smith, K.M., 2003. Molecular mechanisms of bacterial quorum sensing as a new drug target. Curr. Opin. Chem. Biol. 7 (5), 586–591.

Suneby, E.G., Herndon, L.R., Schneider, T.L., 2017. *Pseudomonas aeruginosa* LasR·DNA binding is directly inhibited by quorum sensing antagonists. ACS Infect. Dis. 3 (3), 183–189.

Taylor, M.W., Schupp, P.J., Baillie, H.J., Charlton, T.S., De Nys, R., Kjelleberg, S., Steinberg, P.D., 2004. Evidence for acyl Homoserine lactone signal production in Bacteria associated with marine sponges. Appl. Environ. Microbiol. 70, 4387–4389.

Thiel, V., Kunze, B., Verma, P., Wagner-Döbler, I., Schulz, S., 2009. New structural variants of homoserine lactones in bacteria. Chembiochem 10, 1861–1868.

Tomasz, A., 1965. Control of the competent state in Pneumococcus by a hormone-like cell product: an example for a new type of regulatory mechanism in bacteria. Nature 208, 155–159.

Uroz, S., Chhabra, S.R., Cámara, M., Williams, P., Oger, P., Dessaux, Y., 2005. N-acylhomoserine lactone quorum-sensing molecules are modified and degraded by *Rhodococcus erythropolis* W2 by both amidolytic and novel oxidoreductase activities. Microbiology 151, 3313–3322.

Van Mooy, B.A.S., Hmelo, L.R., Sofen, L.E., Campagna, S.R., May, A.L., Dyhrman, S.T., Heithoff, A., Webb, E.A., Momper, L., Mincer, T.J., 2012. Quorum sensing control of phosphorus acquisition in *Trichodesmium* consortia. ISME J. 6, 422–429.

Wagner-Döbler, I., Thiel, V., Eberl, L., Allgaier, M., Bodor, A., Meyer, S., Ebner, S., Hennig, A., Pukall, R., Schulz, S., 2005. Discovery of complex mixtures of novel long-chain quorum sensing signals in free-living and host-associated marine alphaproteobacteria. Chembiochem 6, 2195–2206.

Winzer, K., Hardie, K.R., Williams, P., 2002. Bacterial cell-to-cell communication: Sorry, can't talk now—gone to lunch! Curr. Opin. Microbiol. 5, 216–222.

Wu, H., Song, Z., Hentzer, M., Andersen, J.B., Molin, S., Givskov, M., Høiby, N., 2004. Synthetic furanones inhibit quorum-sensing and enhance bacterial clearance in *Pseudomonas aeruginosa* lung infection in mice. J. Antimicrob. Chemother. 53 (6), 1054–1061.

Xavier, K.B., Miller, S.T., Lu, W., Kim, J.H., Rabinowitz, J., Pelczer, I., Semmelhack, M.F., Bassler, B.L., 2007. Phosphorylation and processing of the quorum-sensing molecule autoinducer-2 in enteric bacteria. ACS Chem. Biol. 2, 128–136.

Zhou, L., Zhang, L., Cámara, M., He, Y., 2017. The DSF family of quorum sensing signals: diversity, biosynthesis, and turnover. Trends Microbiol. 25, 293–303.

Ziesche, L., Bruns, H., Dogs, M., Wolter, L., Mann, F., Wagner-Döbler, I., Brinkhoff, T., Schulz, S., 2015. Homoserine lactones, methyl Oligohydroxybutyrates, and other extracellular metabolites of macroalgae-associated Bacteria of the *Roseobacter* clade: Identification and functions. Chembiochem 16, 2094–2107.

Zou, Y., Nair, S.K., 2009. Molecular basis for the recognition of structurally distinct autoinducer mimics by the *Pseudomonas aeruginosa* LasR quorum sensing signaling receptor. Chem. Biol. 16 (9), 961–970.

CHAPTER 2

Analytical Approaches for the Identification of Quorum Sensing Molecules

Adele Cutignano

National Research Council of Italy—Institute of Biomolecular Chemistry, Pozzuoli, Italy

1 Introduction

Quorum sensing (QS) is a widely conserved cell-to-cell communication strategy of the bacterial world (Camilli, 2006; Hmelo, 2017; Taga and Bassler, 2003; Waters and Bassler, 2005). Small chemicals called autoinducers (AIs) are produced and released in the local environment, diffusing through the cellular membrane into adjacent cells and regulating gene expression in a cell-density/concentration dependent manner. Quorum sensing is involved in several biological activities, such as biofilm formation, bioluminescence, sporulation, motility, genetic exchange, and the production of antibiotics and virulence factor. In particular, QS is a key process well documented in pathogens and implicated in the onset of bacterial pathogenicity; hence, it has been increasingly considered as an alternative target for antibiotic treatments.

Both Gram-positive and Gram-negative bacteria rely on chemical signals to coordinate population behavior, even though the QS molecules they produce belong to different chemical classes (Hawver et al., 2016) (Fig. 1): typically, they are *N*-acyl-L-homoserine lactones (AHLs) in most Gram-negative species (AI-1) (Fuqua et al., 1994; Fuqua and Greenberg, 2002; Papenfort and Bassler, 2016) and short autoinducing peptides (AIPs) in Gram-positive bacteria (Bofinger et al., 2017; Kleerebezem et al., 1997; Sturme et al., 2002; Verbeke et al., 2017).

Each bacterial species produces and senses a specific pattern of small AIs. The AHLs produced by Gram-negative bacteria from different species may vary for length and unsaturation degree of the *N*-acyl side chain, usually ranging from 4 to 18 carbon atoms, with only one report of a 19 carbons chain length (Doberva et al., 2017); carbon 3 can carry a hydroxyl- or oxo-group (Fig. 1). However, specific molecular entities from surrounding species can be mutually recognized, and a pool of molecules derived by (*S*)-4,5-dihydroxy-2,3-pentanedione (DPD) and equilibrium-connected, furanosyl boron diesters (AI-2) are unique, non-species-specific chemicals produced by both Gram-positive and Gram-negative bacteria

Quorum Sensing. https://doi.org/10.1016/B978-0-12-814905-8.00002-2

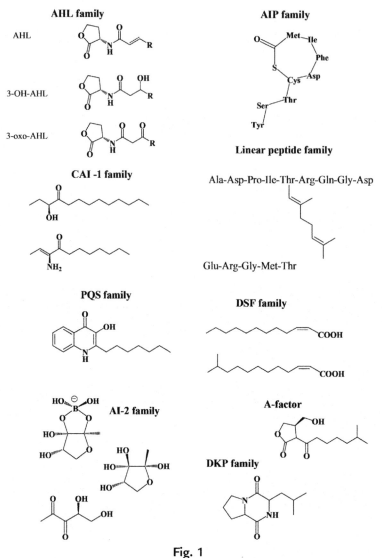

Fig. 1
Representative structures and chemical classes of autoinducers.

and known to mediate interspecies communication, thus constituting a sort of universal QS signal (Bassler et al., 1997; Schauder et al., 2001; Vendeville et al., 2005) (Fig. 1). Besides these three groups of autoinducers (AI-1, AIPs, and AI-2), minor QS molecules have been described from bacterial species and are reported in Fig. 1 (Bofinger et al., 2017; Deng et al., 2011; He and Zhang, 2008; Pesci et al., 1999; Winans, 2011). The knowledge of microbiome-host crosstalk is extending as well, with its clinical significance becoming more and more evident. A complex communication network keeps symbiotic bacteria, pathogen agents, and their host intertwined, thus orchestrating the fight between defense and attack against invading microbes (Curtis and Sperandio, 2011; Goswami, 2017; Sperandio et al., 2003; Williams, 2007); one example is the epinephrine/norepinephrine/autoinducer-3 (AI-3) signaling system, which regulates *Escherichia coli* EHEC pathogenicity in its mammalian host (Curtis and Sperandio, 2011; Reading et al., 2007; Sifri, 2008; Sperandio et al., 2003). For this group of type-3 autoinducers, an aromatic nature has been hypothesized (Reading and Sperandio, 2006), but no structural confirmation has been reported to date in the literature; thus it has to be considered undetermined yet.

The common trait of all these situations is that the signal molecules are produced at a very low concentration (typically nM range) and their detection requires highly sensitive techniques. The development of biosensors has allowed fast screening of microorganisms for the productions of autoinducers; however, for specific identification and quantification of these bacterial metabolites, the gold choice is in the form of mass spectrometry-based methodologies, coupled with either gas chromatography (GC-MS) or liquid chromatography (LC-MS/MS). Sample extraction, cleanup, and preconcentration can greatly improve autoinducer identification from complex matrices, while also allowing reliable quantification. The present chapter will highlight the common strategies adopted for studying these compounds with a special focus on AHLs, and future trends of the research based on state-of-the-art technologies.

2 Sample Preparation

One of the bottlenecks when tackling very low abundance natural metabolites is the choice of the correct protocol to clean up and enrich the biological sample before analysis; this step is demanding to avoid compound overlooking due to a matter of dilution or to other components present in the cell culture supernatants (extracellular products or constituents of the growth media) or in other different environmental matrices, that can mask or compromise proper identification of the target metabolites (Cutignano et al., 2015). On the other hand, direct analysis of the media containing QS molecules is prevented by the presence of salts and macromolecules (e.g., proteins and polysaccharides) that can clog the analytical column of a sequential step. Different strategies can be adopted, including liquid-liquid extraction (LLE) and/or solid phase extraction (SPE) (Sitnikov et al., 2016).

2.1 Liquid-Liquid Extraction (LLE)

LLE represents the most immediate, traditional, and relatively simple method to obtain an enriched fraction of small QS molecules. The aqueous phase constituted by the supernatant of the culture medium or by environmental water samples is extracted by immiscible organic solvents, typically ethyl acetate or dichloromethane (Feng et al., 2014; Kumari et al., 2008;

Ortori et al., 2007; Shaw et al., 1997; Wang et al., 2017). At this stage, an internal standard should be added before further manipulation to correct for recovery and instrumental bias in view of quantitative analysis. Once the extract has been evaporated to dryness, it can be redissolved in a small solvent volume (i.e., Acetonitrile, 1–2 mL) and directly analyzed, for example by thin layer chromatography (TLC); however, it is often submitted to further chromatographic steps for preconcentration and cleanup, which are typically carried out by solid phase extraction (SPE) chromatography (Andrade-Eiroa et al., 2016).

2.2 Solid Phase Extraction (SPE)

SPE is a technique widely and increasingly adopted over the last two decades, especially before quantitative analysis, to increase the concentration of trace level analytes. Two great advantages have contributed to its broad application: the availability of prepacked cartridges with a variety of chemically different sorbents, thus extending its field of use to an ample range of chemically diverse substances; and the possibility to carry out this step by an automatic process, thus avoiding variability due to human operators. Furthermore, SPE grants high recovery rates, which is particularly significant when handling chemical components occurring in trace amount in biological matrices, for quantitative purposes. In fact, SPE can gain sensitivity of detection at least of two up to tenfold with respect to LLE (Schupp et al., 2005).

An array of sorbents have been tested for extraction of quorum sensing AHLs by SPE chromatography, including silica, octadecyl (RP-18), alumina, diol, phenyl, ion exchange, polymeric styrene-divinylbenzene (PSDVB), primary-secondary amines (PAS), and polyamide (PA)-based resins (Li et al., 2006; Schupp et al., 2005; Wang et al., 2017). The best results were obtained with reversed phase columns, while normal phase materials and sorbents with ion exchange activity were not advisable since a high fraction of the loaded amount was always retained (Li et al., 2006). In general, no single sorbent was adequate to recover and identify all compounds. In fact, Schupp et al. (2005) reported on an excellent recovery with polyamide (DPA) resin only for short-chain and part of medium-chain AHLs, while for AHLs with chain length from C_6 to C_{12}, the recovery was higher than 90% with RP-18 and modified PSDVB-based resins (HLB) with poorer results with C_4-HSL and C_{14}-HSL (Li et al., 2006; Wang et al., 2017).

Accordingly, different protocols have been reported for SPE elution (Fekete et al., 2007; Gould et al., 2006; Li et al., 2006; Schupp et al., 2005; Wang et al., 2017). Typically, silica-based column are eluted with sequential mixtures of apolar solvents (i.e., *n*-hexane with increasing amounts of ethyl acetate), and AHLs have been detected in the fraction eluting with *n*-hexane/ethyl acetate 65:35 (Schupp et al., 2005); on the other hand, an optimized protocol on end-capped octadecyl columns relies on initial methanol/water washing (15:85, *v/v*) and *n*-hexane/isopropanol (25:75, *v/v*) elution of AHLs (Fekete et al., 2007; Li et al., 2006). In 2017, Wang and coworkers described fractionation onto a HLB adsorbent, which combines lipophilic and hydrophilic balanced interactions. After washing steps with methanol/water (5:95, *v/v*), target compounds were eluted with excellent recoveries with methanol containing 2% acetic acid.

3 Detection, Identification, and Quantification of QS Molecules

3.1 Analytical Approaches for AHL Detection

3.1.1 Biosensors

The mechanism underpinning the cellular response to AHL autoinducers was recognized more than 20 years ago in *Vibrio fischeri* and is mediated by two proteins of the LuxI-LuxR family (and other homologs proteins)—namely, a LuxI-AHL synthase and a LuxR-AHL transcriptional activator (Fuqua et al., 1994) (Fig. 2).

LuxR-type proteins are cytoplasmic receptors that detect and bind AHLs produced by a cognate LuxI-type synthase, and associate with short DNA sequences, called *lux*-boxes, triggering the expression of hundreds of (up to 600) genes. Examples of AHL-mediated cell-cell signaling are those involved in autoinduction of bioluminescence in the marine bacteria *Photobacterium (Vibrio) fischeri* and *Vibrio harveyi* (Fuqua et al., 1994) and in regulation of pigment violacein production in the bacterium *Chromobacterium violaceum* (McClean et al., 1997). The development of bacterial biosensors not producing AHLs but able to recognize and respond in the presence of AHLs boosted the discovery of a large number of AHL-based QS systems

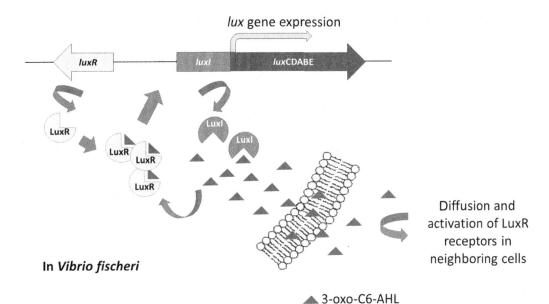

Fig. 2
Schematic representation of the LuxI/LuxR QS system in *V. fischeri. Modified from Galloway et al., 2011.*

in Gram-negative bacteria. These biosensors can be obtained by cloning a functional LuxR-type protein along with a suitable promoter, for example the promoter of the cognate *luxI* synthase, which positively regulates the transcription of a reporter gene, e.g., violacein production, bioluminescence, or galactosidase (Steindler and Venturi, 2007). In other words, the recognition event is coupled to a transducer element that converts it into a readable and eventually measurable output, by using genetically engineered bacteria or alternatively by exploiting constitutively expressed reporters. LuxR proteins preferentially bind AHLs produced by cognate LuxI synthase, but under some extent respond to closely related AHLs analogs exhibiting different types of side chains; thus each biosensor is tailor-made and can be applied to screen for the production of a narrow range of AHL family of autoinducers. On the other hand, by using a panel of biosensors, a broader spectrum of AHLs can be sensed. Cell-based biosensors may be used in different ways. One is plate cross-streaking of the bacterial producer and the biosensor strain: the closest point between the two strains will show the most intense response. Another one is thin layer chromatography (TLC) overlay. Extracts and/or chromatographic fractions obtained from the bioproducer are loaded onto TLC plate along with chemical standards and allowed to resolve into the molecular components in a chromatographic chamber. Detection of the AHLs can be successively appreciated by TLC-overlay with an agar suspension of the biosensor strain. The chromatographic retardation factor (Rf) of the resulting detected spots, compared with AHL-standards, can help in tentatively assigning the chain length and substitution of the autoinducer (McClean et al., 1997; Shaw et al., 1997). Moreover, the intensity of the reporter readout may be interpreted in a quantitative assay. Although easy and robust, these cell-based biosensing systems present few drawbacks, including slow response, chemical interference from bacterial cell components, and batch-to-batch variability. For this reason, in the last two decades alternative tools have been proposed to detect AHL based on plasmids containing genes encoding the green fluorescent protein gfp (Andersen et al., 2001); these sensors, which harness the biological machinery inside cells, harbor the great advantage that they can be used at single-cell level for the detection of AHL production. Very recently, a "cell-free" form of biosensors, where the engineered DNA circuit and cellular machinery are free-floating in a solution, was developed and proposed to test clinical samples for *Pseudomonas aeruginosa* (Wen et al., 2017). In the presented study, quorum sensing molecules occurring in sputum of patients with cystic fibrosis were quantitatively measured at nanomolar level, with results comparable with those obtained with other sensitive techniques based on tandem mass spectrometry, but at possibly lower costs.

3.1.2 Radiolabeled assays

Biosynthesis of acyl homoserine lactone backbone is deputed to two categories of lactone synthases: acyl-CoA utilizing synthase and acyl-ACP utilizing synthase. However, both of them make use of *S*-adenosyl-L-methionine (SAM), which therefore is a conserved substrate (Fig. 3).

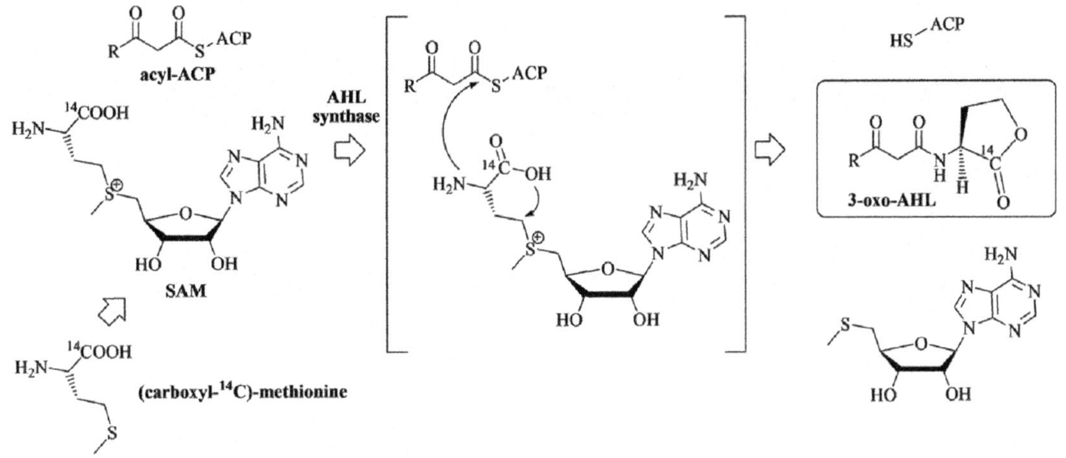

Fig. 3

Biosynthetic scheme of AHLs underlying the use of (carboxyl-^{14}C)-methionine for radiolabeled assays.

^{14}C-AHL radiolabel assay is based on the uptake of [1-^{14}C]-L-methionine by living cells and conversion into the radiolabeled *S*-adenosyl methionine. This latter metabolic intermediate is in turn incorporated into AHL signaling metabolites by an AHL synthase enzyme. Once extracted by organic solvents, radioactivity associated with AHLs can be easily revealed by the means of a scintillator detector, without any bias for specific chain length or oxidation state. Radiolabeled assays were in the past applied to AHL detection in biofilm bacteria: in fact, the very low amount of biological material available and the number of tests required to detect the different types of AHLs prevented the use of bioassays in this context (Greenberg and Parsek, 2001; Jones et al., 1993; Schaefer et al., 2001, 2002). However, despite its simplicity and sensitivity, this approach is currently being replaced by faster and safer mass spectrometric-based methodologies.

3.1.3 Colorimetric assays

In an attempt to develop a method cheap and easy to use, Yang et al. (2006) proposed a colorimetric assay by modifying a previous procedure used for the analysis of ester compounds. The method was originally proposed for lactone compounds and lactonase activity detection and was scaled down to 96-well plate size, requiring minute amount of sample (20–50 µL) and exhibiting a detection limit of 1 nmol of lactone molecule. The colorimetric assay is based on the conversion of the carboxylic esters/lactones into the corresponding hydroxamic acids; the latter, in the presence of ferric ions, form red to purple complexes that can be measured spectrophotometrically. Although colorimetric methods are very sensitive, the scarce specificity of the above suggested for AHL detection may lead to the overestimation or misidentification of the QS molecules.

3.2 Analytical Approaches for AHL Identification and Quantification

3.2.1 Gas chromatography-mass spectrometry (GC-MS)

Gas chromatography coupled to mass spectrometry (GC-MS) is a hyphenated analytical technique that combines the high resolution power of gas chromatographic separation with sensitive detection distinctive of a mass spectrometer. It is suitable for separation of volatile and thermally stable molecules that may be identified by the interpretation of mass spectra obtained from each ion species detected in virtue of its mass-to-charge (m/z) ratio value. Dealing with known molecules, the availability of standards and/or spectral databases allows fast identification of unknown by comparison of retention times and fragmentation patterns. In fact, GC-MS is based on a hard ionization technique that uses electron impact (EI) or chemical ionization (CI) for the generation of a molecular ion, which is detected and registered along with fragment ions originating in source from the cleavage of chemical bonds. Hence, GC-MS represents a reliable and very sensitive technique to identify and quantify *N*-acyl-L-homoserine lactones in bacterial extracts (Osorno et al., 2012). All compounds of the C_n-AHL series exhibited a typical pattern of fragmentation, with the presence of diagnostic ions at m/z 143 and 102 arising from the cleavage of the lactone ring (Thiel et al., 2009a) (Fig. 4). The mass spectra of 3-hydroxy-AHLs are instead dominated by the fragment at m/z 172 resulting from α-cleavage next to the hydroxyl group (Wagner-Döbler et al., 2005) (Fig. 4). Therefore, when the mass spectrometric analyzer is a quadrupole-type, as often encountered in GC-MS platforms, the most appropriate experimental design will contemplate the acquisition of spectral data in both Full-Scan and Selected Ion Monitoring (SIM) mode by using the

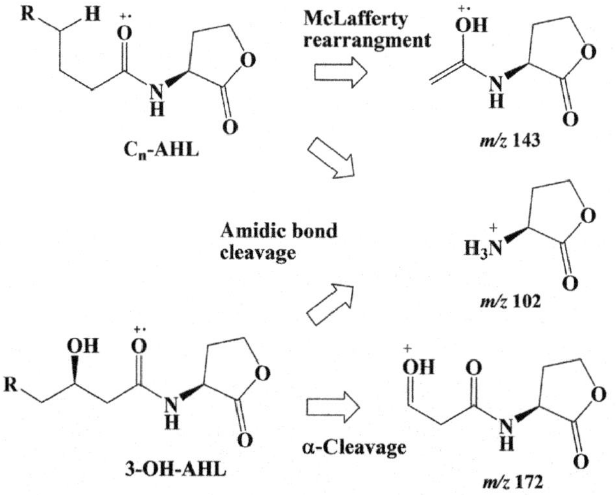

Fig. 4

Schematic electron impact fragmentation of AHL molecules generating diagnostic ions at m/z 143 and 102 for saturated, unsubstituted C_n-AHL and at m/z 172 and 102 for 3-OH-AHLs.

prominent fragment at m/z 143 and/or 172. This latter acquisition mode exploits a unique electronic feature of quadrupole-based mass analyzer that works as a mass filter. Ions formed in the source are accumulated in the analyzer, and only those presenting a defined m/z ratio are selected and allowed to reach the detector. A peak is showed only when the preset m/z ion is above a sensitivity threshold, but since other ions are not monitored, all the acquisition time is dedicated to listed ions, with substantial improvement of signal-to-noise ratio; this fact, together with fast scan rate, provides a hundredfold increase in sensitivity in SIM rather than in full scan acquisition mode.

Methods have been reported for specific GC-MS analysis of labile 3-oxo-AHLs after their conversion into the corresponding, more stable pentafluorobenzyloxime (PFBO) derivatives (Charlton et al., 2000); however, in a few protocols, all AHLs were analyzed without employing any chemical derivatization and monitoring the most abundant and diagnostic fragments in SIM-mode (Cataldi et al., 2004, 2007; Celio et al., 2006; Chi et al., 2017; Pomini and Marsaioli, 2008; Ran et al., 2016; Rani et al., 2011). Branched AHLs show mass spectra similar to those of the corresponding linear homoserine lactones, although they exhibit elution time shorter than unbranched counterparts. However, their identification can be accomplished through GC-MS analysis by using an empirical model that was developed to calculate the position of the methyl branch on a long alkyl chain in different lipids by using a retention index value (Schulz, 2001). According to this model, for the first time $isoC_9$-HSL and 3-OH-$isoC_9$-HSL were identified in *Aeromonas culicicola*. Furthermore, a close inspection of higher mass region revealed the presence of a $[M-43]^+$ ion, diagnostic of an alkyl chain methyl branched at position ω-1 (Thiel et al., 2009a). Chromatographic separations were usually carried out on fused silica capillary column 20–30 cm long by using Helium as carrier gas and temperature gradients from 100 to 300°C. The quantification was in some cases performed by using C_7-HSL as internal standard (Cataldi et al., 2007), but the occurrence of odd-chain AHLs has been documented in few bacterial species (Dickschat, 2010), thus resulting in a definitively more suitable deuterated analog as an internal standard reference (Gould et al., 2006). The field of application ranged from ecological studies on algicidal activity of marine bacteria *Ponticoccus* sp. associated with microalgae (Chi et al., 2017), to biomedical screening for the identification of *Pseudomonas aeruginosa* AHL-marker in sputum samples (Rani et al., 2011), to chemical characterization of novel structural variants of homoserine lactones (Thiel et al., 2009a).

GC-MS analysis has been also used to assign absolute configuration of chiral carbons in AHLs. All molecules of this group share a stereogenic center represented by the C-3 of the lactone ring, while 3-OH-AHL derivatives exhibit an additional asymmetric center at the hydroxylated position of the alkyl side chain. Chiral analysis was addressed by GC with flame ionization detector (FID) analysis on a chiral column containing nonpolar β-cyclodextrin, by comparison with synthetic standards (Pomini and Marsaioli, 2008; Malik et al., 2009). Interestingly, D-homoserine acyl lactones were found along with L-derivatives in the *Burkholderia cepacia* LA3 strain (Malik et al., 2009) by using a novel single drop microextraction technique, suggesting that chiral analytical technologies are needed to differentiate enantiomers with putatively different biological activities. On the other hand, the absolute configuration at C-3′ was determined by acidic methanolysis of the 3-OH-AHL and release of the corresponding fatty acid methyl ester; chromatographic analysis on the same chiral GC phase by comparison with racemic and enantiomerically pure synthetic standards was carried out to assign carbon stereogenicity of the natural compounds (Thiel et al., 2009a). Double bond localization in mono- and less common di-unsaturated analogues was accomplished by GC-MS analysis of their dimethylsulfide derivatives (DMDS) (Wagner-Döbler et al., 2005; Thiel et al., 2009a): the preferential α-cleavage between the methyltio groups allows to assign the double bond position straightforward (Fig. 5). Olefin geometry of the distal unsaturation was assigned by comparison with *E/Z* standards prepared by selective synthesis, assuming a *trans* configuration at C-2 as commonly found in similar acids of biological origin.

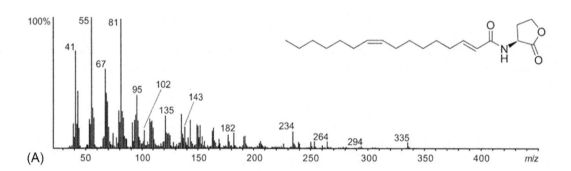

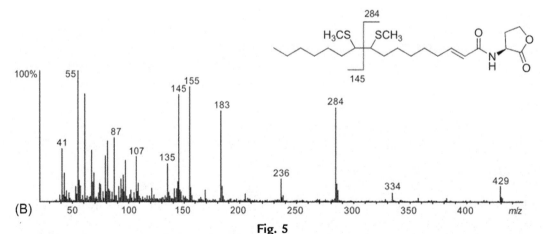

Fig. 5

EI-MS spectra of: A) $C_{16:2}$-AHL, and B) DMDS adduct of $C_{16:2}$-AHL for internal double bond localization. *Modified from Thiel et al. 2009a.*

3.2.2 Liquid chromatography-mass spectrometry (LC-MS)

Compared with other analytical techniques, liquid chromatography online to mass spectrometry (LC-MS) has proven to be the most reliable and sensitive approach for QS molecule detection, identification, and quantitative measurements. Under the general definition of LC-MS methodologies are encompassed different and peculiar technological solutions, which offer different degrees of sensitivity, mass resolution, and accuracy. Thus they may meet and satisfy different scientific requests according to the field of applicability. The two hyphenated analytical blocks (i.e., the chromatographic system and the mass spectrometric instrument) can be configured, tailored, and interfaced for a target analysis according to the output one is expected to receive. Up to 2004, the hyphenation of liquid chromatography to mass spectrometers was realized only by means of high performance liquid chromatography (HPLC) systems. HPLC is a well-known and largely diffused technique for the separation of a complex mixture into its individual components, in virtue of their different affinity between a mobile, liquid phase over a stationary, solid phase and measured as time of elution. The solid phase is typically packed in steel columns of different length and diameter, and since its first application now permits to resolve an impressive range of analytes belonging to many chemical classes. Although many technological modifications have been introduced, all stationary phases can be ascribed to one of these three types of interaction: (1) size exclusion, (2) ion exchange, and (3) adsorption chromatography. The latter can be carried out in the normal phase (NP) or the in reversed phase (RP). In NP chromatography, the stationary bed is higher polar silica or silica-based compositions, and the eluent phase is predominantly constituted by nonpolar solvent (e.g., *n*-hexane). It is appropriate for the separation of lipophilic compounds. Conversely, RP chromatography relies on a nonpolar stationary phase eluted by a mixture of polar solvents, including methanol, acetonitrile, and water. This phase suitable for separation of polar to low polar compounds was typically chosen for AHL analysis, by using methanol/water or acetonitrile/water mobile phases in isocratic or gradient elution programs. Like LC component, MS part is available in several different technological solutions too. The component that mostly affects the pairing of LC with MS is the interface, which is the place where molecules are transformed in ions. Typically, in LC-MS the interface is heated and allows ionization to occur in an electric field at atmospheric pressure (API). This can be obtained in two ways—by chemical ionization (APCI) and by electrospray ionization (ESI). ESI is by far the most widespread mode, being suitable for a large variety of chemical substances, including acyl homoserine lactones. It is a soft ionization, which means that only the molecular ion is formed in source; hence, on one hand, the molecular mass and formula (in accurate mass measurements) is easily calculated, but on the other hand, to obtain structural information it is necessary to take a further step of mass fragmentation (tandem MS or MS/MS), which generally occurs in a distinct place and at a different time during the mass spectrometric analysis, although it is possible to force in-source fragmentation as well (CID).

The core of a mass spectrometer is constituted by the analyzer. The most common analyzers are those based on quadrupole, time of flight, and ion-trap. Without going into deeper technical description, for most of targeted and quantitative approaches and in all cases when sensitivity is the crucial point such as for QS detection, quadrupole-based analyzers are the preferred choice, especially when three quadrupole units are coupled in series (QqQ, triple quadrupole). Furthermore, QqQ offers high speed of analysis due to its extremely fast scan, although mass resolution is generally low (one mass unit). The best LC-MS coupling was obtained with the introduction in 2004 of ultra performance liquid chromatography (UPLC). UPLC technology, based on sub-2 μm particles, paved the way to a new era in chromatographic separations, allowing shortening time of analytical runs, but still retaining optimal peak resolution. An example of UPLC application in QS analysis was provided by the work of Fekete et al. (2007). In the strategy adopted for AHL analysis, they included a prepurification step of the bacterial supernatant by RP-18 SPE cartridges, followed by UPLC analysis on a BEH C18 column eluted with a mixture of water and acetonitrile. The gradient used was very fast, from 10% to 100% ACN in only 1 min, holding at 100% ACN in isocratic elution; nevertheless, the resolution among homologs member of saturated and unsubstituted AHLs was appreciable. Furthermore, the retention times for most AHLs were predicted because of their theoretical log$P$ values. However, to confirm AHL identity in a bacterial extract, they followed an integrated offline approach resulting in additional FTICR-MS (Fourier-transform ion cyclotron resonance-mass spectrometry) measurements of the protonated $[M+H]^+$ molecular ions. FT-ICRMS is an ultra high-resolution mass spectrometry technique that can grant resolution at 1,000,000 FWHM (full width at half maximum) and mass resolution below 0.5 ppm. This means that known and unknown metabolites can be straightforward chemically characterized in complex mixtures; FTICR-MS was therefore used to identify the major AHL product of the bacterium *Acidovorax sp.* N35 (e.g., 3-OH-C$_{10}$-AHL). However, since the coupling with FTICR-MS is not technically feasible due to different scan rates, they developed an at-line coupling for a rapid screening based on chip for a nano-electrospray ionization (Li et al., 2007). The same UPLC method, however, was successfully applied by using an easier-to-manage online platform UPLC-QqQ (Abbamondi et al., 2016), exploiting the MRM acquisition scan mode for AHL detection. Multiple reaction monitoring (MRM) is one of the most powerful tandem mass-based approaches for the identification of target metabolites with high level of sensitivity and specificity. In an MRM analysis, data acquisition is triggered only when a selected fragmentation occurs. Thus the first and the third quadrupoles work as mass filters, selecting the molecular ion and the expected product ion, respectively; the second, intermediate quadrupole serves as collision chamber, and is the space where a controlled fragmentation of the molecular, parent ion takes place. Only when a product ion is generated from a parent ion as previously set is a signal acquired and registered by the detector; this behavior, together with the fast scanning properties of the quadrupole analyzer, is reflected in an impressive increase of selectivity and sensitivity for the monitored ion pair transitions.

In the paper of Abbamondi et al. (2016), the diagnostic transitions for AHL analysis were those generating the homoserine lactone fragment at *m/z* 102, and accordingly homologous series of synthetic standard derivatives, including unsubstituted, hydroxyl, and oxo-AHLs were processed in the UPLC conditions reported by Fekete et al. (2007) (Fig. 6, Grauso and Cutignano, unpublished data).

Based on a combined approach relying on the use of biosensors and mass spectrometry, AHL-related QS activity was detected in the culture media of the halophilic bacterium *Halomonas smyrnensis* AAD6 by TLC overlay and unequivocally identified and ascribed to an unusual long-chain C_{16}-HSL by UPLC-MS/MS (MRM) analysis (Abbamondi et al., 2016).

According to the same basic principle, triple quadrupole mass analyses in MRM acquisition mode were employed in several other research works, by using similar UPLC chromatographic supports from different manufacturers but variable column dimensions and chromatographic run lengths (Chan et al., 2014, 2016) or by traditional HPLC on RP-18 (Kumari et al., 2008; Morin et al., 2003; Tan et al., 2014) or more polar phases, such as Atlantis T3 (Sun et al., 2018).

An alternative to triple quadrupole comes in the form of hybrid configurations such as quadrupole-linear ion trap mass spectrometers (QqQLIT). These instruments offer the benefit to combine in a single LC run the advantages of both conventional triple quadrupole and linear ion trap (LIT) analyzers. Therefore, in addition to usual scan modes of triple quadrupole, including MRM but also product ion (PI), precursor ion (PC), and neutral loss (NL) scan modes, the high ion storage capacity, scan rate, and resolution of linear ion trap are exploited, thus allowing getting structural information even on unknown compounds. In fact, when the third quadrupoles are operating as a LIT, it is possible to acquire high quality product ion spectra useful for structural identification. This HPLC-QqQLIT technique was applied to determine the

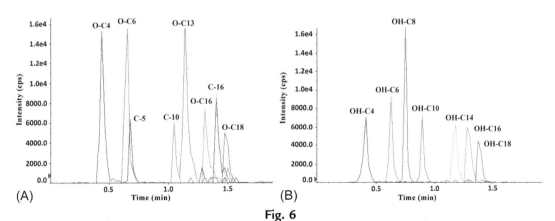

Fig. 6
Representative UPLC-ESI⁺-MS/MS (MRM) profiles of standard AHLs by using a BEH-C18 column with water/acetonitrile gradient. (A) Pool of five oxo-AHLs and three C-AHLs; (B) pool of seven OH-AHLs.

profile and the relative abundance of 24 AHLs in *Yersinia pseudotuberculosis* that therefore resulted the bacterium with the broadest spectrum of AHLs known to be produced by a single species (Ortori et al., 2007).

3.2.3 Capillary electrophoresis-mass spectrometry (CE-MS)

Capillary electrophoresis (CE) on uncoated fused silica capillaries coupled with mass spectrometry is an analytical technique that offers a few attractive advantages: separation efficiency, detection sensitivity, and chemical identification. A major drawback is, however, represented by the compatibility of buffer solutions required for electrophoretic separation with the mass spectrometric ionization process by ESI. In fact, electrophoresis separations are intrinsically grounded on differential migration of charged species in an electrical field, and as such, the buffer medium used for chromatographic separation has great influence on pH stability and difference in mobility. On the other hand, common buffers used in CE (e.g., phosphate and TRIS buffers) are not compatible with ESI that imposes volatile buffers only. In addition, a high molarity generally improves buffering capacity of the system, but again ESI works better with buffer concentration at about 10 mM. To further complicate things, AHLs do not exhibit chargeable groups, being the secondary amine stabilized by the amidic linkage. Thus a way to overcome this aspect was proposed by means of micellar electrokinetic chromatography (MEKC) (Frommberger et al., 2003). Best results were obtained by using sodium dodecyl sulfate (SDS) 10 mM for micelles formation with partial filling of the capillary with the micellar solution (PF-MEKC). Two AHLs were detected in *Burkholderia cepacia* in a relative short time analysis (20 min), but no quantitative data could be obtained with the proposed analytical tool. Technical improvements were then achieved analyzing AHLs in *Pseudomonas fluorescens* culture supernatants by CE-MS after basic hydrolysis of the lactone ring. Separations were performed in 20 mM ammonium carbonate buffer to preserve MS interface, and quantitative measurements were obtained via C7-HSL internal standard and sample preconcentration by anion exchange-SPE (Frommberger et al., 2005).

3.2.4 Matrix-assisted laser desorption ionization-mass spectrometry (MALDI-MS)

MALDI-MS-based methods have been traditionally applied to the identification of biomolecules (peptides/proteins, oligo/polysaccharides, nucleic acids), which are revealed as molecular ions with no fragmentation occurring during the ionization process. From a technical point of view, the sample is mixed with a defined amount of a matrix [e.g., α-cyano-4-hydroxycinnamic acid (α-CHCA) and 2,5-dihydroxybenzoic acid (DHB)] and allowed to co-crystallize onto a plate. Ionization takes place by irradiation with a laser light that quickly heats the matrix, which in turn vaporizes together with the sample. It is commonly accepted that MALDI was originally not designed for the ionization of small molecules (below 500 Da) because of the interference due to peaks from matrix. To address this issue, several methods have been reported (Zhang et al., 2010). Nevertheless, 3-oxo-AHL molecules were identified by MALDI-TOF-MS as such in bacterial extracts of *Pantoea agglomerans* (Jiang et al., 2015) or quantified after derivatization with Girard's reagent in *P. aeruginosa* (Kim et al., 2015).

3.2.5 Nuclear magnetic resonance spectroscopy

Nuclear magnetic resonance (NMR) is the primary technique used to elucidate organic structures. Unfortunately, its detection limit is at mM scale; thus it is in a far higher concentration range than MS-based techniques. The consequence is that it requires likely high volumes of bacterial culture for AHL extraction and purification before the acquisition of NMR spectra of the isolated pure compound (Pearson et al., 1994). Although fundamental at the beginning of the story to structurally characterize QS signals and to recognize novel molecular skeletons (Doberva et al., 2017; Shen et al., 2016), its role moves to a secondary level in analytical approaches aiming at identify and quantify known autoinducers.

3.3 Analytical Approaches for Identification and Quantification of Other QS Molecules

Autoinducing peptides (AIPs) from virulent Gram positive *Staphylococcus aureus* control the secretion of enterotoxins belonging to the family of superantigens: these proteic toxins trigger abnormal activation of T cells, producing a series of clinical symptoms ranging from atopic dermatitis to toxic shock syndromes. Notably, AIPs from one group cross-inhibit gene expression in strains from other groups, thus a method for detection and differentiation of AIPs resulted of clinical relevance in infection and expected to be of diagnostic utility. So, in a study carried out by Kalkum et al. (2003), a multistage mass spectrometry method was developed with a matrix-assisted laser desorption ionization (MALDI)-quadrupole ion trap instrument, implementing a traditional approach to study peptide molecules relying on MALDI ion sources (Chaurand et al., 1999). In a first case, it was used following a hypothesis-driven approach. In fact, AIPs are synthesized as large propeptides and successively hydrolyzed to mature QS-peptides during excretion (Sturme et al., 2002); thus the knowledge of the DNA sequence allows for the prediction of all the possible aminoacidic sequences of shorter peptides originating from the cleavage of the biosynthesized precursor; the m/z values of these putative peptides can be searched by target screening in the bacterial supernatants and listed for MS^2 (MS/MS) and MS^3 (MS/MS/MS) confirmation experiments. In a second case, it was applied to assign the unknown structure of AIP of *Staphylococcus intermedius* as a nonapeptide lactone. In both cases, the ability of a MALDI-ion trap to isolate specific ions from noisy matrices and convert them into fragments registered with lower background interference by using a tiny amount of samples in a short time was fully exploited. Nevertheless, the limit of conventional MALDI mass spectrometry is in its qualitative information for compound identification, being not as reliable as other mass spectrometric platforms for quantitative purposes. To overcome this aspect, recently it has been proposed for peptide analysis the application of a hybrid configuration constituted by a combination of quadrupole or linear ion trap and orbitrap mass analyzers online with UPLC to ensure short time (10 min or less) for the chromatographic runs

(Junio et al., 2013; Olson et al., 2014; Todd et al., 2016). Therefore, a series of putative mature AIPs from *Staphylococcus epidermidis* and methicillin-resistant *Staphylococcus aureus* (MRSA) strains were predicted based on the gene sequence of the putative longer precursor peptide (*agrD* gene), and then their corresponding *m/z* ions, matching the exact mass, were searched in the active fractions, which were selected by bioassay reporter strain. Using this strategy, different AIPs were detected and their aminoacidic sequence established by tandem MS experiments and by synthesis, as well (Olson et al., 2014). This methodology was used to monitor the time-dependent production of AIP-I from *S. aureus* and to compare its levels in different strains (Junio et al., 2013) as well as to assess the inhibition of AIP production by quantitative measurement of the peptide in the bacterial filtrate under different antibiotic treatments (Todd et al., 2016).

Small cyclic dipeptides (diketopiperazines, DKPs) with the ability to activate a LuxR-based AHL biosensor have been reported from Gram negative bacteria, especially from the marine environment (Abbamondi et al., 2014; Holden et al., 1999; Tommonaro et al., 2012). Their molecular identification can be performed by GC-MS analysis carried out on underivatized molecules and confirmed by chemical synthesis, as reported by Gu et al. (2013).

GC-MS and HPLC-MS/MS on a triple quadrupole platform were both exploited to direct or postderivatization quantitation of AI-2 signals. Considering the chemical nature of this class of autoinducers, constituted by a mixture of interconverting borated and unborated molecules, the identification of the active components of the pool has been always difficult. Besides their low abundance, one of the main complications was the hydrophilic nature of these compounds that prevented both the use of organic solvent extraction for sample concentration and chemical reaction in lipophilic media to convert them into less polar derivatives. Hence, alternative strategies were developed to derivatize dihydroxy pentanedione (DPD) by reaction with 1,2-phenylenediamine in aqueous buffer to obtain its (3-methylquinoxalin-2-yl)ethane-1,2-diol derivative (De Keersmaeckert et al., 2005). This product was originally identified directly by HPLC-MS/MS (Hauck et al., 2003) or, after further derivatization of the diol moieties with *N*-methyl-*N*-(trimethylsilyl)trifluoroacetamide (MSTFA), by GC-MS (Thiel et al., 2009b). This latter step was necessary to make the quinoxaline diol derivative volatile and suitable for GC-MS analysis. Applying this procedure and by using an isotopically labeled standard analog, a quantitative methodology was developed and validated in *Vibrio harveyi* BB152 and *Streptococcus mutans* UA159, exhibiting a sensitivity comparable to biosensor systems. Recently, improved methods for postderivatization AI-2 quantitative determination by HPLC coupled to tandem mass spectrometry on QqQ have been also proposed (Campagna et al., 2009; Xu et al., 2017).

QS molecules of the quinolone family (alkyl quinolone, AQ), including 4-hydroxy-2-alkylquinoline and their *N*-oxides, have been identified by both GC-electron capture MS and LC-MS (Lépine et al., 2004; Taylor et al., 1995; Vial et al., 2008). Due to the thermal instability and poor volatility in the experimental conditions, GC-MS methodology required conversion of the natural compounds into *bis*-trifluoromethylbenzoyl derivatives, which have long been used to identify and quantify a range of hydroxylated and nitrogenous compounds, and for *N*-oxides, further reduction to the parent hydroxyquinoline. A linear relationship was disclosed between the alkyl chain length and the retention time, which was convenient when searching for low abundance compounds. As usual, reported LC-MS analytical methods did not require any derivatization, instead. Both approaches have been used for AQs quantitation with deuterated internal standards. Interestingly, a simultaneous identification of components of both AQ and AHLs families in *P. aeruginosa* was developed by a targeted HPLC-ESI-MS/MS (MRM) procedure (Ortori et al., 2011). Undoubtedly, this is an advantage when interest lies in a broad array of QS signals, offering a clear-cut profile of the bacterial signaling response. Just recently, a miniaturized platform for AQs identification and quantification was proposed, relying on dispersive liquid-liquid microextraction and MALDI-MS with ionic liquid matrices (ILM) (Leipert et al., 2018). This latter constitutes an alternative matrix for MALDI sample preparations that allows one to overcome the inhomogeneous distribution of analytes and matrices in conventional solid samples, thus resulting in more reliable quantitative measurements (Zabet-Moghaddam et al., 2004). Although it is not able to detect AHLs, it is presented as a fast, reproducible, and straightforward method for simultaneous identification of diverse signaling molecules, toxins, and virulence factors (e.g., AQs and pyocyanin) in clinical samples and validated on cystic fibrosis patients' biological samples as a tool for basic research or diagnostic purposes.

4 Concluding Remarks

Quorum sensing (QS) communication in bacteria is one of the most intriguing discoveries in the past two decades. Despite the simple chemistry that drives the molecular message among microorganisms and between these and the host, a large analytical effort is required for the detection, identification, and measurement of the levels of all these different low-molecular-weight metabolites. The tremendous progress over the past years in technological platforms, especially in the field of mass spectrometry, has opened the doors to a new era in chemical and biochemical studies for a deeper understanding of the mechanisms underlying organism interactions and their regulation. In fact, QS elicits a global metabolic readjustment inducing secondary metabolites biosynthesis normally encoded by silent gene clusters that may ultimately lead to the discovery of novel natural products. MS-based metabolomics as an analytical approach is a way to monitor these changes under the influence of different molecular QS signals, in both intra- and interspecies communication. Few studies have been reported so far, but more are expected to address this topic (Davenport et al., 2015; Tang et al., 2018). The importance of QS-mediated metabolic network regulation lies in the fact that it can cross the boundaries of the microbial realm and involve interkingdom cross-talk, shaping both nonpathogenic and pathogenic bacteria-host interactions. A deep knowledge on how host cells respond to external cues like QS molecules can pave the way to designing novel cell-based therapeutics (McNerney and Styczynski, 2018). In general, a deeper and wider identification of the chemical basis of QS signaling not only in bacteria but also in other microorganisms (e.g., fungi and microalgae) promises to have a huge impact in ecology, agriculture, and medicine.

Glossary

Capillary electrophoresis (CE) is an analytical technique relying on narrow-bore (20–200 μm ID) capillaries to separate with high efficiency both small and large molecules. Separations are achieved by application of high voltages, which promote electroosmotic and electrophoretic flow of buffer solutions and ionic species within the capillary.

Electrospray ionization (ESI) is a "soft" ionization technique used in mass spectrometry to produce molecular ions by nebulizing a liquid under high voltages without or with little fragmentation.

Fourier-transform ion cyclotron resonance-mass spectrometry (FTICR-MS) is a mass analyzer based on the cyclotron frequency of the ions placed in a fixed magnetic field. It is characterized by high resolution and mass accuracy of 1–2 ppm.

Gas chromatography-mass spectrometry (GC-MS) is a hyphenated analytical technique that combines the high resolving power of gas chromatography with sensitive detection of mass spectrometry. It is suitable for separation of volatile and thermally stable molecules detected by virtue of their mass-to-charge (*m/z*) ratio value.

High performance liquid chromatography (HPLC) is an analytical technique used to separate, identify, and quantify single components in complex mixtures. It requires a pumping apparatus to allow the solvent containing the sample to pass through a column packed with an adsorbent material. Different solvent compositions (mobile phase) and adsorbent materials (stationary phase) allow separation of different components, eluting from the column at different times according to their chemical properties and specific interaction with the stationary phase.

Liquid chromatography-mass spectrometry (LC-MS) is an analytical technique where liquid chromatography is online to mass spectrometry. It represents one of the most reliable and sensitive approaches for molecule detection, identification, and quantitative measurements. Under the general definition of LC-MS methodologies are encompassed different technological solutions, which offer different degrees of sensitivity, mass resolution, and accuracy. When liquid chromatography is coupled to a multistage mass analyzer (LC-MS/MS), it allows the detection of product ions generated by collision-induced fragmentation.

Matrix-assisted laser desorption/ionization (MALDI) is a "soft" ionization technique used in mass spectrometry to obtain ions from macromolecules with minimal fragmentation. It uses a pulsed laser to irradiate the sample mixed with a suitable matrix.

Mass spectrometry (MS) is a powerful analytical technique used to measure the mass to charge (m/z) ratio of ions. Its basic instrumental configuration encompasses an ion source, an analyzer, and a detector system. It is used to identify unknown compounds within a sample, to elucidate the structure and chemical properties of different molecules, and to quantify known materials.

Multiple reaction monitoring (MRM) is a multistage experimental design used with a triple quadrupole (QqQ) mass analyzer to specifically detect and simultaneously quantify hundreds of target analytes in mixture with high sensitivity. It requires a preset of precursor and fragment ions linked in a fragmentation reaction to detect and register a signal.

Nuclear magnetic resonance spectroscopy (NMR) is a spectroscopic technique that exploits the magnetic properties of some nuclei (i.e., ^{1}H and ^{13}C), allowing to gather structural information on organic molecules by measuring the absorption of electromagnetic radiation when a sample is placed in a strong magnetic field.

Selected ion monitoring (SIM) is a single-stage mass experiment in which one or more ions with specific mass-to-charge (m/z) ratios are monitored. In the SIM acquisition mode, the ions formed in the source are accumulated in the analyzer, which acts as a filter and allows only those with the selected m/z ratio to be transferred to the detector.

Solid phase extraction (SPE) is a sample preparation process by which compounds in a mixture are dissolved or suspended in a liquid phase and separated from each other by using an adsorbent material, according to their physical and chemical properties. It is used to extract or to concentrate and purify samples for analysis from a wide variety of matrices, including biological fluids (urine, blood), bacterial cultural media, vegetal and animal tissues, and soil.

Thin layer chromatography (TLC) is a traditional chromatographic technique used to separate nonvolatile compounds in mixture. It is carried out on a glass plate or aluminium foil covered with a stationary phase (typically silica), loaded at a base point with the sample and developed in a chromatographic chamber containing an elution mixture (mobile phase). The different retardation factor (Rf)—that is, the distance traveled with respect to the solvent—of the various components depends on their chemical properties and affinity for the mobile phase.

Triple Quadrupole (QqQ) is a mass analyzer configured with three quadrupoles in series. It is suitable for highly sensitive and targeted quali-quantitative analysis of metabolites in mixtures through the application of different experimental designs, including MRM.

Ultra Performance Liquid Chromatography (UPLC/UHPLC) is a modern technique that improves performance of HPLC with respect to speed, resolution, and sensitivity. It relies on columns packed with particle less than 2 μm in diameter and requires higher pressure (15–18 kpsi) than HPLC. It is particularly suitable for coupling with modern mass spectrometers.

Abbreviations

AHL	*N*-acyl-L-homoserine lactones
AIP	autoinducing peptide
AQ	alkyl quinolone
CE	capillary electrophoresis
ESI	electrospray ionization
FTICR-MS	Fourier-transform ion cyclotron resonance-mass spectrometry
FWHM	full width at half maximum
GC-FID	gas chromatography-flame ionization detector
GC-MS	gas chromatography-mass spectrometry
HPLC	high performance liquid chromatography
HSL	homoserine lactone
LC-MS	liquid chromatography-mass spectrometry
LC-MS/MS	liquid chromatography-tandem mass spectrometry
LLE	liquid/liquid extraction
MALDI	matrix assisted laser desorption/ionization
MRM	multiple reaction monitoring
MS	mass spectrometry
NMR	nuclear magnetic resonance
QqQ	triple quadrupole
QqQLIT	triple quadrupole-linear ion trap
QS	quorum sensing
Rf	retention factor
SIM	selected ion monitoring
SPE	solid phase extraction
TOF	time of flight
TLC	thin layer chromatography
UPLC	ultra performance liquid chromatography

References

Abbamondi, G.R., De Rosa, S., Iodice, C., Tommonaro, G., 2014. Cyclic dipeptides produced by marine sponge-associated Bacteria as quorum sensing signals. Nat. Prod. Commun. 9 (2), 229–232.

Abbamondi, G.R., Suner, S., Cutignano, A., Grauso, L., Nicolaus, B., Toksoy Oner, E., Tommonaro, G., 2016. Identification of *N*-Hexadecanoyl-L-homoserine lactone (C16-AHL) as signal molecule in halophilic bacterium *Halomonas smyrnensis* AAD6. Ann. Microbiol. 66(3).

Andersen, J.B.O., Heydorn, A., Hentzer, M., Eberl, L.E.O., Geisenberger, O., Molin, R.E.N., Givskov, M., Christensen, B.B.A.K., 2001. Gfp-based *N*-acyl Homoserine-lactone sensor systems for detection of bacterial communication. Appl. Environ. Microbiol. 67 (2), 575–585.

Andrade-Eiroa, A., Canle, M., Leroy-Cancellieri, V., Cerdà, V., 2016. Solid-phase extraction of organic compounds: a critical review. Part ii. TrAC: Trends Anal. Chem. 80, 655–667.

Bassler, B.L., Greenberg, E.P., Stevens, A.M., 1997. Cross-species induction of luminescence in the quorum-sensing bacterium *Vibrio harveyi*. J. Bacteriol. 179 (12), 4043–4045.

Bofinger, M.R., De Sousa, L.S., Fontes, J.E.N., Marsaioli, A.J., 2017. Diketopiperazines as cross-communication quorum-sensing signals between *Cronobacter sakazakii* and *Bacillus cereus*. ACS Omega 2 (3), 1003–1008.

Camilli, A., 2006. Bacterial small-molecule signaling pathways. Science 311 (5764), 1113–1116.

Campagna, S.R., Gooding, J.R., May, A.L., 2009. Direct quantitation of the quorum sensing signal, Autoinducer-2, in clinically relevant samples by liquid chromatography-tandem mass spectrometry. Anal. Chem. 81 (15), 6374–6381.

Cataldi, T.R.I., Bianco, G., Frommberger, M., Schmitt-Kopplin, P., 2004. Direct analysis of selected N-acyl-L-homoserine lactones by gas chromatography/mass spectrometry. Rapid Commun. Mass Spectrom. 18 (12), 1341–1344.

Cataldi, T.R.I., Bianco, G., Palazzo, L., Quaranta, V., 2007. Occurrence of *N*-acyl-L-homoserine lactones in extracts of some gram-negative bacteria evaluated by gas chromatography-mass spectrometry. Anal. Biochem. 361 (2), 226–235.

Celio, S., Troxler, H., Durka, S.S., Chládek, J., Wildhaber, J.H., Sennhauser, F.H., Heizmann, C.W., Moeller, A., 2006. Free 3-nitrotyrosine in exhaled breath condensates of children fails as a marker for oxidative stress in stable cystic fibrosis and asthma. Nitric Oxide Biol. Chem. 15 (3), 226–232.

Chan, K.-G., Cheng, H.J., Chen, J.W., Yin, W.-F., Ngeow, Y.F., 2014. Tandem mass spectrometry detection of quorum sensing activity in multidrug resistant clinical isolate *Acinetobacter baumannii*. Sci. World J. 891041.

Chan, X.-Y., How, K.-Y., Yin, W.-F., Chan, K.-G., 2016. *N*-acyl Homoserine lactone-mediated quorum sensing in *Aeromonas veronii* biovar *sobria* strain 159: Identification of LuxRI homologs. Front. Cell. Infect. Microbiol. 6, 1–6.

Charlton, T.S., de Nys, R., Netting, A., Kumar, N., Hentzer, M., Givskov, M., Kjelleberg, S., 2000. A novel and sensitive method for the quantification of *N*-3-oxoacyl homoserine lactones using gas chromatography-mass spectrometry: application to a model bacterial biofilm. Environ. Microbiol. 2 (5), 530–541.

Chaurand, P., Luetzenkirchen, F., Spengler, B., 1999. Peptide and protein identification by matrix-assisted laser desorption ionization (MALDI) and MALDI-post-source decay time-of-flight mass spectrometry. J. Am. Soc. Mass Spectrom. 10 (2), 91–103.

Chi, W., Zheng, L., He, C., Han, B., Zheng, M., Gao, W., Sun, C., Zhou, G., Gao, X., 2017. Quorum sensing of microalgae associated marine *Ponticoccus* sp. PD-2 and its algicidal function regulation. AMB Express Springer Berlin Heidelberg, 7 (1).

Curtis, M.M., Sperandio, V., 2011. A complex relationship: the interaction among symbiotic microbes, invading pathogens, and their mammalian host. Mucosal Immunol. 4 (2), 133–138.

Cutignano, A., Nuzzo, G., Ianora, A., Luongo, E., Romano, G., Gallo, C., Sansone, C., Aprea, S., Mancini, F., D'Oro, U., Fontana, A., 2015. Development and application of a novel SPE-method for bioassay-guided fractionation of marine extracts. Mar. Drugs 13 (9), 5736–5749.

Davenport, P.W., Griffin, J.L., Welch, M., 2015. Quorum sensing is accompanied by global metabolic changes in the opportunistic human pathogen *Pseudomonas aeruginosa*. J. Bacteriol. 197 (12), 2072–2082.

De Keersmaeckert, S.C.J., Varszegi, C., Van Boxel, N., Habel, L.W., Metzger, K., Daniels, R., Marchal, K., De Vos, D., Vanderleyden, J., 2005. Chemical synthesis of (S)-4,5-dihydroxy-2,3-pentanedione, a bacterial signal molecule precursor, and validation of its activity in *Salmonella typhimurium*. J. Biol. Chem. 280 (20), 19563–19568.

Deng, Y., Wu, J., Tao, F., Zhang, L. H., 2011. Listening to a new language: DSF-based quorum sensing in gram-negative bacteria. Chem. Rev. 111 (1), 160–179.

Dickschat, J.S., 2010. Quorum sensing and bacterial biofilms. Nat. Prod. Rep. 27 (3), 343–369.

Doberva, M., Stien, D., Sorres, J., Hue, N., Sanchez-Ferandin, S., Eparvier, V., Ferandin, Y., Lebaron, P., Lami, R., 2017. Large diversity and original structures of acyl-Homoserine lactones in strain MOLA 401, a marine Rhodobacteraceae bacterium. Front. Microbiol. 8, 1–10.

Fekete, A., Frommberger, M., Rothballer, M., Li, X., Englmann, M., Fekete, J., Hartmann, A., Eberl, L., Schmitt-Kopplin, P., 2007. Identification of bacterial *N*-acylhomoserine lactones (AHLs) with a combination of ultra-performance liquid chromatography (UPLC), ultra-high-resolution mass spectrometry, and in-situ biosensors. Anal. Bioanal. Chem. 387 (2), 455–467.

Feng, H., Ding, Y., Wang, M., Zhou, G., Zheng, X., He, H., Zhang, X., Shen, D., Shentu, J., 2014. Where are signal molecules likely to be located in anaerobic granular sludge? Water Res. 50, 1–9.

Frommberger, M., Schmitt-Kopplin, P., Menzinger, F., Albrecht, V., Schmid, M., Eberl, L., Hartmann, A., Kettrup, A., 2003. Analysis of N-acyl-L-homoserine lactones produced by *Burkholderia cepacia* with partial filling micellar electrokinetic chromatography—electrospray ionization-ion trap mass spectrometry. Electrophoresis 24 (17), 3067–3074.

Frommberger, M., Hertkorn, N., Englmann, M., Jakoby, S., Hartmann, A., Kettrup, A., Schmitt-Kopplin, P., 2005. Analysis of N-acylhomoserine lactones after alkaline hydrolysis and anion-exchange solid-phase extraction by capillary zone electrophoresis-mass spectrometry. Electrophoresis 26 (7–8), 1523–1532.

Fuqua, C., Greenberg, E.P., 2002. Listening in on bacteria: acyl-homoserine lactone signalling. Nat. Rev. Mol. Cell Biol. 3 (9), 85–695.

Fuqua, W.C., Winans, S.C., Greenberg, E.P., 1994. Quorum sensing in Bacteria: the LuxR-LuxI family of cell density-responsive transcriptional regulators. J. Bacteriol. 176 (2), 269–275.

Galloway, W.R.J.D., Hodgkinson, J.T., Bowden, S.D., Welch, M., Spring, D.R., 2011. Quorum sensing in gram-negative bacteria: small-molecule modulation of AHL and AI-2 quorum sensing pathways. Chem. Rev. 111 (1), 28–67.

Goswami, J., 2017. Quorum sensing by super bugs and their resistance to antibiotics, a short review. Global J. Pharm. Pharmaceut. Sci. 3 (3), 1–7.

Gould, T.A., Herman, J., Krank, J., Murphy, R.C., Churchill, M.E.A., 2006. Specificity of acyl-homoserine lactone synthases examined by mass spectrometry. J. Bacteriol. 188 (2), 773–783.

Gu, Q., Fu, L., Wang, Y., Lin, J., 2013. Identification and characterization of extracellular cyclic dipeptides as quorum-sensing signal molecules from *Shewanella baltica*, the specific spoilage organism of *Pseudosciaena crocea* during 4°C storage. J. Agric. Food Chem. 61 (47), 11645–11652.

Hauck, T., Hübner, Y., Brühlmann, F., Schwab, W., 2003. Alternative pathway for the formation of 4,5-dihydroxy-2,3-pentanedione, the proposed precursor of 4-hydroxy-5-methyl-3(2H)-furanone as well as autoinducer-2, and its detection as natural constituent of tomato fruit. Biochim. Biophys. Acta Gen. Subj. 1623 (2–3), 109–119.

Hawver, L.A., Jung, S.A., Ng, W.L., 2016. Specificity and complexity in bacterial quorum-sensing systems. FEMS Microbiol. Rev. 40 (5), 738–752.

He, Y.W., Zhang, L.H., 2008. Quorum sensing and virulence regulation in *Xanthomonas campestris*. FEMS Microbiol. Rev. 32 (5), 842–857.

Hmelo, L.R., 2017. Quorum sensing in marine microbial environments. Annu. Rev. Mar. Sci. 9 (1), 257–281.

Holden, M.T.G., Ram Chhabra, S., de Nys, R., Stead, P., Bainton, N.J., Hill, P.J., Manefield, M., Kumar, N., Labatte, M., England, D., Rice, S., Givskov, M., Salmond, G.P.C., Stewart, G.S.A.B., Bycroft, B.W., Kjelleberg, S., Williams, P., 1999. Quorum-sensing cross talk: Isolation and chemical characterization of cyclic dipeptides from *Pseudomonas aeruginosa* and other gram-negative bacteria. Mol. Microbiol. 33 (6), 1254–1266.

Jiang, J., Wu, S., Wang, J., Feng, Y., 2015. AHL-type quorum sensing and its regulation on symplasmata formation in *Pantoea agglomerans* YS19. J. Basic Microbiol. 55 (5), 607–616.

Jones, S., Yu, B., Bainton, N.J., Birdsall, M., Bycroft, B.W., Chhabra, S.R., Cox, A.J.R., Winson, M.K., Salmond, G.P.C., Stewart, G.S., Williams, P., 1993. The lux autoinducer regulates the production of exoenzyme virulence determinants in *Erwinia carotovora* and *Pseudomonas aeruginosa*. EMBO J. 12 (6), 2477–2482.

Junio, H.A., Todd, D.A., Ettefagh, K.A., Ehrmann, B.M., Kavanaugh, J.S., Horswill, A.R., Cech, N.B., 2013. Quantitative analysis of autoinducing peptide I (AIP-I) from *Staphylococcus aureus* cultures using ultrahigh performance liquid chromatography—high resolving power mass spectrometry. J. Chromatogr. B 930, 7–12.

Kalkum, M., Lyon, G.J., Chait, B.T., 2003. Detection of secreted peptides by using hypothesis-driven multistage mass spectrometry. Proc. Natl. Acad. Sci. 100 (5), 2795–2800.

Kim, Y.-W., Sung, C., Lee, S., Kim, K.-J., Yang, Y.-H., Kim, B.-G., Lee, Y.K., Ryu, H.W., Kim, Y.-G., 2015. MALDI-MS-based quantitative analysis for ketone containing Homoserine lactones in Pseudomonas aeruginosa. Anal. Chem. 87 (2), 858–863.

Kleerebezem, M., Quadri, L.E., Kuipers, O.P., de Vos, W.M., 1997. Quorum sensing by peptide pheromones and two-component signal-transduction systems in gram-positive bacteria. Mol. Microbiol. 24 (5), 895–904.

Kumari, A., Pasini, P., Daunert, S., 2008. Detection of bacterial quorum sensing *N*-acyl homoserine lactones in clinical samples. Anal. Bioanal. Chem. 391 (5), 1619–1627.

Leipert, J., Bobis, I., Schubert, S., Fickenscher, H., Leippe, M., Tholey, A., Clark, K.D., 2018. Miniaturized dispersive liquid-liquid microextraction and MALDI MS using ionic liquid matrices for the detection of bacterial communication molecules and virulence factors. Anal. Bioanal. Chem. 410, 4337–4748.

Lépine, F., Milot, S., Déziel, E., He, J., Rahme, L.G., 2004. Electrospray/mass spectrometric identification and analysis of 4-hydroxy-2-alkylquinolines (HAQs) produced by *Pseudomonas aeruginosa*. J. Am. Soc. Mass Spectrom. 15 (6), 862–869.

Li, X., Fekete, A., Englmann, M., Götz, C., Rothballer, M., Frommberger, M., Buddrus, K., Fekete, J., Cai, C., Schröder, P., Hartmann, A., Chen, G., Schmitt-Kopplin, P., 2006. Development and application of a method for the analysis of *N*-acylhomoserine lactones by solid-phase extraction and ultra high pressure liquid chromatography. J. Chromatogr. A 1134 (1–2), 186–193.

Li, X., Fekete, A., Englmann, M., Frommberger, M., Lv, S., Chen, G., Schmitt-Kopplin, P., 2007. At-line coupling of UPLC to chip-electrospray-FTICR-MS. Anal. Bioanal. Chem. 389 (5), 1439–1446.

Malik, A.K., Fekete, A., Gebefuegi, I., Rothballer, M., Schmitt-Kopplin, P., 2009. Single drop microextraction of homoserine lactones based quorum sensing signal molecules, and the separation of their enantiomers using gas chromatography mass spectrometry in the presence of biological matrices. Microchim. Acta 166 (1–2), 101–107.

McClean, K.H., Winson, M.K., Fish, L., Taylor, A., Chhabra, S.R., Camara, M., Daykin, M., John, H., Swift, S., Bycroft, B.W., Stewart, G.S.a.B., Williams, P., 1997. Quorum sensing and *Chrornobacteriurn violaceum*: exploitation of violacein production and inhibition for the detection of *N*-acyl homoserine lactones. Microbiology 143, 3703–3711.

McNerney, M.P., Styczynski, M.P., 2018. Small molecule signaling, regulation, and potential applications in cellular therapeutics. Wiley interdisciplinary reviews: systems. Biol. Med. 10(2), e1405.

Morin, D., Grasland, B., Vallee-Rehel, K., Dufau, C., Haras, D., 2003. On-line high-performance liquid chromatography-mass spectrometric detection and quantification of *N*-acylhomoserine lactones, quorum sensing signal molecules, in the presence of biological matrices. J. Chromatogr. A 1002 (1–2), 79–92.

Olson, M.E., Todd, D.A., Schaeffer, C.R., Paharik, A.E., Van Dyke, M.J., Büttner, H., Dunman, P.M., Rohde, H., Cech, N.B., Fey, P.D., Horswill, A.R., 2014. *Staphylococcus epidermidis* agr quorum-sensing system: signal identification, cross talk, and importance in colonization. J. Bacteriol. 196 (19), 3482–3493.

Ortori, C.A., Atkinson, S., Chhabra, S.R., Cámara, M., Williams, P., Barrett, D.A., 2007. Comprehensive profiling of *N*-acylhomoserine lactones produced by *Yersinia pseudotuberculosis* using liquid chromatography coupled to hybrid quadrupole-linear ion trap mass spectrometry. Anal. Bioanal. Chem. 387 (2), 497–511.

Ortori, C.A., Dubern, J.-F., Chhabra, S.R., Cámara, M., Hardie, K., Williams, P., Barrett, D.A., 2011. Simultaneous quantitative profiling of *N*-acyl-L-homoserine lactone and 2-alkyl-4(1H)-quinolone families of quorum-sensing signaling molecules using LC-MS/MS. Anal. Bioanal. Chem. 399 (2), 839–850.

Osorno, O., Castellanos, L., A., F., ArvaloFerro, C., 2012. Gas Chromathography as a tool in quorum sensing studies, in Salih, B. (ed.) Gas Chromatography—Biochemicals, Narcotics and Essential Oils. InTech, 67–96.

Papenfort, K., Bassler, B.L., 2016. Quorum sensing signal-response systems in gram-negative bacteria. Nat. Rev. Microbiol. 14 (9), 576–588.

Pearson, J.P., Gray, K.M., Passador, L., Tucker, K.D., Eberhard, A., Iglewski, B.H., Greenberg, E.P., 1994. Structure of the autoinducer required for expression of *Pseudomonas aeruginosa* virulence genes. Proc. Natl. Acad. Sci. 91 (1), 197–201.

Pesci, E.C., Milbank, J.B.J., Pearson, J.P., McKnight, S., Kende, A.S., Greenberg, E.P., Iglewski, B.H., 1999. Quinolone signaling in the cell-to-cell communication system of *Pseudomonas aeruginosa*. Proc. Natl. Acad. Sci. 96 (20), 11229–11234.

Pomini, A.M., Marsaioli, A.J., 2008. Absolute configuration and antimicrobial activity of acylhomoserine lactones. J. Nat. Prod. 71 (6), 1032–1036.

Ran, T., Zhou, C.-i., Xu, L.-w., Geng, M.-m., Tan, Z.-l., Tang, S.-x., Wang, M., Han, X.-f., Kang, J.-h., 2016. Initial detection of the quorum sensing autoinducer activity in the rumen of goats in vivo and in vitro. J. Integr. Agric. 15 (10), 2343–2352.

Rani, S., Kumar, A., Malik, A.K., Schmitt-Kopplin, P., 2011. Occurrence of *N*-acyl Homoserine lactones in extracts of bacterial strain of *Pseudomonas aeruginosa* and in sputum sample evaluated by gas chromatography–mass spectrometry. Am. J. Anal. Chem. 2 (2), 294–302.

Reading, N.C., Sperandio, V., 2006. Quorum sensing: the many languages of bacteria. FEMS Microbiol. Lett. 254 (1), 1–11.

Reading, N.C., Torres, A.G., Kendall, M.M., Hughes, D.T., Yamamoto, K., Sperandio, V., 2007. A novel two-component signaling system that activates transcription of an enterohemorrhagic *Escherichia coli* effector involved in remodeling of host actin. J. Bacteriol. 189 (6), 2468–2476.

Schaefer, A.L., Greenberg, E.P., Parsek, M.R., 2001. Acylated homoserine lactone detection in *Pseudomonas aeruginosa* biofilms by radiolabel assay. In: Methods in Enzymology. Academic Press, pp. 41–47.

Schaefer, A.L., Taylor, T.A., Beatty, J.T., Greenberg, E.P., 2002. Long-chain acyl-Homoserine lactone quorum-sensing regulation of *Rhodobacter capsulatus* gene transfer agent production. J. Bacteriol. 184 (23), 6515–6521.

Schauder, S., Shokat, K., Surette, M.G., Bassler, B.L., 2001. The LuxS family of bacterial autoinducers: Biosynthesis of a novel quorum-sensing signal molecule. Mol. Microbiol. 41 (2), 463–476.

Schulz, S., 2001. Composition of the silk lipids of the spider *Nephila clavipes*. Lipids 36 (6), 637–647.

Schupp, P.J., Charlton, T.S., Taylor, M.W., Kjelleberg, S., Steinberg, P.D., 2005. Use of solid-phase extraction to enable enhanced detection of acyl homoserine lactones (AHLs) in environmental samples. Anal. Bioanal. Chem. 383 (1), 132–137.

Shaw, P.D., Ping, G., Daly, S.L., Cha, C., Cronan, J.E., Rinehart, K.L., Farrand, S.K., 1997. Detecting and characterizing *N*-acyl-homoserine lactone signal molecules by thin-layer chromatography. Proc. Natl. Acad. Sci. 94 (12), 6036–6041.

Shen, Q., Gao, J., Liu, J., Liu, S., Liu, Z., Wang, Y., Guo, B., Zhuang, X., Zhuang, G., 2016. A new acyl-homoserine lactone molecule generated by *Nitrobacter winogradskyi*. Sci. Rep. 6 (1), 22903.

Sifri, C.D., 2008. Healthcare epidemiology: Quorum sensing: Bacteria talk sense. Clin. Infect. Dis. 47 (8), 1070–1076.

Sitnikov, D.G., Monnin, C.S., Vuckovic, D., 2016. Systematic assessment of seven solvent and solid-phase extraction methods for metabolomics analysis of human plasma by LC-MS. Sci. Rep. 6, 1–11.

Sperandio, V., Torres, A.G., Jarvis, B., Nataro, J.P., Kaper, J.B., 2003. Bacteria-host communication: The language of hormones. Proc. Natl. Acad. Sci. 100 (15), 8951–8956.

Steindler, L., Venturi, V., 2007. Detection of quorum-sensing *N*-acyl homoserine lactone signal molecules by bacterial biosensors. FEMS Microbiol. Lett. 266 (1), 1–9.

Sturme, M.H.J., Kleerebezem, M., Nakayama, J., Akkermans, D.L., Vaughan, E.E., de Vos, W.M., 2002. Cell to cell communication by autoinducing peptides in gram- positive bacteria. Antonie Van Leeuwenhoek 81 (1–4), 233–243.

Sun, Y., He, K., Yin, Q., Echigo, S., Wu, G., Guan, Y., 2018. Determination of quorum-sensing signal substances in water and solid phases of activated sludge systems using liquid chromatography-mass spectrometry. J. Environ. Sci. 69, 85–94.

Taga, M.E., Bassler, B.L., 2003. Chemical communication among bacteria. Proc. Natl. Acad. Sci. 100 (Suppl. 2), 14549–14554.

Tan, C.H., Koh, K.S., Xie, C., Tay, M., Zhou, Y., Williams, R., Ng, W.J., Rice, S.A., Kjelleberg, S., 2014. The role of quorum sensing signalling in EPS production and the assembly of a sludge community into aerobic granules. ISME J. 8 (6), 1186–1197.

Tang, X., Guo, Y., Wu, S., Chen, L., Tao, H., Liu, S., 2018. Metabolomics uncovers the regulatory pathway of acyl-homoserine lactones-based quorum sensing in anammox consortia. Environ. Sci. Technol. 52 (4), 2206–2216.

Taylor, G.W., Machan, Z.A., Mehmet, S., Cole, P.J., Wilson, R., 1995. Rapid identification of 4-hydroxy-2-alkylquinolines produced by *Pseudomonas aeruginosa* using gas chromatography-electron-capture mass spectrometry. J. Chromatogr. B Biomed. Sci. Appl. 664 (2), 458–462.

Thiel, V., Kunze, B., Verma, P., Wagner-Döbler, I., Schulz, S., 2009a. New structural variants of homoserine lactones in bacteria. Chembiochem 10 (11), 1861–1868.

Thiel, V., Vilchez, R., Sztajer, H., Wagner-Döbler, I., Schulz, S., 2009b. Identification, quantification, and determination of the absolute configuration of the bacterial quorum-sensing signal autoinducer-2 by gas chromatography-mass spectrometry. Chembiochem 10 (3), 479–485.

Todd, D.A., Zich, D.B., Ettefagh, K.A., Kavanaugh, J.S., Horswill, A.R., Cech, N.B., 2016. Hybrid quadrupole-Orbitrap mass spectrometry for quantitative measurement of quorum sensing inhibition. J. Microbiol. Methods 127, 89–94 (d).

Tommonaro, G., Abbamondi, G.R., Iodice, C., Tait, K., De Rosa, S., 2012. Diketopiperazines produced by the halophilic archaeon, *Haloterrigena hispanica*, activate AHL bioreporters. Microb. Ecol. 63 (3), 490–495.

Vendeville, A., Winzer, K., Heurlier, K., Tang, C.M., Hardie, K.R., 2005. Making "sense" of metabolism: Autoinducer-2, LuxS and pathogenic bacteria. Nat. Rev. Microbiol. 3 (5), 383–396.

Verbeke, F., De Craemer, S., Debunne, N., Janssens, Y., Wynendaele, E., Van de Wiele, C., De Spiegeleer, B., 2017. Peptides as quorum sensing molecules: measurement techniques and obtained levels in vitro and in vivo. Front. Neurosci. 11, 1–18.

Vial, L., Lépine, F., Milot, S., Groleau, M.C., Dekimpe, V., Woods, D.E., Déziel, E., 2008. *Burkholderia pseudomallei*, *B. thailandensis*, and *B. ambifaria* produce 4-hydroxy-2-alkylquinoline analogues with a methyl group at the 3 position that is required for quorum-sensing regulation. J. Bacteriol. 190 (15), 5339–5352.

Wagner-Döbler, I., Thiel, V., Eberl, L., Allgaier, M., Bodor, A., Meyer, S., Ebner, S., Hennig, A., Pukall, R., Schulz, S., 2005. Discovery of complex mixtures of novel long-chain quorum sensing signals in free-living and host-associated marine Alphaproteobacteria. Chembiochem 6 (12), 2195–2206.

Wang, J., Ding, L., Li, K., Schmieder, W., Geng, J., Xu, K., Zhang, Y., Ren, H., 2017. Development of an extraction method and LC–MS analysis for N-acylated-L-homoserine lactones (AHLs) in wastewater treatment biofilms. J. Chromatogr. B Anal. Technol. Biomed. Life Sci. 1041–1042, 37–44.

Waters, C.M., Bassler, B.L., 2005. Quorum sensing: cell-to-cell communication in Bacteria. Annu. Rev. Cell Dev. Biol. 21 (1), 319–346.

Wen, K.Y., Cameron, L., Chappell, J., Jensen, K., Bell, D.J., Kelwick, R., Kopniczky, M., Davies, J.C., Filloux, A., Freemont, P.S., 2017. A cell-free biosensor for detecting quorum sensing molecules in *P. aeruginosa*-infected respiratory samples. ACS Synth. Biol. 6 (12), 2293–2301.

Williams, P., 2007. Quorum sensing, communication and cross-kingdom signalling in the bacterial world. Microbiology 153 (12), 3923–3938.

Winans, S.C., 2011. A new family of quorum sensing pheromones synthesized using S-adenosylmethionine and acyl-CoAs. Mol. Microbiol. 79 (6), 1403–1406.

Xu, F., Song, X., Cai, P., Sheng, G., Yu, H., 2017. Quantitative determination of AI-2 quorum-sensing signal of bacteria using high performance liquid chromatography-tandem mass spectrometry. J. Environ. Sci. 52, 204–209.

Yang, Y.-H., Lee, T.-H., Kim, J.H., Kim, E.J., Joo, H.-S., Lee, C.-S., Kim, B.-G., 2006. High-throughput detection method of quorum-sensing molecules by colorimetry and its applications. Anal. Biochem. 356 (2), 297–299.

Zabet-Moghaddam, M., Heinzle, E., Tholey, A., 2004. Qualitative and quantitative analysis of low molecular weight compounds by ultraviolet matrix-assisted laser desorption/ionization mass spectrometry using ionic liquid matrices. Rapid Commun. Mass Spectrom. 18 (2), 141–148.

Zhang, S., Liu, J., Chen, Y., Xiong, S., Wang, G., Chen, J., Yang, G., 2010. A novel strategy for MALDI-TOF MS analysis of small molecules. J. Am. Soc. Mass Spectrom. 21 (1), 154–160.

CHAPTER 3

Quorum Sensing in Marine Biofilms and Environments

Raphaël Lami

Sorbonne Université, CNRS, Laboratoire de Biodiversité et Biotechnologies Microbiennes, LBBM, Banyuls-sur-Mer, France

1 Introduction

Quorum sensing is suspected to lead to spectacular phenomenon in marine waters. For example, an impressive "Milky sea" of $15,400\,km^2$ in the Arabian Sea has been attributed to bioluminescent *Vibrio harveyi*. These cells were suspected to express quorum sensing to glow and bloom in response to a phytoplankton bloom (Miller et al., 2005). However, since its discovery in the 1970s, quorum sensing has been evidenced not only in this type of spectacular bioluminescence but also in many different types of bacterial activities and in phylogenetically very diverse marine cells. This review aims to synthesize the current knowledge about quorum sensing mechanisms in marine biofilms and environments. Maintaining a specific focus on marine waters, we will review the wide diversity of chemical compounds involved in marine quorum sensing, the wide range of biological functions of quorum sensing, and the large number of ecological niches where these processes occur. The last section of this chapter will review applications, targeting the marine environment as a source of quorum sensing-inhibiting compounds and enzymes or focusing on the manipulations of marine bacterial physiologies using quorum quenching-based strategies.

2 The Discovery of and Growing Interest in Quorum Sensing in Marine Environments

2.1 1970s: Quorum Sensing was Discovered in the Marine Environment—The Vibrio-squid Model

The concept of quorum sensing referring to a population density-based physiological response of bacterial cells was introduced in the 1990s (Fuqua et al., 1994). However, most of the observations that led to the elaboration of this concept were acquired from experiments conducted by marine scientists during the 1970s. During this decade, a lot of data were collected on *Vibrio fischeri* strains that were able to colonize the light organ of the Hawaiian bobtail squid *Euprymna scolopes*, where they produce bioluminescence (Greenberg et al., 1979; Nealson et al., 1970). In particular, a density-dependent phenotype was originally noted in this symbiotic bacterial community. In the surrounding seawater, these cells are free living and scarce, and do not produce light. However, they are able to bioluminesce when they reach high concentrations, similar to in lab cultures or when they colonize the light organ of the squid.

Interestingly, the cell abundances of *V. fischeri* within the squid follow a circadian pattern. At night, *V. fischeri* are present at high concentrations (10^{10}–10^{11} cells mL^{-1}) and emit a diffusible factor, also named an autoinducer (AI), associated with the production of light. At the end of the night, most of the bacterial cells are expulsed from the light organ, leading to a dramatic reduction in bacterial concentration and in the diffusible factor. During the day, the concentrations of *V. fischeri* that have not been expulsed are very low, the diffusible factor is not produced, and the squid do not bioluminesce. However, this remaining population of *Vibrio* grow steadily under favorable conditions within the squid throughout the day and again reach at night a cell abundance that is sufficient to produce bioluminescence. This bacteria-squid association constitutes a bacteria-animal symbiosis. The squid relies on *Vibrio*'s light to escape predators or hunt preys. In return, the squid provide host and nutrients to the *Vibrio* (Graf and Ruby, 1998).

Since the first observations in the 1970s, this original system of bioluminescence regulation has been fully chemically and genetically described. The diffusible signal was identified in 1981 as an acyl-homoserine lactone (AHL) and described as 3-oxo-hexanoyl-homoserine lactone (3-oxo-C6-HSL) (Eberhard et al., 1981). The genetic cluster involved in this phenomenon was then characterized as a bidirectionally transcribed operon with eight genes, named *luxA-E*, *luxG*, *luxI*, and *luxR*. The LuxA and LuxB proteins are the two subunits of luciferase, the enzyme responsible for light production. The LuxC-D-E proteins are involved in the synthesis of the luciferase substrates, while LuxG is a flavin reductase. However, in quorum sensing research, most of the interest is focused on LuxI and LuxR proteins. LuxI is the AI synthase responsible for AIs production, while LuxR is the receptor of this diffusible signal. When the AIs reach a threshold concentration in the nearby environment of bacterial cells (reflecting the increase in cell abundances), they bind to the LuxR receptors, which act as transcription factors and activate the expression of all *lux* genes. The diffusible signal is designated an AI as it promotes its own production through the autoinduction of *luxI* (Fig. 1) (Eberhard et al., 1981; Engebrecht et al., 1983).

2.2 1990–2010s: A Growing Interest in the Study of Quorum Sensing in Marine Environments

After these initial discoveries and subsequent full elucidation of the genetic system of quorum sensing, the study of this mechanism garnered little interest from the scientific community for more than a decade. Likely, quorum sensing appeared then to be a kind of regulation specialized

Low cell density

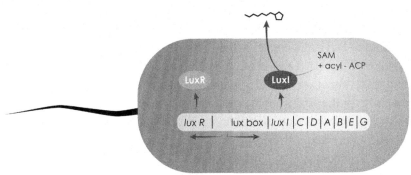

High cell density

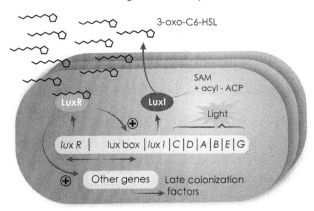

Fig. 1

A schematic representation of the first discovered *luxI/luxR*-based quorum sensing system in the model species *Vibrio fischeri*, producing 3-oxo-C6-HSL.

for bioluminescence. This interest was renewed in the 1990s with the development of DNA sequencing methods and the discovery of a large diversity of *luxI* and *luxR* homologs in many different types of bacteria. Little by little, it appeared that the *luxI-luxR* model developed for *V. fischeri* was also relevant to a wide diversity of bacterial strains. These observations led to the establishment of the quorum sensing concept in 1994 (Fuqua et al., 1994).

However, despite the full characterization of the environmental *Vibrio*-squid model, most of the scientific effort in the field of quorum sensing was put forth in the 1990s and focused on strains with a medical or agronomic interest. Researchers devoted little attention to these mechanisms in the field of environmental sciences before 2005. An important reason for this interest in the medical field, among others, is that an increasing number of links were established during the 1990s between virulence and quorum sensing in pathogenic bacteria, such as in *Staphylococcus* strains and *Pseudomonas aeruginosa*. It was only in the following decade that work began to be published about bacteria in the field of environmental sciences, including those isolated from the marine waters. In 1998, one of the first reports of AIs present in the natural environment aquatic was published under the title "Quorum Sensing Autoinducers: Do They Play a Role in Natural Environments?," which revealed some early interest in naturally occurring aquatic biofilms (McLean et al., 1997). It was then hypothesized in 2001 that quorum sensing might function in marine particle-attached bacteria (Kiørboe, 2001). In 2002, Gram et al. reported for the first time the production of AHLs within *Roseobacter* and *Marinobacter* strains isolated from marine snow. Since then, a growing number of reports had focused on the nature and role of quorum sensing in marine bacteria, and large sets of culture-dependent (Rasmussen et al., 2014; Wagner-Döbler et al., 2005) and culture-independent (Doberva et al., 2015; Muras et al., 2018a) studies have highlighted the importance of quorum sensing mechanisms in marine biofilms and environments.

3 Overview of Prokaryotic Diversity and Compounds Involved in Quorum Sensing in Marine Environments and Biofilms

3.1 Experimental Approaches to Characterize AIs in the Marine Environment

One important difficulty in characterizing AIs from the marine environment is their low concentration, which is at picomolar levels. Only in a few cases have researchers been able to identify AIs directly on samples collected in situ, such as in marine snow (Jatt et al., 2015), phycosphere of phytoplankton cells (Bachofen and Schenk, 1998; Van Mooy et al., 2012), mucus covering cnidarians (Ransome et al., 2014), or microbial mats (Decho et al., 2009). However, such direct measurements of AIs concentrations remain rare, and indirect detection has generally been used, such as that described in the pioneering paper of Gram et al. (2002). In this workflow, the first step consists of the isolation of bacterial strains from the studied environment, such as a phytoplankton bloom (Bachofen and Schenk, 1998), algae cultures, microbial mats, or marine snow (Gram et al., 2002; Schaefer et al., 2008; Wagner-Döbler et al., 2005). Then, the supernatant of the isolated cells is added to a culture of whole-cell biosensors, such as *Escherichia coli* JB523, *Chromobacterium violaceum* CV026, or *V. harveyi* JMH612. These biosensors are very diverse and present large variation of sensibilities to detect various types of AHLs. They are genetically modified organisms that are able to produce a signal in the presence of the AIs eventually emitted in the supernatant of the screened strains. For example, *C. violaceum* produces the purple pigment violacein, and *E. coli* JB523 emits a GFP signal in presence of AHLs (Steindler and Venturi, 2006).

The next step in quorum sensing compounds characterization relies on the tools used in the field of natural substances chemistry. Pioneering studies used thin-layer liquid chromatography (TLC) (Gram et al., 2002). However, more recent approaches are usually based on liquid chromatography coupled with mass spectrometry (LC-MS) (Schaefer et al., 2008), gas chromatography coupled with MS (GC-MS) (Wagner-Döbler et al., 2005), and MS/MS approaches (Van Mooy et al., 2012). In some cases, these analyses are preceded by a microfractionation step, which allows a better separation and concentration of the extracted compounds (Doberva et al., 2017). When focusing on AHLs characterization, the position of double bonds on the side chain of AHLs can be determined by additional derivatization steps using dimethyl disulfide (Neumann et al., 2013). Definitive characterization can sometimes be achieved by 1D and 2D nuclear magnetic resonance analyses, depending on whether the purity and concentrations of targeted compounds are sufficient to allow such analyses.

3.2 AHLs, or Autoinducer-1 (AI-1)

In marine environments, the production of AHLs is mostly due to Gram-negative bacteria and is found in many diverse types of marine *Alpha-*, *Beta-*, and *Gamma-Proteobacteria*. AI-1 is produced by LuxI family enzymes and detected by LuxR family receptors (Engebrecht et al., 1983). In addition, some AHLs are produced by *ainS*-like genes (Gilson et al., 1995), and in some cases by *hdtS*-like genes (Laue et al., 2000; Rivas et al., 2007). The AHLs are formed from *S*-adenosylmethionine and a fatty acid residue, which are the AIs synthases substrates. Also, it has been evidenced that some bacteria harbor "*luxR* orphans," meaning that these organisms are able to catch AHLs in the marine environment without producing them. Thus they might save the energetic cost of AHL production, but they are able to "sense" chemical dialogues in their nearby environment and adapt their physiology using such "spying" system. In addition, some "*luxI* orphans" have also been found in some marine bacterial genomes (Cude and Buchan, 2013). The ecological roles and importance of LuxI and LuxR orphans in marine waters remains to be studied.

AHLs are composed of a lactone ring attached to a fatty acid residue (the acyl side chain) with an amide bound (Fig. 2). These AHLs present many types of structural variants that differ in length, with 4–19 carbons (Doberva et al., 2017) in the acyl side chain that can be saturated or unsaturated. These compounds also differ in the substitution that occurs at the C3 position (hydrogen, hydroxyl, or carbonyl group) and sometimes on other carbons in the acyl side chain. Additionally, a few AHLs with branched acyl side chains have been described, but to the best of our knowledge, not yet in marine strains. Other authors have also reported

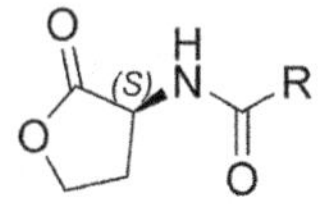

Fig. 2

The general chemical structure of AHLs. The lactone ring moiety is linked to a fatty acid residue (R) with an amide bound. The acyl side chain is highly variable in terms of length, oxidization state, and the presence of hydrogen, hydroxyl, or carbonyl groups; this variability provides to the AHL signal its specificity.

AHLs presenting side chains with aromatic acid residues (*p*-coumaric acid or cinnamic acid). This is the case of *p*-coumaroyl-HSL, which has been characterized in *Rhodopseudomonas palustris* but also discovered in the marine bacterium *Silicibacter pomeroyi* DSS-3 (Schaefer et al., 2008).

Most *Rhodobacteraceae* species (a major marine bacterial group involved in quorum sensing) produce long-chain AHLs with additional modifications (Fig. 3) (Cude and Buchan, 2013). The strain *Rhodobacter sphaeroides*, which is phylogenetically very closed to various marine strains, synthesizes C14:1-HSL. The marine *Dinoroseobacter shibae* emits C18:2-HSL, C18:1-HSL, and traces of C16-HSL, C15-HSL, and C14-HSL (Neumann et al., 2013; Patzelt et al., 2013; Wagner-Döbler et al., 2005). The sponge symbiont *Ruegeria* sp. emits 3-OH-C14-HSL, 3-OH-C14:1-HSL, and 3-OH-C12-HSL (Zan et al., 2012). More recent studies using

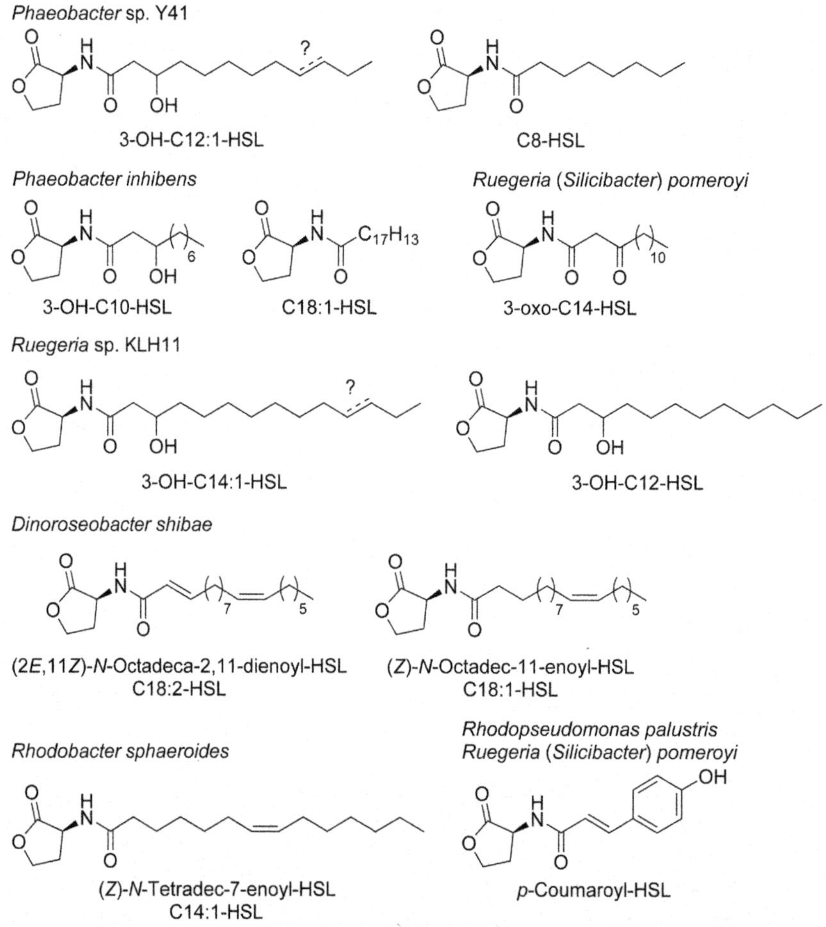

Fig. 3

A snapshot of AHL diversity in marine *Rhodobacteraceae*.

UHPLC-HRMS/MS approaches have revealed a much broader diversity of AHL compounds in *Rhodobacteraceae* strains. The strain MOLA401 (which should be classified as *Palleronia rufa*; Barnier C, pers. communication, submitted manuscript) emits a large diversity of long-chain AHLs, with 20 different putative types of these compounds, including one with 19 carbons in the acyl side chain (Doberva et al., 2017). The strain *Paracoccus* sp. Ss63 from the sponge *Sarcotragus* emits long-chain AHLs, including 12 saturated and 4 unsaturated putative ones (Saurav et al., 2016b). Clearly, the real extent of AHLs diversity in marine *Rhodobacteraceae* strains remains an open question that still requires further investigation.

Vibrionaceae strains are found in many types of marine environments and hosts and are frequently pathogens of fishes, mollusks, and corals (*Vibrio anguillarum*, *Vibrio vulnificus*, *V. harveyi*) or symbionts of squids (*Aliivibrio fischeri*, *Vibrio pomeroyi*, *Vibrio aesturianus*). The first quorum sensing signal was discovered in the marine squid associated *V. fischeri* (see earlier in this chapter). Since this discovery, a few *Vibrio* have been fully described as prokaryotic models for the study of quorum sensing. Nevertheless, the extent of AHL signals diversity across *Vibrio* species is far from fully explored. The screening of *Vibrio* collections for AHL production has revealed contradictory data, with between 9% and 85% of strains testing positive in these biotests (Garcia-Aljaro et al., 2012; Girard et al., 2017; Purohit et al., 2013), making it difficult to determine the real extent of AHL types distribution across this genus. Nevertheless, a few studies have attempted to characterize the chemical diversity of AHLs in diverse *Vibrio* strains. In general, the lengths of acyl side chains of AHL were found shorter than those detected in *Rhodobacteraceae* strains: C4-HSL to C12-HSL in diverse *Vibrio* strains (Fig. 4) (Purohit et al., 2013; Rasmussen et al., 2014) and C10-HSL to C14-HSL in *Vibrio tasmaniensis* LGP32 (Girard et al., 2017).

An AHL-based quorum sensing system has also been found in some marine *Mesorhizobium* (Krick et al., 2007), *Bacteroidetes* (Huang et al., 2008), and *Cyanobacteria* (Sharif et al., 2008). In addition, some evidence of AHL production has been detected in aquatic Archaea (Paggi et al., 2003), as well as in the marine Gram positive bacteria, *Exiguobacterium* (Biswa and Doble, 2013). By contrast, some important groups of marine bacteria do not appear to produce AHLs, such as members of the SAR11 group, which dominate many types marine prokaryotic communities (R. Lami, unpublished data based on biosensors *P. putida* F117 and *E. coli* MT102).

3.3 Autoinducer-2 (AI-2)

4,5-Dihydroxy-2,3-pentanedione (DPD) is the precursor of AI-2 and is synthesized by the LuxS enzyme. The enzyme LuxS, which converts *S*-ribosyl homocysteine to homocysteine and DPD, catalyzes the key step in AI-2 biosynthesis. The AI-2 is found in many different types of both Gram-negative and Gram-positive bacteria; thus, this AI may serve as an interspecies signal (Surette et al., 1999). In the presence of boron, DPD leads to (2*S*,4*S*)-2 methyl-2,3,3,4-tetrahydroxytetrahydrofuran-borate (S-THMF-borate). In the marine environment, this

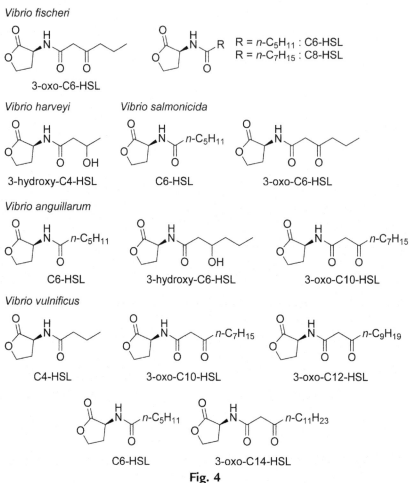

Fig. 4

A snapshot of AHL diversity in marine *Vibrionaceae*.

compound is used as AI-2 in *Vibrio* (Chen et al., 2002). In the absence of boron, DPD leads to (2*R*,4*S*)-2-methyl-2,3,3,4-tetrahydroxytetrahydrofuran (R-THMF), which is the AI-2 signaling compound in enteric bacteria. The presence or absence of borate in local environments may shift the spontaneous nonenzymatic rearrangements between the two known forms of AI-2 (Miller et al., 2004). However, some researchers have raised doubt about the signaling function of AI-2, as DPD is also a side product of the activated methyl cycle (Rezzonico and Duffy, 2008). Thus, AI-2 is at the same time a signal, a cue, or a waste product depending of the bacterial species which produce or receive it. How metabolic waste products can evolve to become signals remains an important and open question that should be addressed in the field of chemical communication (Whiteley et al., 2017).

The presence of DPD in the marine environment has been confirmed by direct measurements in natural samples (Van Mooy et al., 2012). AI-2 signaling is frequently utilized by *Vibrio* species, as confirmed by culture-based studies (Yang et al., 2011). The examination of the phylogenetic diversity of *luxS*-translated sequences in the Global Ocean Sampling database suggests that *Shewanella*-related species constitute a major group using AI-2 signaling in marine environments (Doberva et al., 2015), as also observed among cultivated strains (Bodor et al., 2008). *Sulfurovum lithotrophicum* and *Caminibacter mediatlanticus* are *Epsilonproteobacteria* and AI-2 producers that grow within biofilms colonizing deep-sea hydrothermal environments. These bacterial strains were found to express AI-2-based quorum sensing, and *luxS* transcripts were also found directly from RNA extracts and subsequent RT-qPCR performed on deep sea biofilms (Perez-Rodriguez et al., 2015). Despite these few studies, the real extent of bacterial diversity involved in AI-2 signaling in the marine environment, as well as the ecological role of this AI, require further investigation.

3.4 Other AIs

In marine environments, *V. harveyi* synthesizes the AI (Z)-3-aminoundec-2-en-4-one. In fact, a chemically very similar compound was first discovered in *Vibrio cholerae*, which synthesizes (S)-3-hydroxytriecan-4-one, which gives the name to this family of AIs: CAI-1 or Cholera AI-1 (Fig. 5). The enzyme that synthesizes CAI-1 belongs to the CqsA-family, which is found in all *Vibrio*. Whether CAI-1 plays an ecological role in natural marine communities remains to be evaluated in future experiments (Higgins et al., 2007; Kelly et al., 2009; Ng et al., 2011; Wei et al., 2011).

There are many other AIs that were discovered in marine strains and for which little data have yet to be published. 3,5-Dimethylpyrazin-2-ol (DPO) has been demonstrated to be an AI in *V. cholerae* (Fig. 5), activating the expression of vqmR, encoding small regulatory RNAs

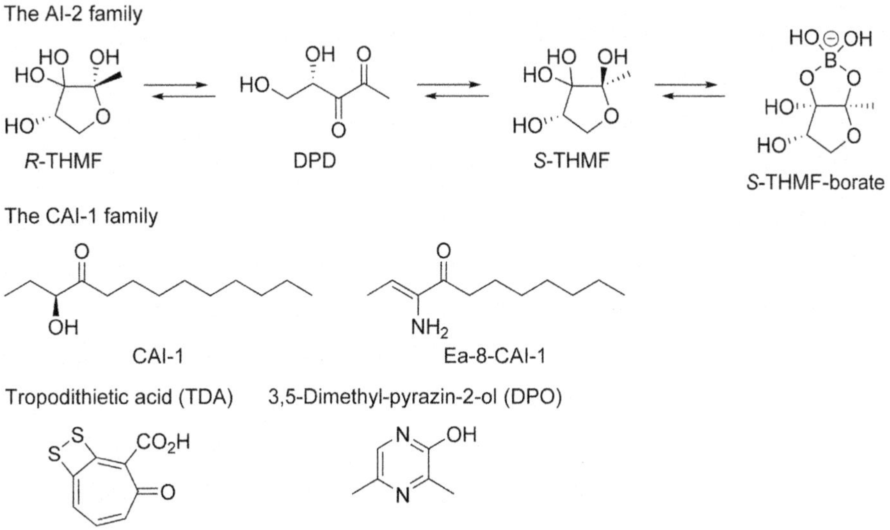

Fig. 5
The AI-2 and CAI-1 autoinducer families and the structure of the autoinducers TDA and DPO.

(Papenfort et al., 2017). In addition, tropodithietic acid (TDA) has been described as an AI in *Rhodobacteraceae* (Geng and Belas, 2010). Collectively, these observations clearly reveal that quorum sensing AIs are not limited to AI-1 and AI-2 families in the marine environment. There is probably a much broader diversity of quorum sensing semiochemicals that remains to be discovered and probably play important ecological roles in association with the dynamics of marine microbial communities.

3.5 Diffusion of AIs in the Marine Environment

Once produced by marine bacteria, AIs passively diffuse through the marine environment to reach their target. It appears that the signaling compounds cannot diffuse over "calling distances" longer than 10–100 µm (Gantner et al., 2006), in accordance with their thermodynamic and chemical characteristics, like solubility, diffusivity, dispersion potential, or chemical stability (Harder et al., 2014). According to Fick's law, short-chain AHLs will diffuse more rapidly than long-chain AHLs. The molecular weight of AHLs and their solubility are positively correlated with their mobility, but other characteristics modify this general rule. On one hand, the presence of hydroxy- or oxo-substitutions along their acyl side chain can increase their solubility and thus their capacity to passively diffuse (Doberva et al., 2017). On the other hand, long-chain AHLs with high molecular weight are also predicted to adsorb to organic surfaces (Decho et al., 2011; Harder et al., 2014).

Many other factors must be considered to differentiate AHL diffusion processes from one bacterium to another one. Abiotic processes can lead to a rapid base hydrolysis of the lactone ring, especially under alkaline conditions, and converts AHLs in inactive γ-hydroxy carboxylates. However, this process is reversible under acidic conditions (Tait et al., 2005; Yates et al., 2002). Additionally, it is worth noting that AHLs that present an oxo-substitution on their 3-carbon can be subjected to a spontaneous Claisen condensation, forming tetramic acids, which inhibit their capacity to transmit information (Kaufmann et al., 2005). The half-life of 3-oxo-C6-HSL has been established in days according to the following formula: $1/(1 \times 10^7 \times [OH^-])$ (Schaefer et al., 2000). However, much slower degradation rates of 3-oxo-C6-HSL have been experimentally reported in artificial marine waters (0.094 vs $0.26\,h^{-1}$ according to the formula), suggesting that AHLs can diffuse over longer distances than initially hypothesized (Hmelo and Van Mooy, 2009). In the same report, slower degradation rates than those previously published for unsubstituted AHLs were also reported. In addition, a few reports have noted that short-chain AHLs present shorter half-lives than those of long-chain AHLs, suggesting that long-chain AHLs might diffuse over longer distances than do short-chain AHLs (Decho et al., 2011).

AHL diffusion can be limited not only by the effects of abiotic factors but also by those of biotic factors. For example, it is well known that many bacteria or eukaryotes emit quorum quenching enzymes or chemical compounds that can degrade affect AIs diffusion and reception. Hmelo and Van Mooy (2009) reported that C6-HSL, 3-oxo-C6-HSL, and 3-oxo-C8-HSL have degradation rates that are respectively 54%, 23%, and 57% higher in natural seawater than in artificial seawater, and these authors attributed this increase to the occurrence of quorum quenching processes in the marine waters, for which more details are provided in the last section of this chapter.

Overall, the diffusion through marine environments of AHLs appears to be a complex phenomenon that is dependent on many diverse abiotic and biotic variables. One important last factor to consider is that these infochemicals are usually released in complex tridimensional biofilms, made of complex exopolymers given density and thickness to these biological architectures (Harder et al., 2014). Thus some authors tend to consider the chromatographic movement rather than the passive diffusion movement as the best concept to model the travel of AHLs through marine microniches (Decho et al., 2011). The architecture of biofilms can also provoke the sequestration AHLs in microniches, leading to locally important concentrations of AIs that can activate quorum sensing processes, even with few cells in the local environment (Charlton et al., 2000).

3.6 V. harveyi *(and the* Harveyi *clade), a Model for the Study of Quorum Sensing in Marine Bacteria*

V. harveyi is a major pathogen in the marine environment and is responsible for many diseases observed in farmed oysters, mollusks, and shrimps, leading to substantial economic loss. *V. harveyi* possesses a complex quorum sensing system with three interdependent channels: AI-1, AI-2, and CAI-1-based quorum sensing (Fig. 6). In all cases, AIs bind to their cognate membrane receptors and thus activate an intracellular phosphorylation/dephosphorylation signal transduction cascade. The key protein in this transduction cascade is LuxO, a response regulator that integrates the signals from the three channels presenting various levels of phosphorylation (Lilley and Bassler, 2000). At low cell densities, when phosphorylated, LuxO activates through five quorum regulatory sRNAs (Qrr RNAs) the protein AphA which regulates many genes, in particular those involved in the expression of most of the identified virulence factors, as well as biofilm formation (see below in this chapter for detailed discussion).

At low cell densities, Qrr RNAs inhibit the transcription of the master regulator LuxR (Lenz et al., 2004). However, when high concentrations of quorum sensing signals are present, LuxO is dephosphorylated and the expression of *lux* genes is induced. More precisely, the concentration of LuxR in the cytoplasm is dependent on the concentrations of these five different types of small regulatory RNAs, which are correlated with the level of phosphorylation of LuxO. At such high cell densities, the production of AphA is inhibited and LuxR-dependent genes are activated, like those involved in bioluminescence production.

In Section 5 of this chapter, we'll review the diversity of quorum sensing dependent phenotypes and functions regulated by quorum sensing. One important fact to keep in mind is this

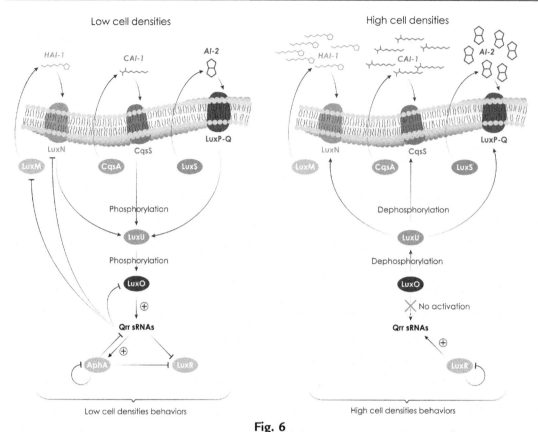

Fig. 6
Quorum sensing signaling pathways in the model marine strain *V. harveyi* and their mode of functioning at low and high cell densities.

opposition between the quorum sensing master regulator AphA, which coordinates low cell densities behaviors, and the quorum sensing master regulator LuxR, which controls high cell densities behaviors. In this sense, the regulation of *V. harveyi* quorum sensing dependent phenotypes is complex. For example, among the virulence factors, quorum sensing positively regulates metalloproteases and extracellular toxins at high cell densities. By contrast, at such high cell densities, quorum sensing negatively regulates chitinases, phospholipases, siderophores, and type III secretion system. Such opposition probably reflects the need of different types of virulence factors at the different stages of infection (Natrah et al., 2011b).

4 Surface Biofilms and Holobionts: The Diversity of Microbial Habitats Enabling Quorum Sensing in the Marine Environment

Quorum sensing occurs at high bacterial cell concentrations. Thus it appears unlikely that free-living marine bacterial cells will perform quorum sensing. However, there are many (micro)-niches and habitats in the marine environment where bacterial cells are organized in highly structured biofilms, where they reach sufficient abundances to induce quorum sensing-based dialogues. The biofilms are organized in a gel-like exo-macropolymeric matrix, a three-dimensional architecture that includes a micrometer-scale spatial organization and efficiently concentrates chemical compounds (Decho et al., 2011; McLean et al., 1997). Also, prokaryotes colonize marine plants, vertebrates and invertebrates, and the recognition of such importance of the bacterial microbiota for other organisms leads to the concept of holobiont. Again, an holobiont provides possible bacterial concentrations and the occurrence of prokaryotic interactions, including quorum sensing (Teplitski et al., 2016). Some examples will be described in this section.

4.1 Phycosphere of (Micro-)algae and Phyllosphere of Marine Plants

The phycosphere corresponds to the immediate region of surrounding algae. It describes a microbial habitat that is deeply shaped by algae. In this microenvironment, bacterial abundances can reach 10^8–10^{11} cells mL^{-1} (Rolland et al., 2016), enabling the occurrence of quorum sensing. The pioneering study published by Bachofen and Schenk (1998) revealed the production of AHLs within a cyanobacterial phytoplankton bloom in concentrations that can reach 10 mg L^{-1} (Bachofen and Schenk, 1998). Since then, many reports about quorum sensing processes in the phycosphere of (micro)algae have been published (Rolland et al., 2016). A recent experiment using transcriptomics approaches noted an increase in *luxI* and *luxR* gene expression within *Ruegeria pomeroyi* co-cultured with the microalgae *Alexandrium tamarense*. This work revealed putative important links between microalgae growth and quorum sensing-dependent physiological regulations within *Rhodobacteraceae* strains associated with some microalgae (Landa et al., 2017).

Many strains capable of quorum sensing were isolated from the phycosphere of micro- and macro-algae, and the compounds that they produce and their involvement in cell-to-cell signaling were elucidated. For example, strains affiliated with different species isolated in diverse phycospheres were revealed as AHL producers, such as *D. shibae*, *Hoeflea phototrophica*, and *Roseovarius mucosus* producing C18:1-HSL and C14:1-HSL (Wagner-Döbler et al., 2005). Some authors have also reported the capacity of cyanobacteria to produce AHLs, particularly the toxin producer *Microcystis aeruginosa* (Zhai et al., 2012) as well as the epilithic *Gloeothece* PCC6909 (Sharif et al., 2008) that emits C8-HSL.

There are not only AHLs that are involved in quorum sensing signaling within the phycosphere but also a large diversity of compounds, of which many fewer observations have been made. For example, few AI-2 producers were isolated from the phycosphere of phytoplankton. However, recent reports noted some *Vibrio* that were isolated in the phycosphere of *Trichodesmium* and were able to produce this AI (Van Mooy et al., 2012). A potential role for AI-2 in the control of the algaecide activity against the dinoflagellate *Gymnodinium catenatum* has also been hypothesized (Skerratt et al., 2002). TDA has been characterized as an AI in many *Rhodobacterales* species (Geng and Belas, 2010) and is frequently associated with microalgae, similar to the bacterial strains affiliated with the genera *Phaeobacter*, *Silicibacter*, and *Ruegeria* (Bruhn et al., 2005; Geng et al., 2008; Porsby et al., 2008).

A recent paper investigated the prevalence of quorum sensing in the microbiota associated with the Mediterranean seagrass *Posidonia oceanica*, a marine angiosperm that plays a key role in coastal ecology. To address this question, 60 strains from *P. oceanica* leaves and rhizomes were isolated, including epiphytic and endophytic strains. From this collection, the authors of this study were able to detect 6 strains able to emit 8 different types of AHLs and 19 strains able to activate the AI-2 biosensors. These results reveal the importance of examining quorum sensing relationships also within the microbiota of marine angiosperms to better understand the complex relationships that may exist between these microbes and their host (Blanchet et al., 2017).

4.2 Biofilm Around Sinking Marine Snow Particles

Marine snow includes particles that are larger than 0.5 mm and are sinking in the deep ocean. These aggregates, which are composed of organic and inorganic carbon, have been recognized to play a major role in the oceanic and global carbon cycles. Important consortia of marine bacteria colonize marine snow, and their metabolic activities largely contribute to shape the composition of these aggregates. For example, these communities express large sets of hydrolytic enzymes that are involved in the degradation of particulate organic carbon. These bacterial consortia are structured in a biofilm, and marine snow constitutes hotspots of high bacterial concentrations in the oceans, with 10^8–10^9 cells mL^{-1}, which are two to four orders of magnitude compared with the ambient seawater (Simon et al., 2002).

A first report in 2002 by Gram et al. (2002) revealed the production of AHLs among 4 of 43 bacterial strains isolated from marine snow. They highlighted in these strains the production of C6-HSL and C8-HSL, and identified them as affiliated with the genera *Roseobacter* and *Marinobacter*. This pioneering study revealed that quorum sensing regulations occur in particle-attached marine microbial communities. This paper then hypothesized that quorum sensing might regulate bacterial hydrolytic enzymes activities, biofilm formation, or antibiotic production in such marine niches. Other interesting pieces of information were then published by Hmelo et al. (2011). In this study, the authors were able to detect AHLs directly on aggregates collected from the Pacific Ocean near Vancouver Island. In addition, they were able to prove that the addition of exogenous AHLs into flasks containing sinking marine snow aggregates increased the activity of bacterial hydrolytic enzymes. In a similar approach, Jatt et al. (2015) were also able to detect in situ the production of AHLs and revealed the presence of 3-oxo-C6-HSL and C8-HSL in marine snow aggregates collected from China's marginal seas. Among diverse isolates which were detected as AHLs producers, the authors reported the isolation of a marine strain affiliated with the species *Pantoea ananatis* able to produce six different types of AHLs. In their experiments, they revealed that the addition of exogenous AHLs to the culture media increased the extracellular hydrolytic enzymes activities. In particular, they noticed an upregulation of alkaline phosphatases activities. Collectively, these data reveal that quorum sensing might play a crucial role in the regulation of the activities of bacteria colonizing marine snow. In this sense, quorum sensing might be a crucial process regulating the carbon cycle in the ocean; many more studies coupled with in situ measurements are required to support this hypothesis.

4.3 Marine Microbial Mats and Other Types of Subtidal Biofilms

Very little work has been conducted about quorum sensing in marine microbial mats and diverse types of subtidal biofilms. These biofilms are one of the earliest forms of life known on Earth (3.4–3.5 Gy B. P.) and have been identified in fossils such as stromatolites. These microbial mats are highly diverse in terms of microbial community composition and functions, and are organized in very well structured, layered assemblages. Thus they present important cell abundances that are fully compatible with the occurrence of quorum sensing regulations.

The pioneering study published by McLean et al. (1997) was the first report of AHLs in natural environments and focused on submerged biofilms. While this study did not focus on marine biofilms, it provided a lot of interesting data for aquatic environments. These researchers revealed the absence of AHLs on rocks not coated with natural biofilms and their presence on those covered with biofilms. A few years later, a report published by Decho et al. (2009) revealed very interesting patterns of AHLs regulation in marine mats. First, they were able to collect and characterize AHLs directly on marine mat samples collected in situ and detected both long- and short-chain AHLs (C4- C6- oxo-C6-, C7-, C8-, oxo-C8-, C10-, C12-, and C14-HSLs). Even more interestingly, the authors revealed that short-chain AHLs were significantly less present during the day. One possible interpretation is that diel variations in pH may alter short-chain AHLs when pH is above 8.2 during the day, due to photosynthetic processes. The development of subtidal biofilms appears to be very dynamic, and it has been shown that with the variation in community composition, the pattern of emitted AHLs is also modified. Many AHL producers were isolated in such biofilms (Huang et al., 2008). In particular, AHL-producing *Vibrio* spp. appear be pioneer species in these biofilms (Huang et al., 2009).

Many studies have demonstrated that the presence of a bacterial biofilm favors the settlement on surfaces of invertebrate larvae. In particular, a series of interesting papers have highlighted that the AHLs produced in marine biofilms are detected by diverse types of larger organisms to guide their settlement (Hadfield, 2011; Hadfield and Paul, 2001; Wieczorek and Todd, 1998). Among the chemical attractants, AHLs were found to play an important role for the zoospores of the macroalga *Ulva*. The detection of AHLs by these spores leads to a modification of their swimming behavior and the provocation of chemokinesis, which favor their settlement on the biofilm. These data revealed that AHLs produced in bacterial biofilms can also be sensed by eukaryotes, and these studies pointed out that AHLs also act as interkingdom chemical signals (Joint et al., 2002; Tait et al., 2005). In a similar vein, it has been shown that AHLs favor the settlement of cyprid larvae of *Balanus improvisus* in bacterial biofilms (Tait and Havenhand, 2013).

4.4 Corals and Other Cnidarian-associated Communities

Many cnidarians species harbor bacterial-associated communities in which members communicate using quorum sensing signals. For example, *Anemonia viridis* and the *Gorgonacea Eunicella verrucosa* harbor very diverse associated bacterial communities, including quorum sensing AIs producers (Ransome et al., 2014). However, most of the work on cnidarian species in the field of quorum sensing has focused on corals, especially in the context of pathogenicity (i.e., black band disease, white band disease, etc.).

Coral species harbors important and dense associated microbial communities that are 10–1000-fold more concentrated compared with ambient seawater (Rosenberg et al., 2007). A study conducted in 2011 reports that 30% of bacteria in these consortia are capable of quorum sensing. These include a *Vibrio* strain that has been shown to emit 3-OH-C10-HSL detected using TLC (Golberg et al., 2011). Similarly, in another study published in 2010, a total of 29 *Vibrio* isolates were collected in different healthy or diseased corals (samples collected from mucus, tissues, surrounding waters) and were tested for their ability to produce quorum sensing compounds (Tait et al., 2010). The authors found that all isolated *Vibrio* were able to emit AI-2, and 17 were found capable of AHL synthesis, including strains isolated from both healthy and diseased animals. Interestingly, in the same study, the authors reported that temperature can inhibit AHL production in the pathogen *V. harveyi*. Collectively, these data suggest that quorum sensing may play a role in coral diseases and *Vibrio* infections, but the potential mechanisms behind this hypothesis remain to be elucidated (Hmelo et al., 2011).

More pieces of information have recently come from researchers studying the white and black band diseases, which are polymicrobial illnesses threatening corals. Some strains capable of quorum sensing were isolated within the bacterial consortia responsible for the coral affections (Zimmer et al., 2014). Interestingly, it was then shown that the exposure of *Acropora cervicornis* to C6-HSL converts its healthy microbiome into one responsible for white band disease (Meyer et al., 2016). Moreover, the addition of quorum sensing-inhibiting compounds dramatically modified the composition of the infected coral microbiota, as revealed by 16SrRNA gene sequencing (Certner and Vollmer, 2015). Also, the inoculation to *Aiptasia pallida* of coral commensals presenting a quorum quenching activity inhibited the capacity of *Serratia marcescens* to degrade polyps (Alagely et al., 2011). Similarly, some authors studied the black band disease and pointed the production by associated cyanobacteria of lyngbic acid, a strong antiquorum sensing compound that can selectively inhibit cell-to-cell communication in some bacterial species (Meyer et al., 2016). Collectively, these results acquired from studying two different coral diseases underline the importance of quorum sensing in the regulation of coral-associated microbiota in either healthy or diseased animals (Hmelo et al., 2011; Teplitski et al., 2016).

4.5 Sponge-associated Communities

The sponge (phylum *Porifera*) microbiome is complex and diverse. It has been found that some sponges are poorly colonized by bacteria, while some others are not (Hentschel et al., 2003). However, bacteria colonizing sponge tissues can represent up to 35% of the sponge biomass (Vacelet and Donadey, 1977). Again, in this type of micro-niche, bacteria can frequently reach a sufficient abundance to establish a quorum sensing-based communication, and diverse sponge symbionts able to produce AIs are found (Mohamed et al., 2008; Taylor et al., 2004). In addition, some reports suggest that this type of communication between sponge bacterial symbionts is frequent, as 77% of studied Australian sponges are able to activate AHL-based biosensors (Taylor et al., 2004) and 46% of sponge species collected in the Mediterranean and Red Sea (Britstein et al., 2017). The chemical diversity of AHLs involved in these microbiota has been elucidated and was found to be very diverse, including short and long acyl side chains and between 6 and 18 carbons (Saurav et al., 2017). For example, C6-HSL, C7-HSL and 3-oxo-C12-HSL were detected in the Celtic sea sponge *Suberites domuncula* (Gardères et al., 2012), and an increasing number of AHLs from sponges are now being characterized (Bose et al., 2017; Britstein et al., 2016). The strain *Ruegeria* sp. KLH11, isolated from *Mycale laxissima*, has been developed as a model to better understand the cellular effects of quorum sensing and to better characterize the relationship between quorum sensing expression and the traits of the bacterial sponge symbionts. These studies revealed the presence of two pairs of quorum sensing *luxI/R* genes and an orphan *luxI* gene that control biofilm formation (negatively regulated) and flagella-based motility (positively regulated) in this KLH11 strain model (Zan et al., 2011, 2013, 2015). Such regulation might limit the bacterial aggregation within the sponge, and favors its dispersion and release in the environment. Quorum sensing-dependent relationships between sponges and their symbionts appear to be very complex. For example, bacterial 3-oxo-C12-HSL modifies gene expression in *S. domuncula* and inhibits its innate immune system (Gardères et al., 2014), and some sponge compounds are able to interfere with quorum sensing signaling (Costantino et al., 2017).

5 Diversity of Prokaryotic Functions Regulated by Quorum Sensing in Marine Environments

A large diversity of prokaryotic functions is regulated by quorum sensing in marine environments and include among others the production of bioluminescence, the formation and inhibition of biofilms, pigment synthesis, and many more. A proposition of synthetic view to summarize this diversity of functions is presented on Fig. 7.

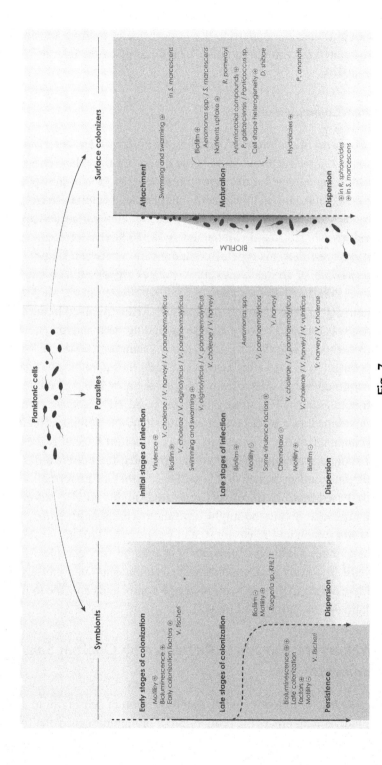

Fig. 7

A synthetic picture that attempts to synthesize and order the diversity of quorum sensing-regulated functions in marine environments and biofilms. *V.* = *Vibrio*; *S.* = *Serratia*; *R. pomeroyi* = *Ruegeria pomeroyi*; *R. sphaeroides* = *Rhodobacter sphaeroides*.

5.1 Bioluminescence and Pigment Production

Pioneering papers about quorum sensing described the details of bioluminescence regulation in the historical model *V. fischeri*. These details and references are provided to the reader in the first section of this chapter, and here is a reminder that bioluminescence is an important marine bacterial trait regulated by quorum sensing.

Pigment production (and in particular the purple violacein) is known to be regulated by quorum sensing (McClean et al., 1997). In the marine environment, a quorum sensing-dependent production of violacein has been reported in the marine strain *Pseudoalteromonas ulvae* TC14 (Ayé et al., 2015). Interestingly, in planktonic conditions, the strain does not appear to produce AHLs; nevertheless, the production of violacein is upregulated by the addition of C6-, C12-, 3-oxo-C8-, and 3-oxo-C12-HSL and downregulated by 3-oxo-C6-HSL. In sessile conditions, the 3-oxo-C8-HSL was found to upregulate the emission of the purple pigment.

5.2 Motility

In the initial stages of colonization, or at low or average cell densities, motility is frequently activated by quorum sensing. For example, motility has been found upregulated in the *V. fischeri* colonizing the squid *E. scolopes* (Lupp and Ruby, 2005; Lupp et al., 2003). The AphA quorum sensing master regulator is involved in low cell densities motility behaviors, like in *Vibrio alginolyticus* (Gu et al., 2016) and *Vibrio parahaemolyticus* (Wang et al., 2013). Similar results were observed with *V. harveyi* (van Kessel et al., 2013). However, in *V. cholerae*, the upregulation of motility is indirectly controlled by quorum sensing and depends on the intracellular c-di-GMP pool (Rutherford et al., 2011). At high cell densities, the master regulators LuxR/HapR are also able to control motility in some *Vibrio* species, like in the marine pathogen *V. vulnificus* (Lee et al., 2007). By contrast, OpaR negatively regulates motility in *V. parahaemolyticus* and *V. fischeri* (Lupp and Ruby, 2005). Interestingly, LuxR also activates motility in *V. cholerae* at high cell densities, probably to promote cell detachment and biofilm dispersion (Reidl and Klose, 2002).

5.3 Initiation of Biofilm Formation, Maturation, and Dispersion

Pioneer studies in the medical field revealed a positive relationship between quorum sensing expression and biofilm formation (Davies et al., 1998). However, the relationship between biofilm formation and emission of quorum sensing compounds is more complex. For example, among *Rhodobacteraceae*, *Silicibacter lacuscaerulensis*, and *S. pomeroyi* have quorum sensing genes, but they do not have the same capacities for surface colonization (Slightom and Buchan, 2009). One possible explanation for such complexity is that biofilm formation, maturation, and dispersion rely on diverse and multiple steps that are under the control of multiple regulatory and signaling pathways that are frequently interconnected and can act differently before being integrated at the cellular level. The in-depth genetic analysis conducted on the aquatic pathogens *V. cholerae* and *V. harveyi* sheds light on such complexity. In these cells, the sensory transduction pathways of quorum sensing, chemotaxis, and two-component signaling are connected to each other and to the c-di-GMP intracellular pool, as well as small RNA-mediating signaling pathways. In these strains, it has been experimentally demonstrated that this network of signaling pathways is associated with biofilm formation (Hunter and Keener, 2014; Srivastava et al., 2011; Srivastava and Waters, 2012; Svenningsen et al., 2009; Tu and Bassler, 2007).

Another layer of explanation is that biofilm formation, surface attachment, biofilm maturation and dispersion require different types of regulation according to the life strategies of bacteria. Quorum sensing has been found to be involved in surface attachment in *S. marcescens* (Labbate et al., 2007), an opportunist pathogen that can sometimes be found in marine environments (Alagely et al., 2011). In some *Vibrios* species, at low cell densities, the AphA master regulator also controls biofilm formation. This has been observed in *V. cholerae*. (Yang et al., 2010), *V. parahaemolyticus* (Wang et al., 2013), *V. alginolyticus* (Gu et al., 2016).

At high cell densities, quorum sensing LuxR-type proteins repress biofilm production in *V. cholerae* (Waters et al., 2008). However, the opposite pattern is observed in *V. parahaemolyticus* and *V. vulnificus* (Lee et al., 2007; Yildiz and Visick, 2009), and *V. anguillarum* (Croxatto et al., 2002). Many bacteria also positively regulate biofilm production and maturation with quorum sensing. This is the case in the genus *Aeromonas* (Lynch et al., 2002), which includes fish pathogens. In maturating biofilms, other functions are activated by quorum sensing, such as the production of antimicrobial compounds or the maintenance of cell shape heterogeneity (see later in this chapter).

The dispersion of biofilm is also under the control of quorum sensing in many marine bacteria. In the sponge symbiont *Ruegeria* sp. KLH11 (Zan et al., 2012), cell motility is activated by quorum sensing and favors dissemination. At the same time, biofilm production is inhibited by quorum sensing at high cell densities, which probably also favor the dispersion of *Ruegeria* sp. KLH11 cells from its host (Zan et al., 2012). This is also probably why biofilm production is negatively regulated at high cell densities in the pathogen *V. cholerae* (Zhu and Mekalanos, 2003). The production of hydrolases may favor such dynamics. It has been demonstrated that the expression of some of these enzymes is positively regulated by quorum sensing, like in *P. ananatis* (Jatt et al., 2015).

Collectively, these data emphasize the roles of quorum sensing at each stage of biofilm development, regardless of the type of marine surfaces on which they grow: crucial in the establishment, maturation, and dispersion of marine biofilms. In addition, similar biofilm-dependent phenotypes are positively or negatively regulated by quorum sensing in diverse strains, reflecting diverse types of life strategies and adaptations of bacterial cells.

5.4 Virulence

Virulence factors are extracellular products released by bacterial cells and are involved in pathogenesis. Their production represents an important metabolic cost for the cell and thus is strictly regulated (Yang and Defoirdt, 2015). The relationships among virulence and expression of quorum sensing compounds are complex: it has been shown that virulence factors can be either positively or negatively controlled by quorum sensing, either at low cell density or high cell densities. Such complex regulation is interpreted as a need to differently tune the production factors at the different stages of infection (Natrah et al., 2011b). For example, in *V. harveyi*, both the low cell density master regulator AphA and the high cell density master regulator LuxR-type control the production of the type III secretion system 1 (T3SS1). Such tuning of T3SS1 allows a well-controlled peak of expression in between the low cell density state and the high cell density state (van Kessel et al., 2013).

At high cell densities, a positive regulation of caseinase, gelatinase, and other types of proteases has been reported in *V. harveyi* (Mok et al., 2003; Natrah et al., 2011b), *V. alginolyticus* (Rui et al., 2008), and *V. vulnificus* (Shao and Hor, 2001). Similarly, the production of metalloproteases (Mok et al., 2003) and extracellular toxins (Manefield et al., 2000) appears positively regulated by quorum sensing, as well as flagellar motility, which is frequently considered an important virulence factor (Yang and Defoirdt, 2015). By contrast, downregulation of chitinase A (Defoirdt et al., 2010), siderophore production (Lilley and Bassler, 2000) by quorum sensing has been evidenced in *V. harveyi*. Lipases and hemolysin are virulence factors whose production appears independent of quorum sensing expression (Natrah et al., 2011b).

5.5 Emission of Colonization Factors

In marine bacteria, quorum sensing regulates many colonization factors in *V. fischeri* colonizing Hawaiian bobtail squid *E. scolopes*. The production of quorum sensing mutants, combined with the examination of microarray data collected from both mutant and wild-type cultures, provided interesting data. Thus it has been demonstrated that *ainS* gene expression favors the initiation of squid colonization and, among others, positively regulates cell motility and exopolysaccharide production (Lupp and Ruby, 2005). At this step of colonization, *lux* genes are not fully expressed. By contrast, *lux* gene expression is involved in the production of late colonization factors and in the persistence of cells in the crypts of the squid's light organ. Collectively, these data suggest that the *ainS*-dependent quorum sensing system operates at moderate cell densities, while the lux-dependent system activity peaks at very high cells densities, such as those observed in the crypts of the light organ.

5.6 Horizontal Gene Transfer

The existence of quorum sensing-dependent horizontal gene transfer has been demonstrated in *V. cholerae* cultured in biofilm composed of mixed *Vibrio* species. These processes are under the control of the gene *comEA*, whose transcription was found to be induced in the presence of AI-1 and AI-2. As the experimental setup was based on mixed biofilms, it is probable that such horizontal transfer in *V. cholerae* occurs also through AIs emitted by other *Vibrio* species (Antonova and Hammer, 2011). It has also been shown in *V. cholerae* that the type IV secretion system, which permits cell lysis and extracellular DNA acquisition, is positively regulated by quorum sensing (Papenfort and Bassler, 2016).

The importance of quorum sensing in gene transfer has also been demonstrated in *Rhodobacter capsulatus*, a model bacterium in microbiology whose close relatives have been detected in marine environments. Interestingly, when the AIs synthase gene *gtaI* was muted with a Sp^r cassette, researchers observed a reduction in gene transfer agent production, which was restored by the addition of C16-HSL (Schaefer et al., 2002). Overall, all these data suggest a great importance of quorum sensing in the control of gene transfer, raising concerns about possible major ecological and evolutionary impacts. These questions remains fully open for future studies.

5.7 Nutrient Acquisition

One of the earliest hypotheses to explain the role of quorum sensing in bacterial cells is that cooperation may favor the acquisition of nutrients. In this sense, the extracellular hydrolytic enzymes emitted by diverse bacteria are public goods that will benefit the whole community. This hypothesis is not specifically adapted to the marine environment, with a population-level benefit from the secretion of proteases shown in *P. aeruginosa* cultures (Darch et al., 2012).

In the marine environment, this hypothesis has been explored mostly with bacteria colonizing the phycosphere of microalgae or marine snow. In a paper published in 2012, it has been shown that the epibionts of *Trichodesmium* rely on quorum sensing to upregulate their phosphate acquisition by the production of alkaline phosphatases. It appeared that AHLs are involved in this process, while AI-2 production led to a decrease in phosphate uptake (Van Mooy et al., 2012). In a similar vein, *R. pomeroyi* overproduces 3-oxo-C14-HSL when cultured in the presence of dimethylsulfoniopropionate (DMSP) as an energy source. Interestingly, this observation combined with the record of a drastic modification of the cell metabolome suggests that *Ruegeria* switches to a cooperative lifestyle when cultured with algal DMSP provided as a source of sulfur (Johnson et al., 2016).

Another series of papers revealed interesting links between the production of chemical compounds by decaying algae and the expression of quorum sensing. Indeed, it has been shown that *p*-coumaric acid, a product of algal lignin degradation released by decaying phytoplankton cells, is also the precursor of *p*-coumaroyl-HSL involved in *R. palustris* quorum sensing, as well as in the marine *S. pomeroyi* DSS-3 strain (Schaefer et al., 2008). In this sense, the production of semiochemicals associated with the release of phytoplanktonic compounds might convey some information about nutrients and exogenously supplied substrates availability within the phycosphere (Buchan et al., 2014; Schaefer et al., 2008). Interestingly, the algae symbiont *D. shibae* controls flagellar biosynthesis with quorum sensing (Patzelt et al., 2013), which potentially enables chemotaxis to microalgae and thus favors the acquisition of nutrients.

The coordination of marine bacteria for nutrient acquisition probably has broader consequences at a global scale, although this hypothesis remains poorly explored. Nonetheless, the reports of Hmelo et al. (2011) and Jatt et al. (2015) support similar conclusions. Hmelo et al. (2011) collected and incubated marine sinking particles sampled near Vancouver Island, noticing an increase in hydrolytic enzyme activity when adding commercially available AHLs. Similarly, Jatt et al. (2015) observed an enhancement of alkaline phosphatase activity when adding C10-AHL to a *P. ananatis* culture initially isolated from marine snow. Collectively, these data suggest that quorum sensing may regulate marine snow mineralization kinetics, yet many more observations are required to better characterize and conceptualize the biogeochemical implications of quorum sensing expression in seawater.

5.8 Production of Antimicrobial Compounds

Quorum sensing is involved in the regulation of many compounds that act on cell density and on community dynamics, such as antimicrobial molecules (Wood and Pierson, 1996). In the marine environment, such observations have been made on many strains producing algaecidal compounds. Among those, TDA is an antimicrobial compound that also induces its own synthesis and thus also acts as an AI (Berger et al., 2011; Bruhn et al., 2005; Geng et al., 2008; Porsby et al., 2008). Interestingly, the production of TDA is also dependent on AHLs in diverse *Roseobacter* species (Berger et al., 2011; Rao et al., 2007; Thole et al., 2012). The production of TDA regulates interesting community traits in the association between *Phaeobacter gallaeciensis* BS107 and the microalgae *Emiliania huxleyi*. The bacteria provides growth inducers to the alga, such as auxins during bloom conditions, and produce antibiotics including TDA probably to target algal pathogens (Geng et al., 2008; Thiel et al., 2010). In this symbiotic relationship, *P. gallaeciensis* receives DMSP produced by the microalgae as a sulfur source (González et al., 1999; Newton et al., 2010).

Different types of algaecides whose production could be under the control of quorum sensing compounds have been identified from marine bacteria (Nakashima et al., 2006; Paul and Pohnert, 2011; Skerratt et al., 2002). For example, *Kordia algicida* displays algaecide activity due to the emission of proteases, and some experimental data suggest that this expression is quorum sensing-regulated (Paul and Pohnert, 2011). In addition, Skerratt et al. (2002) suggested a potential role of AI-2 in the algaecide activity against the dinoflagellate *G. catenatum*. More recently, it has been shown that the strain *Ponticoccus* sp. PD-2 isolated from the phycosphere of the microalga *Prorocentrum donghaiense* regulates its algicidal activity using quorum sensing. This strain produces 3-oxo-C8-HSL and 3-oxo-C10-HSL using two quorum sensing networks (*zlaI/R* and *zlbI/R*). The inhibition of quorum sensing activity using β-cyclodextrin reduced the algicical activity of more than 50% (Chi et al., 2017).

5.9 Induction and Maintenance of Phenotypic Heterogeneity within a Cell Population

Interestingly, a few reports have suggested that quorum sensing could be involved in the maintenance of population heterogeneity, which is hypothesized to be a survival strategy under fluctuating environmental conditions. It has been suggested that such phenotypic heterogeneity might increase the fitness of the whole population by enhancing survival in a subpopulation when facing changing conditions. Such observations were made initially on *V. harveyi* (now reclassified as *Vibrio campbellii*). The authors of this study revealed that in a bioluminescent population of cells, only 69% were effectively emitting light, and that the strength of the bioluminescence was different between cells. The key LuxO protein in the quorum sensing signaling pathway (see earlier in this chapter for more details) seems to also play a role in these mechanisms, as the culture of a *luxO* mutant is composed of only glowing cells (Anetzberger et al., 2009).

Interesting data linking quorum sensing and phenotypic heterogeneity were also collected, coupling genetic and transcriptomic studies conducted on the marine *D. shibae* (Patzelt et al., 2013). In these experiments, the authors revealed that the lack of AHL production in a *luxI* mutant strain modified the transcription of 344 genes involved in the regulation of many different physiological activities, including cell division, flagellar biosynthesis, and sigma factor synthesis. Interestingly, this *luxI* mutant culture presented a single ovoid phenotype of *Dinoroseobacter* cells. By contrast, the wild-type phenotype, as well as the mutant culture amended with addition of C18-AHL at saturating concentrations, included more cell morphologies with ovoid, rod-shaped, and very elongated cells. In marine bacterial populations, the existence of such a morphological heterogeneity might be an ecological advantage. During phytoplankton blooms, grazing is intense and has been demonstrated to be cell shape-dependent. Thus a bacterial population might enhance its fitness by allowing a proportion of the population to stochastically vary its phenotype and ensuring the survival of at least a fraction of the population during environmental fluctuations (Acar et al., 2008), such as over a seasonal marine plankton bloom. Such observations are not specific to the marine environment and remain a subject of interrogation in diverse fields of microbiology (Grote et al., 2015).

6 Sourcing and Applying Quorum Sensing Inhibitors in the Marine Environment: Aquaculture Protection and Innovation in Green Antifouling

6.1 A model story: The Discovery of Antifouling Compounds in the Red Marine Alga Delisea pulchra

The first quorum sensing-inhibiting compound was discovered in seawater. This pioneering research is frequently cited as a model process to identify antiquorum sensing compounds with potential economic applications. This research story begins with a naturalist observation of *D. pulchra*, a small red benthic macroalgae whose fronds measure a few centimeters. First, these fronds have the particularity of being poorly colonized by different types of organisms. Unlike most other types of marine plants and algae, bacterial abundances on fronds surfaces are significantly lower than those on other seaweed surfaces (Maximilien et al., 1998). Then, it is important to notice that on the east coast of Australia, these algae are subjected to seasonal bleachings (Campbell et al., 2011) that appear in summer when the temperature rises. Interestingly, *D. pulchra* is able to produce halogenated furanones, whose concentrations are reduced during these episodes of bleaching (Campbell et al., 2011).

Interestingly, it has been observed that halogenated furanones share structural similarities with AHLs, and it has been hypothesized that they could interfere with quorum sensing signaling pathways (Givskov et al., 1996). These halogenated furanones are able to modify many different bacterial properties, such as swarming in *Serratia liquefaciens* (Givskov et al., 1996) and *Proteus mirabilis* (Gram et al., 1996), exoenzyme production in diverse bacteria (Kjelleberg et al., 1997), as well as bioluminescence and pigment production (Givskov et al., 1996; Kjelleberg et al., 1997). The molecular and genetic mechanisms underlying the furanones' mode of action were then investigated, and it was demonstrated that these compounds bind to the LuxR receptors and increase their degradation rates turnover (Manefield et al., 2000, 2002).

Then the relationships between *Delisea*'s bleaching events, temperature increase, and bacterial quorum sensing were also investigated. The incubation of *D. pulchra* sporelings in water containing natural bacterioplankton increased their susceptibility to be subjected to bleaching compared to algae immerged in sterile seawater, leading the authors to involve some role for bacteria in bleaching events. Also, the absence of halogenated furanones increased the susceptibility of *D. pulcha* to bleachings (Campbell et al., 2011), revealing the importance of quorum quenching to limit the effects of this disease. One of the responsible bacteria has been isolated and identified as a *Nautella* sp., which penetrates algae's tissues (Case et al., 2011). The virulence mechanisms remain to be elucidated (Harder et al., 2014); it appears that these bacteria could provoke at least part of the observed summer bleachings and that the production of halogenated furanones allows *D. pulchra* to control pathogens.

Natural or synthetic halogenated furanones were then tested with some success against common pathogens (Fig. 8). It has been shown that these compounds inhibit quorum sensing

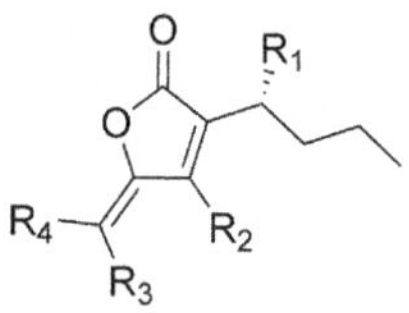

	R$_1$	R$_2$	R$_3$	R$_4$
Furanone 1	H	Br	Br	Br
Furanone 2	H	Br	H	Br
Furanone 3	OAc	Br	H	Br
Furanone 4	OH	Br	H	Br
Furanone 5	OAc	Br	H	I
Furanone 6	H	H	Br	Br

Fig. 8

Some of the halogenated furanones produced by *Delisea pulchra*. *Reproduced from Manefield et al. (1999). Compound (6) is synthetic.*

processes in *P. aeruginosa* biofilms (Hentzer et al., 2002), as well as carbapenem antibiotic synthesis and exoenzyme virulence factor production in *Erwinia carotovora*, a plant pathogen (Manefield et al., 2001). Quorum sensing-inhibiting compounds are usually suspected to avoid any apparition of resistance in targeted bacteria. However, this paradigm has to be revisited with recent data and observations. Some *P. aeruginosa* strains were found to be resistant to some of the synthetic halogenated furanones intensively used in laboratories (García-Contreras et al., 2013). The resistance mechanism appears to be linked to the capacity of the strains to increase the efflux of the furanones out of cells (Maeda et al., 2012).

6.2 Diversity of Quorum Quenching Compounds Isolated from Marine Organisms

A large panel of quorum quenching compounds has been isolated from a wide range of marine organisms (Fig. 9). These include bacteria, cyanobacteria, fungi, marine sponges, marine algae, cnidarians, and Bryozoa. A recently published review fully details the current catalog of quorum sensing-inhibiting compounds isolated from seawater (Saurav et al., 2017) and reports

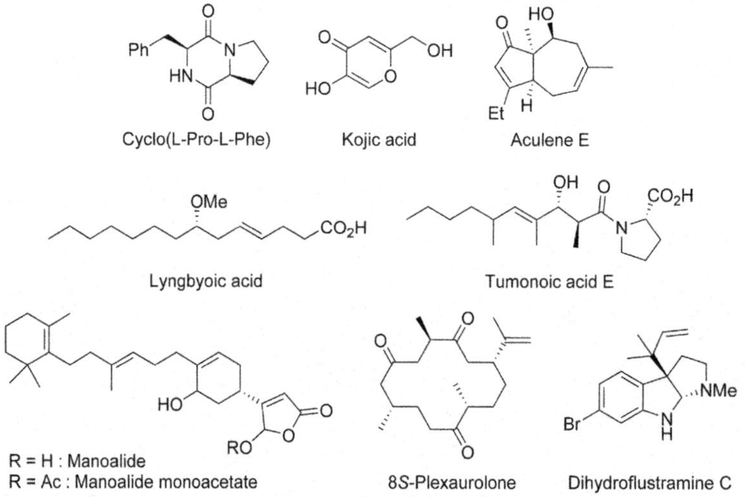

Fig. 9

A snapshot of the diversity of quorum sensing-inhibiting compounds.

the existence of 70 marine-derived molecules presenting such activity. This review reveals the wide spectrum of marine organisms and chemical compounds capable of such biological activity. In bacteria, compounds as diverse as phenethylamides, cyclic dipeptides, tyrosol, and tyrosol acetate have been found to present quorum quenching activity (Abed et al., 2013; Martínez-Matamoros et al., 2016; Teasdale et al., 2011). In marine fungi, aculenes C, D, E; penicitor; aspergillumarins A, B; meleagrin; and kojic acid were found to be quorum sensing inhibitors (Dobretsov et al., 2011; Kong et al., 2017; Li et al., 2003). In cyanobacteria, malyngamide C, 8-epi-malyngamide C, malyngolide, and lyngbyic acid showed antiquorum sensing properties, as well as tumonoic acids; honaucins A, B, C; pitinoic acid; and microcolins A, B (Choi et al., 2012; Clark et al., 2008; Dobretsov et al., 2011; Kwan et al., 2010, 2011; Montaser et al., 2013). Such biological activity was also reported in manoalides isolated from sponges (Montaser et al., 2013), cembranoids isolated from cnidarians (Tello et al., 2012), and brominated alkaloids from Bryozoa (Peters et al., 2003).

6.3 Diversity of Marine Quorum Quenching Enzymes

A large diversity of bacteria is able to produce quorum quenching enzymes, and the use of these enzymes to fight quorum sensing-dependent traits in bacteria is very promising. These enzymes act extracellularly to degrade AIs and can be used in catalytic quantities (Bzdrenga et al., 2017). They are able to disrupt the quorum sensing signal in the nearby environment of bacterial cells and thus do not have to penetrate bacterial cells, facilitating their mode of action and effects. Although various types of enzymes present such activity (Romero et al., 2015), two major types of quorum quenching enzymes have been described: lactonases (which hydrolyze the lactone ring of AHLs) and acylases (which cleave the amide bond linking the lactone cycles to the acyl side chains in AHLs) (Grandclement et al., 2016). Many studies have tried to use quorum quenching enzymes to disrupt the virulence of pathogens or biofilm formation. Specifically, quorum quenching enzymes have been tested against plant pathogens and against biofilm formation on membrane bioreactors and medical devices (Grandclement et al., 2016).

Marine bacteria are an important resource for prospecting new quorum quenching enzymes. A report published by Romero et al. (2011) screened 166 marine strains and identified 24 that were able to significantly degrade AHLs (Romero et al., 2011). It has recently been shown that *Alteromonas stellipolaris*, a bacterium selected among 450 strains isolated from a bivalve hatchery, has a very efficient lactonase activity (Torres et al., 2016). Similarly, enzymatic quorum quenching activities were also detected in corals (Golberg et al., 2013; Tait et al., 2010) and sponges (Saurav et al., 2016a). In the same perspective, metagenomic studies revealed a large diversity of quorum quenching genes in marine microbial communities (Muras et al., 2018a; Romero et al., 2012).

Unsurprisingly, a few research groups attempted to express, characterize, and test the potential of quorum quenching enzymes issued from the marine environment. For example, MomL is a lactonase isolated from the *Flavobacteria Muricauda olearia* Th120, a flounder mucus-derived strain. Interestingly, this enzyme was shown able to significantly reduce the virulence and biofilm formation of *P. aeruginosa* (Tang et al., 2015). Similarly, a *Bacillus licheniformis* strain was isolated from dissected guts collected from Indian white shrimps. This enzyme presents a high resistance to acid conditions (likely as a consequence of the native conditions experienced in the gut), has a large spectrum of AHLs that it can degrade, and appears to reduce *Vibrio* biofilms (Vinoj et al., 2014). Similarly, a wide-spectrum thermostable lactonase has been isolated from the marine bacteria *Tenacibaculum* sp. 20J (Mayer et al., 2015), which present antbiofilm activity against *Streptococcus mutans* (Muras et al., 2018b) and against the fish pathogen *Edwardsiella tarda* (Romero et al., 2014).

6.4 Applications for Biofouling on Marine Underwater Surfaces

The term "biofouling" refers to the biological colonization by organisms on living and nonliving materials. Such colonization of underwater surfaces by marine organisms causes significant problems, such as biocorrosion, increased frictional forces on boat hulls, obstruction of submerged instruments, and devices such as propellers or scientific instruments, and problems in marine aquaculture (Dobretsov et al., 2011). This colonization starts with the development of a bacterial biofilm (microfoulers), which serves as a substrate for the settlement of larger organisms (algae, metazoans: macrofoulers). Thus fighting biofilm development constitutes a current major challenge in many industries, such as underwater electricity production, boat construction, and naval transportation. Most of the antifouling compounds used are toxic biocides, which deeply impact the natural environment since they are released in the seawater. For example, they accumulate along the food chain, leading to a high concentration of toxic substances in marine mammals. They can also directly affect filter-feeding organisms such as oysters (Thomas and Brooks, 2010). One possible strategy to overcome these difficulties is to target key mechanisms responsible for biofilm formation without affecting cell viability to select compounds with low toxicity. In this sense, the application of quorum quenching compounds appears to meet these criteria and seems to be a promising strategy against biofouling. Thus quorum sensing-inhibiting compounds have received growing interest in recent years, especially when considering the interesting results collected on the model macro-algae *D. pulchra* described previously (Rasmussen et al., 2000).

While natural quorum sensing-inhibiting compounds have been widely isolated from a large diversity of living organisms, their effects on marine biofilms remain poorly explored. One reason is that these compounds are produced at low concentrations, which limits their application in the environment. However, synthetic quorum sensing-inhibiting compounds can now be used. In a pioneering study published in 2011 focusing on natural compounds, Dobretsov et al.

screened 78 natural products isolated from marine organisms and terrestrial plants (Dobretsov et al., 2011). Among them, 24% were able to inhibit quorum sensing on the classical reporter strain *C. violaceum* CV017 without causing toxicity. The authors demonstrated that hymenialdisine, demethoxy encecalin, microcolins A and B, and kojic acid (an oxo-pyrone) were involved in these processes, and kojic acid was able to inhibit the formation of biofilms on glass slides. Since then, a few studies have used similar approaches to evaluate the effect of quorum sensing-inhibiting compounds on marine biofilms. For example, some studies have evaluated the potential antifouling effects of three quorum sensing inhibitors (3,4-dibromo-2(5) H-furanone, 4-nitropyridine-*N*-oxide and indole) on the growth of two diatoms that can be found in marine biofilms, *Cylindrotheca* sp. and *Nitzschia closterium*. The authors of this study revealed a significant effect on the biofilm formation of these compounds and reported that the production of the polymeric substance was significantly reduced. 4-Nitropyridine-*N*-oxide appeared to be the most effective compound. In addition, AI-2 based quorum sensing has also been targeted in recent studies to limit biofilm formation. In this vein, it has been reported that the penicillic acid is able to inhibit AI-2 production in *Halomonas pacifica* as well as biofilm formation without affecting bacterial growth and at up to $10\,\mu M$ in concentration (Liaqat et al., 2014).

6.5 Applications in Aquaculture

Aquaculture is another industry in which quorum quenching-based strategies have potentially important applications (Grandclement et al., 2016). The dense breeding conditions along with the growth of this economic sector mean that fish diseases are becoming more prevalent. Prophylactic measures have limited efficacy, while vaccination is effective against some aquatic pathogens but also presents many side effects. The use of antibiotics restrains commercialization of fish culture, as their overuse leads many pathogens to become resistant to these compounds. Thus there is an important need to renew our panel of methods to fight fish pathogens if sustained growth of the aquaculture economy is to continue. Again, quorum sensing is a key mechanism in the activation of virulence in many fish pathogens; thus quorum quenching strategies appear to be a very promising research direction in controlling fish diseases (Defoirdt et al., 2011).

One of the first strategies was based on the use of chemical compounds derived from (micro) algae in aquaculture, to protect fishes, shellfishes, and crustaceans. It has been shown that natural and synthetic furanones were able to protect the shrimp *Artemia franciscana* from *V. harveyi*, *V. campbellii*, and *V. parahaemolyticus* infections (Defoirdt et al., 2006; Givskov et al., 1996). These compounds have also been reported to protect rainbow trout from vibriosis (Rasch et al., 2004). However, these compounds present toxicity for some fishes, such as the rainbow trout. The use of synthetic compounds like brominated thiophenones, which present a much lower toxicity, appears promising (Defoirdt et al., 2012; Rasch et al.,

2004). In the same vein, the potential of diverse microalgae to produce quorum quenching compounds has been evaluated, and *Chlorella saccharophila* has been revealed to present an important potential for use in aquaculture (Natrah et al., 2011a).

Another strategy is the use of quorum quenching enzymes or the provision of probiotics able to deliver such enzymes. The application of these ideas has already achieved some success. For example, carps that were fed quorum quenching AiiA recombinant protein were more resistant to *Aeromonas hydrophila* (Cao et al., 2012). In addition, the microbial communities from the guts of the shrimp *Penaeus vannamei* present AI-1-degrading activity that improved the growth of rotifers in the presence of *V. harveyi*. Similarly, bacterial consortia from the guts of *Dicentrarchus labrax* and *Lates calcarifer* were evaluated to protect prawn (*Macrobrachium rosenbergii*) larviculture (Nhan et al., 2010), and AHL-degrading communities improved the survival of first-feeding turbot larvae (*Scophthalmus maximus* L.) (Tinh et al., 2008). Such selection of quorum quenching-producing strains also gave interesting preliminary data for biocontrol in coral aquaculture. The strain *A. stellipolaris* significantly reduced the pathogenicity of *Vibrio mediterranei* upon the coral *Oculina patagonica* (Torres et al., 2016). A few research teams have also tried to deliver to fishes as probiotics *Bacillus* strains able to produce quorum quenching enzymes. For example, *Bacillus*-related strains isolated from the gut of the fish *Carassius auratus* and emitting quorum quenching enzymes were able to protect zebrafish from infections (Chu et al., 2014). This field of research had led some companies to develop commercial products, such as Biomin (Inzersdorf-Getzersdorf, Austria), which markets the probiotic Aquastar. This product is a *Bacillus* strain encapsulated with food and producing quorum quenching enzymes to be delivered in shrimp cultures (Grandclement et al., 2016).

7 Concluding Remarks

Since its discovery in the 1970s in the marine environment, quorum sensing has been mostly studied in strains of medical or agronomic interest. The roles and importance of quorum sensing among marine strains remain poorly studied, with the notable exception of *Vibrio* strains, as many are important pathogens. Thus a lot of work remains to be done to characterize the diversity of quorum sensing genes, the diversity of quorum sensing compounds, and their biological roles in strains of interest in the field of environmental sciences. In addition, the consequences of quorum sensing expression in biogeochemical cycles have been poorly explored. Another important field of future investigations is the elucidation of quorum sensing roles in the holobionts' microbial balance, as a growing number of studies are revealing the importance of bacteria-eukaryotes (inter)relationships based on quorum sensing compounds emission, perception, and inhibition, including in marine models.

Knowledge of quorum sensing can lead to important biotechnological applications, especially in the marine environment, where aquaculture and antifouling industries can benefit from these technologies. While many interesting results have been published, data concerning larger experiments or applications tested in realistic conditions remain scarce. There are diverse possible explanations for such observations: (i) studies are underway in industrial labs and have not been yet published, (ii) negative results were obtained or difficulties were raised when experimenting on larger scales than in research labs, and (iii) the output of these works is protected by industrial secrets and cannot be released to the scientific community. Also, few quorum quenching agents have been tested for industrial applications, and the catalog of available natural and synthetic compounds needs to be expanded. Whatever the case(s), much more information is required to better evaluate whether quorum quenching strategies can be applied at large scales to solve industrial problems.

Glossary

Claisen condensation Claisen condensation leads to the formation of a carbon-carbon bond from two esters (or one ester and a carbonyl compound). This reaction occurs in the presence of a strong base and leads to a β-keto ester or a β-diketone.

Phytoplankton bloom A phytoplankton bloom is a rapid growth of microalgae in fresh- or seawaters.

Small regulatory RNAs (Small RNAs, sRNAs) 50- to 500-nucleotides RNAs that affect gene expression.

Abbreviations

AHL	acyl-homoserine lactone
AI	autoinducer
CAI	cholerae autoinducer
DPD	4,5-dihydroxy-2,3-pentanedione
GC-MS	gas chromatography-mass spectrometry
LC-MS	liquid chromatography-mass spectrometry
NMR	nuclear magnetic resonance
RT-qPCR	quantitative reverse transcription PCR
TDA	tropodithietic acid
TLC	thin-layer chromatography
UHPLC	ultra-high-performance liquid chromatography

Acknowledgments

The author deeply thanks Carole Petetin and Didier Stien for help in drawing the figures, and Sheree Yau for advices in English grammar and spelling.

References

Abed, R.M., Dobretsov, S., Al-Fori, M., Gunasekera, S.P., Sudesh, K., Paul, V.J., 2013. Quorum-sensing inhibitory compounds from extremophilic microorganisms isolated from a hypersaline cyanobacterial mat. J. Ind. Microbiol. Biotechnol. 40 (7), 759–772.

Acar, M., Mettetal, J.T., Van Oudenaarden, A., 2008. Stochastic switching as a survival strategy in fluctuating environments. Nat. Genet. 40 (4), 471–475.

Alagely, A., Krediet, C.J., Ritchie, K.B., Teplitski, M., 2011. Signaling-mediated cross-talk modulates swarming and biofilm formation in a coral pathogen *Serratia marcescens*. ISME J. 5 (10), 1609–1620.

Anetzberger, C., Pirch, T., Jung, K., 2009. Heterogeneity in quorum sensing-regulated bioluminescence of *Vibrio harveyi*. Mol. Microbiol. 73 (2), 267–277.

Antonova, E.S., Hammer, B.K., 2011. Quorum-sensing autoinducer molecules produced by members of a multispecies biofilm promote horizontal gene transfer to *Vibrio cholerae*. FEMS Microbiol. Lett. 322 (1), 68–76.

Ayé, A.M., Bonnin-Jusserand, M., Brian-Jaisson, F., Ortalo-Magné, A., Culioli, G., Nevry, R.K., Rabah, N., Blache, Y., Molmeret, M., 2015. Modulation of violacein production and phenotypes associated with biofilm by exogenous quorum sensing *N*-acylhomoserine lactones in the marine bacterium *Pseudoalteromonas ulvae* TC14. Microbiology 161 (10), 2039–2051.

Bachofen, R., Schenk, A., 1998. Quorum sensing autoinducers: do they play a role in natural microbial habitats? Microbiol. Res. 153 (1), 61–63.

Berger, M., Neumann, A., Schulz, S., Simon, M., Brinkhoff, T., 2011. Tropodithietic acid production in *Phaeobacter gallaeciensis* is regulated by *N*-acyl homoserine lactone-mediated quorum sensing. J. Bacteriol. 193 (23), 6576–6585.

Biswa, P., Doble, M., 2013. Production of acylated homoserine lactone by gram-positive bacteria isolated from marine water. FEMS Microbiol. Lett. 343 (1), 34–41.

Blanchet, E., Prado, S., Stien, D., Oliveira da Silva, J., Ferandin, Y., Batailler, N., Intertaglia, L., Escargueil, A., Lami, R., 2017. Quorum sensing and quorum quenching in the mediterranean seagrass *Posidonia oceanica* microbiota. Front. Mar. Sci. 4, 218.

Bodor, A., Elxnat, B., Thiel, V., Schulz, S., Wagner-Döbler, I., 2008. Potential for *luxS* related signalling in marine bacteria and production of autoinducer-2 in the genus *Shewanella*. BMC Microbiol. 8 (1), 13.

Bose, U., Ortori, C.A., Sarmad, S., Barrett, D.A., Hewavitharana, A.K., Hodson, M.P., Fuerst, J.A., Shaw, P.N., Boden, R., 2017. Production of *N*-acyl homoserine lactones by the sponge-associated marine actinobacteria *Salinispora arenicola* and *Salinispora pacifica*. FEMS Microbiol. Lett. 364(2), fnx002.

Britstein, M., Devescovi, G., Handley, K.M., Malik, A., Haber, M., Saurav, K., Teta, R., Costantino, V., Burgsdorf, I., Gilbert, J.A., 2016. A new N-acyl homoserine lactone synthase in an uncultured symbiont of the red sea sponge *Theonella swinhoei*. Appl. Environ. Microbiol. 82 (4), 1274–1285.

Britstein, M., Saurav, K., Teta, R., Sala, G.D., Bar-Shalom, R., Stoppelli, N., Zoccarato, L., Costantino, V., Steindler, L., 2017. Identification and chemical characterization of N-acyl-homoserine lactone quorum sensing signals across sponge species and time. FEMS Microbiol. Ecol. 94(2), fix182.

Bruhn, J.B., Nielsen, K.F., Hjelm, M., Hansen, M., Bresciani, J., Schulz, S., Gram, L., 2005. Ecology, inhibitory activity, and morphogenesis of a marine antagonistic bacterium belonging to the *Roseobacter* clade. Appl. Environ. Microbiol. 71 (11), 7263–7270.

Buchan, A., LeCleir, G.R., Gulvik, C.A., Gonzalez, J.M., 2014. Master recyclers: features and functions of bacteria associated with phytoplankton blooms. Nat. Rev. Microbiol. 12 (10), 686–698.

Bzdrenga, J., Daude, D., Remy, B., Jacquet, P., Plener, L., Elias, M., Chabriere, E., 2017. Biotechnological applications of quorum quenching enzymes. Chem. Biol. Interact. 267, 104–115.

Campbell, A.H., Harder, T., Nielsen, S., Kjelleberg, S., Steinberg, P.D., 2011. Climate change and disease: bleaching of a chemically defended seaweed. Glob. Chang. Biol. 17 (9), 2958–2970.

Cao, Y., He, S., Zhou, Z., Zhang, M., Mao, W., Zhang, H., Yao, B., 2012. Orally administered thermostable *N*-acyl homoserine lactonase from *Bacillus* sp. strain AI96 attenuates *Aeromonas hydrophila* infection in zebrafish. Appl. Environ. Microbiol. 78 (6), 1899–1908.

Case, R.J., Longford, S.R., Campbell, A.H., Low, A., Tujula, N., Steinberg, P.D., Kjelleberg, S., 2011. Temperature induced bacterial virulence and bleaching disease in a chemically defended marine macroalga. Environ. Microbiol. 13 (2), 529–537.

Certner, R.H., Vollmer, S.V., 2015. Evidence for autoinduction and quorum sensing in white band disease-causing microbes on *Acropora cervicornis*. Sci. Rep. 5, 11134.

Charlton, T.S., De Nys, R., Netting, A., Kumar, N., Hentzer, M., Givskov, M., Kjelleberg, S., 2000. A novel and sensitive method for the quantification of *N*-3-oxoacyl homoserine lactones using gas chromatography–mass spectrometry: application to a model bacterial biofilm. Environ. Microbiol. 2 (5), 530–541.

Chen, X., Schauder, S., Potier, N., Van Dorsselaer, A., Pelczer, I., Bassler, B.L., Hughson, F.M., 2002. Structural identification of a bacterial quorum-sensing signal containing boron. Nature 415 (6871), 545–549.

Chi, W., Zheng, L., He, C., Han, B., Zheng, M., Gao, W., Sun, C., Zhou, G., Gao, X., 2017. Quorum sensing of microalgae associated marine *Ponticoccus* sp. PD-2 and its algicidal function regulation. AMB Express 7 (1), 59.

Choi, H., Mascuch, S.J., Villa, F.A., Byrum, T., Teasdale, M.E., Smith, J.E., Preskitt, L.B., Rowley, D.C., Gerwick, L., Gerwick, W.H., 2012. Honaucins A-C, potent inhibitors of inflammation and bacterial quorum sensing: synthetic derivatives and structure-activity relationships. Chem. Biol. 19 (5), 589–598.

Chu, W., Zhou, S., Zhu, W., Zhuang, X., 2014. Quorum quenching bacteria *Bacillus* sp. QSI-1 protect zebrafish (*Danio rerio*) from *Aeromonas hydrophila* infection. Sci. Rep. 4, 5446.

Clark, B.R., Engene, N., Teasdale, M.E., Rowley, D.C., Matainaho, T., Valeriote, F.A., Gerwick, W.H., 2008. Natural products chemistry and taxonomy of the marine cyanobacterium *Blennothrix cantharidosmum*. J. Nat. Prod. 71 (9), 1530–1537.

Costantino, V., Della Sala, G., Saurav, K., Teta, R., Bar-Shalom, R., Mangoni, A., Steindler, L., 2017. Plakofuranolactone as a quorum quenching agent from the Indonesian sponge *Plakortis* cf. *lita*. Mar. Drugs 15 (3), 59.

Croxatto, A., Chalker, V.J., Lauritz, J., Jass, J., Hardman, A., Williams, P., Camara, M., Milton, D.L., 2002. VanT, a homologue of *Vibrio harveyi* LuxR, regulates serine, metalloprotease, pigment, and biofilm production in *Vibrio anguillarum*. J. Bacteriol. 184 (6), 1617–1629.

Cude, W.N., Buchan, A., 2013. Acyl-homoserine lactone-based quorum sensing in the *Roseobacter* clade: complex cell-to-cell communication controls multiple physiologies. Front. Microbiol. 4, 336.

Darch, S.E., West, S.A., Winzer, K., Diggle, S.P., 2012. Density-dependent fitness benefits in quorum-sensing bacterial populations. Proc. Natl. Acad. Sci. U. S. A. 109 (21), 8259–8263.

Davies, D.G., Parsek, M.R., Pearson, J.P., Iglewski, B.H., Costerton, J.W., Greenberg, E.P., 1998. The involvement of cell-to-cell signals in the development of a bacterial biofilm. Science 280 (5361), 295–298.

Decho, A.W., Visscher, P.T., Ferry, J., Kawaguchi, T., He, L., Przekop, K.M., Norman, R.S., Reid, R.P., 2009. Autoinducers extracted from microbial mats reveal a surprising diversity of *N*-acylhomoserine lactones (AHLs) and abundance changes that may relate to diel pH. Environ. Microbiol. 11 (2), 409–420.

Decho, A.W., Frey, R.L., Ferry, J.L., 2011. Chemical challenges to bacterial AHL signaling in the environment. Chem. Rev. 111 (1), 86–99.

Defoirdt, T., Crab, R., Wood, T.K., Sorgeloos, P., Verstraete, W., Bossier, P., 2006. Quorum sensing-disrupting brominated furanones protect the gnotobiotic brine shrimp *Artemia franciscana* from pathogenic *Vibrio harveyi*, *Vibrio campbellii*, and *Vibrio parahaemolyticus* isolates. Appl. Environ. Microbiol. 72 (9), 6419–6423.

Defoirdt, T., Darshanee Ruwandeepika, H.A., Karunasagar, I., Boon, N., Bossier, P., 2010. Quorum sensing negatively regulates chitinase in *Vibrio harveyi*. Environ. Microbiol. Rep. 2 (1), 44–49.

Defoirdt, T., Sorgeloos, P., Bossier, P., 2011. Alternatives to antibiotics for the control of bacterial disease in aquaculture. Curr. Opin. Microbiol. 14 (3), 251–258.

Defoirdt, T., Benneche, T., Brackman, G., Coenye, T., Sorgeloos, P., Scheie, A.A., 2012. A quorum sensing-disrupting brominated thiophenone with a promising therapeutic potential to treat luminescent vibriosis. PLoS One. 7(7), e41788.

Doberva, M., Sanchez-Ferandin, S., Toulza, E., Lebaron, P., Lami, R., 2015. Diversity of quorum sensing autoinducer synthases in the Global Ocean sampling metagenomic database. Aquat. Microb. Ecol. 74 (2), 107–119.

Doberva, M., Stien, D., Sorres, J., Hue, N., Sanchez-Ferandin, S., Eparvier, V., Ferandin, Y., Lebaron, P., Lami, R., 2017. Large diversity and original structures of acyl-homoserine lactones in strain MOLA 401, a marine *Rhodobacteraceae* bacterium. Front. Microbiol. 8, 1152.

Dobretsov, S., Teplitski, M., Bayer, M., Gunasekera, S., Proksch, P., Paul, V.J., 2011. Inhibition of marine biofouling by bacterial quorum sensing inhibitors. Biofouling 27 (8), 893–905.

Eberhard, A., Burlingame, A.L., Eberhard, C., Kenyon, G.L., Nealson, K.H., Oppenheimer, N.J., 1981. Structural identification of autoinducer of photobacterium fischeri luciferase. Biochemistry 20 (9), 2444–2449.

Engebrecht, J., Nealson, K., Silverman, M., 1983. Bacterial bioluminescence: isolation and genetic analysis of functions from *Vibrio fischeri*. Cell 32 (3), 773–781.

Fuqua, W.C., Winans, S.C., Greenberg, E.P., 1994. Quorum sensing in bacteria: the LuxR-LuxI family of cell density-responsive transcriptional regulators. J. Bacteriol. 176 (2), 269–275.

Gantner, S., Schmid, M., Durr, C., Schuhegger, R., Steidle, A., Hutzler, P., Langebartels, C., Eberl, L., Hartmann, A., Dazzo, F.B., 2006. *In situ* quantitation of the spatial scale of calling distances and population density-independent N-acylhomoserine lactone-mediated communication by rhizobacteria colonized on plant roots. FEMS Microbiol. Ecol. 56 (2), 188–194.

Garcia-Aljaro, C., Vargas-Cespedes, G.J., Blanch, A.R., 2012. Detection of acylated homoserine lactones produced by *Vibrio* spp. and related species isolated from water and aquatic organisms. J. Appl. Microbiol. 112 (2), 383–389.

García-Contreras, R., Martínez-Vázquez, M., Velázquez Guadarrama, N., Villegas Pañeda, A.G., Hashimoto, T., Maeda, T., Quezada, H., Wood, T.K., 2013. Resistance to the quorum-quenching compounds brominated furanone C-30 and 5-fluorouracil in *Pseudomonas aeruginosa* clinical isolates. Pathog. Dis. 68 (1), 8–11.

Gardères, J., Taupin, L., Saïdin, J.B., Dufour, A., Le Pennec, G., 2012. N-Acyl homoserine lactone production by bacteria within the sponge *Suberites domuncula* (Olivi, 1792) (Porifera, Demospongiae). Mar. Biol. 159 (8), 1685–1692.

Gardères, J., Henry, J., Bernay, B., Ritter, A., Zatylny-Gaudin, C., Wiens, M., Müller, W.E.G., Le Pennec, G., 2014. Cellular effects of bacterial N-3-oxo-dodecanoyl-L-homoserine lactone on the sponge *Suberites domuncula* (Olivi, 1792): insights into an intimate inter-kingdom dialogue. PLoS One. 9(5), e97662.

Geng, H., Belas, R., 2010. Molecular mechanisms underlying *Roseobacter*-phytoplankton symbioses. Curr. Opin. Biotechnol. 21 (3), 332–338.

Geng, H., Bruhn, J.B., Nielsen, K.F., Gram, L., Belas, R., 2008. Genetic dissection of tropodithietic acid biosynthesis by marine *Roseobacters*. Appl. Environ. Microbiol. 74 (5), 1535–1545.

Gilson, L., Kuo, A., Dunlap, P.V., 1995. AinS and a new family of autoinducer synthesis proteins. J. Bacteriol. 177 (23), 6946–6951.

Girard, L., Blanchet, E., Intertaglia, L., Baudart, J., Stien, D., Suzuki, M., Lebaron, P., Lami, R., 2017. Characterization of N-acyl homoserine lactones in *Vibrio tasmaniensis* LGP32 by a biosensor-based UHPLC-HRMS/MS method. Sensors 17 (4), 906.

Givskov, M., de Nys, R., Manefield, M., Gram, L., Maximilien, R., Eberl, L., Molin, S., Steinberg, P.D., Kjelleberg, S., 1996. Eukaryotic interference with homoserine lactone-mediated prokaryotic signalling. J. Bacteriol. 178 (22), 6618–6622.

Golberg, K., Eltzov, E., Shnit-Orland, M., Marks, R.S., Kushmaro, A., 2011. Characterization of quorum sensing signals in coral-associated bacteria. Microb. Ecol. 61 (4), 783–792.

Golberg, K., Pavlov, V., Marks, R.S., Kushmaro, A., 2013. Coral-associated bacteria, quorum sensing disrupters, and the regulation of biofouling. Biofouling 29 (6), 669–682.

González, J.M., Kiene, R.P., Moran, M.A., 1999. Transformation of sulfur compounds by an abundant lineage of marine bacteria in the α-subclass of the class *Proteobacteria*. Appl. Environ. Microbiol. 65 (9), 3810–3819.

Graf, J., Ruby, E.G., 1998. Host-derived amino acids support the proliferation of symbiotic bacteria. Proc. Natl. Acad. Sci. U. S. A. 95 (4), 1818–1822.

Gram, L., de Nys, R., Maximilien, R., Givskov, M., Steinberg, P., Kjelleberg, S., 1996. Inhibitory effects of secondary metabolites from the red alga *Delisea pulchra* on swarming motility of *Proteus mirabilis*. Appl. Environ. Microbiol. 62 (11), 4284–4287.

Gram, L., Grossart, H.P., Schlingloff, A., Kiorboe, T., 2002. Possible quorum sensing in marine snow bacteria: production of acylated homoserine lactones by *Roseobacter* strains isolated from marine snow. Appl. Environ. Microbiol. 68 (8), 4111–4116.

Grandclement, C., Tannieres, M., Morera, S., Dessaux, Y., Faure, D., 2016. Quorum quenching: role in nature and applied developments. FEMS Microbiol. Rev. 40 (1), 86–116.

Greenberg, E., Hastings, J., Ulitzur, S., 1979. Induction of luciferase synthesis in *Beneckea harveyi* by other marine bacteria. Arch. Microbiol. 120 (2), 87–91.

Grote, J., Krysciak, D., Streit, W.R., 2015. Phenotypic heterogeneity, a phenomenon that may explain why quorum sensing does not always result in truly homogenous cell behavior. Appl. Environ. Microbiol. 81 (16), 5280–5289.

Gu, D., Liu, H., Yang, Z., Zhang, Y., Wang, Q., 2016. Chromatin immunoprecipition sequencing technology reveals global regulatory roles of low-cell-density quorum-sensing regulator AphA in the pathogen *Vibrio alginolyticus*. J. Bacteriol. 198 (21), 2985–2999.

Hadfield, M.G., 2011. Biofilms and marine invertebrate larvae: what bacteria produce that larvae use to choose settlement sites. Ann. Rev. Mar. Sci. 3, 453–470.

Hadfield, M.G., Paul, V.J., 2001. Natural chemical cues for settlement and metamorphosis of marine invertebrate larvae. In: McClintock, J.B., Baker, W. (Eds.), Marine Chemical Ecology. In: 2001, CRC Press, Boca Raton, FL, pp. 431–461.

Harder, T., Rice, S.A., Dobretsov, S., Thomas, T., Carre-Mlouka, A., Kjelleberg, S., Steinberg, P.D., McDougald, D., 2014. Bacterial communication systems. In: La Barre, S., Kornprobst, J.M. (Eds.), Oustanding Marine Molecules: Chemistry, Biology, Analysis. Wiley-VCH, Weinheim, pp. 173–187.

Hentschel, U., Fieseler, L., Wehrl, M., Gernert, C., Steinert, M., Hacker, J., Horn, M., 2003. Microbial diversity of marine sponges. In: Müller, W.E.G. (Ed.), Sponges (Porifera). Progress in Molecular and Subcellular Biology. vol. 37.Springer, Berlin, pp. 59–88.

Hentzer, M., Riedel, K., Rasmussen, T.B., Heydorn, A., Andersen, J.B., Parsek, M.R., Rice, S.A., Eberl, L., Molin, S., Hoiby, N., Kjelleberg, S., Givskov, M., 2002. Inhibition of quorum sensing in *Pseudomonas aeruginosa* biofilm bacteria by a halogenated furanone compound. Microbiology 148 (Pt 1), 87–102.

Higgins, D.A., Pomianek, M.E., Kraml, C.M., Taylor, R.K., Semmelhack, M.F., Bassler, B.L., 2007. The major *Vibrio cholerae* autoinducer and its role in virulence factor production. Nature 450 (7171), 883–886.

Hmelo, L., Van Mooy, B.A.S., 2009. Kinetic constraints on acylated homoserine lactone-based quorum sensing in marine environments. Aquat. Microb. Ecol. 54 (2), 127–133.

Hmelo, L.R., Mincer, T.J., Van Mooy, B.A., 2011. Possible influence of bacterial quorum sensing on the hydrolysis of sinking particulate organic carbon in marine environments. Environ. Microbiol. Rep. 3 (6), 682–688.

Huang, Y.-L., Ki, J.-S., Case, R.J., Qian, P.-Y., 2008. Diversity and acyl-homoserine lactone production among subtidal biofilm-forming bacteria. Aquat. Microb. Ecol. 52 (2), 185–193.

Huang, Y.L., Ki, J.S., Lee, O.O., Qian, P.Y., 2009. Evidence for the dynamics of acyl homoserine lactone and AHL-producing bacteria during subtidal biofilm formation. ISME J. 3 (3), 296–304.

Hunter, G.A., Keener, J.P., 2014. Mechanisms underlying the additive and redundant Qrr phenotypes in *Vibrio harveyi* and *Vibrio cholerae*. J. Theor. Biol. 340, 38–49.

Jatt, A.N., Tang, K., Liu, J., Zhang, Z., Zhang, X.H., 2015. Quorum sensing in marine snow and its possible influence on production of extracellular hydrolytic enzymes in marine snow bacterium *Pantoea ananatis* B9. FEMS Microbiol. Ecol. 91 (2), 1–13.

Johnson, W.M., Kido Soule, M.C., Kujawinski, E.B., 2016. Evidence for quorum sensing and differential metabolite production by a marine bacterium in response to DMSP. ISME J. 10 (9), 2304–2316.

Joint, I., Tait, K., Callow, M.E., Callow, J.A., Milton, D., Williams, P., Camara, M., 2002. Cell-to-cell communication across the prokaryote-eukaryote boundary. Science 298 (5596), 1207.

Kaufmann, G.F., Sartorio, R., Lee, S.H., Rogers, C.J., Meijler, M.M., Moss, J.A., Clapham, B., Brogan, A.P., Dickerson, T.J., Janda, K.D., 2005. Revisiting quorum sensing: discovery of additional chemical and biological functions for 3-oxo-*N*-acylhomoserine lactones. Proc. Natl. Acad. Sci. U. S. A. 102 (2), 309–314.

Kelly, R.C., Bolitho, M.E., Higgins, D.A., Lu, W., Ng, W.L., Jeffrey, P.D., Rabinowitz, J.D., Semmelhack, M.F., Hughson, F.M., Bassler, B.L., 2009. The *Vibrio cholerae* quorum-sensing autoinducer CAI-1: analysis of the biosynthetic enzyme CqsA. Nat. Chem. Biol. 5 (12), 891–895.

Kiørboe, T., 2001. Formation and fate of marine snow: small-scale processes with large-scale implications. Sci. Mar. 65 (S2), 57–71.

Kjelleberg, S., Steinberg, P., Givskov, M., Gram, L., Manefield, M., De Nys, R., 1997. Do marine natural products interfere with prokaryotic AHL regulatory systems? Aquat. Microb. Ecol. 13 (1), 85–93.

Kong, F.D., Zhou, L.M., Ma, Q.Y., Huang, S.Z., Wang, P., Dai, H.F., Zhao, Y.X., 2017. Metabolites with Gram-negative bacteria quorum sensing inhibitory activity from the marine animal endogenic fungus *Penicillium* sp. SCS-KFD08. Arch. Pharm. Res. 40 (1), 25–31.

Krick, A., Kehraus, S., Eberl, L., Riedel, K., Anke, H., Kaesler, I., Graeber, I., Szewzyk, U., Konig, G.M., 2007. A marine *Mesorhizobium* sp. produces structurally novel long-chain *N*-acyl-L-homoserine lactones. Appl. Environ. Microbiol. 73 (11), 3587–3594.

Kwan, J.C., Teplitski, M., Gunasekera, S.P., Paul, V.J., Luesch, H., 2010. Isolation and biological evaluation of 8-epi-malyngamide C from the Floridian marine cyanobacterium *Lyngbya majuscula*. J. Nat. Prod. 73 (3), 463–466.

Kwan, J.C., Meickle, T., Ladwa, D., Teplitski, M., Paul, V., Luesch, H., 2011. Lyngbyoic acid, a "tagged" fatty acid from a marine cyanobacterium, disrupts quorum sensing in *Pseudomonas aeruginosa*. Mol. Biosyst. 7 (4), 1205–1216.

Labbate, M., Zhu, H., Thung, L., Bandara, R., Larsen, M.R., Willcox, M.D.P., Givskov, M., Rice, S.A., Kjelleberg, S., 2007. Quorum-sensing regulation of adhesion in *Serratia marcescens* MG1 is surface dependent. J. Bacteriol. 189 (7), 2702–2711.

Landa, M., Burns, A.S., Roth, S.J., Moran, M.A., 2017. Bacterial transcriptome remodeling during sequential co-culture with a marine dinoflagellate and diatom. ISME J. 11 (12), 2677–2690.

Laue, B.E., Jiang, Y., Chhabra, S.R., Jacob, S., Stewart, G.S., Hardman, A., Downie, J.A., O'Gara, F., Williams, P., 2000. The biocontrol strain *Pseudomonas fluorescens* F113 produces the *Rhizobium* small bacteriocin, *N*-(3-hydroxy-7-cis-tetradecenoyl)homoserine lactone, via HdtS, a putative novel *N*-acylhomoserine lactone synthase. Microbiology 146 (Pt 10), 2469–2480.

Lee, J.H., Rhee, J.E., Park, U., Ju, H.M., Lee, B.C., Kim, T.S., Jeong, H.S., Choi, S.H., 2007. Identification and functional analysis of *Vibrio vulnificus* SmcR, a novel global regulator. J. Microbiol. Biotechnol. 17 (2), 325–334.

Lenz, D.H., Mok, K.C., Lilley, B.N., Kulkarni, R.V., Wingreen, N.S., Bassler, B.L., 2004. The small RNA chaperone Hfq and multiple small RNAs control quorum sensing in *Vibrio harveyi* and *Vibrio cholerae*. Cell 118 (1), 69–82.

Li, X., Jeong, J.H., Lee, K.T., Rho, J.R., Choi, H.D., Kang, J.S., Son, B.W., 2003. γ-Pyrone derivatives, kojic acid methyl ethers from a marine-derived fungus *Alternaria* sp. Arch. Pharm. Res. 26 (7), 532–534.

Liaqat, I., Bachmann, R.T., Edyvean, R.G., 2014. Type 2 quorum sensing monitoring, inhibition and biofilm formation in marine microrganisms. Curr. Microbiol. 68 (3), 342–351.

Lilley, B.N., Bassler, B.L., 2000. Regulation of quorum sensing in *Vibrio harveyi* by LuxO and sigma-54. Mol. Microbiol. 36 (4), 940–954.

Lupp, C., Ruby, E.G., 2005. *Vibrio fischeri* uses two quorum-sensing systems for the regulation of early and late colonization factors. J. Bacteriol. 187 (11), 3620–3629.

Lupp, C., Urbanowski, M., Greenberg, E.P., Ruby, E.G., 2003. The *Vibrio fischeri* quorum-sensing systems ain and lux sequentially induce luminescence gene expression and are important for persistence in the squid host. Mol. Microbiol. 50 (1), 319–331.

Lynch, M.J., Swift, S., Kirke, D.F., Keevil, C.W., Dodd, C.E., Williams, P., 2002. The regulation of biofilm development by quorum sensing in *Aeromonas hydrophila*. Environ. Microbiol. 4 (1), 18–28.

Maeda, T., Garcia-Contreras, R., Pu, M., Sheng, L., Garcia, L.R., Tomas, M., Wood, T.K., 2012. Quorum quenching quandary: resistance to antivirulence compounds. ISME J. 6 (3), 493–501.

Manefield, M., de Nys, R., Naresh, K., Roger, R., Givskov, M., Peter, S., Kjelleberg, S., 1999. Evidence that halogenated furanones from Delisea pulchra inhibit acylated homoserine lactone (AHL)-mediated gene expression by displacing the AHL signal from its receptor protein. Microbiology 145 (2), 283–291.

Manefield, M., Harris, L., Rice, S.A., de Nys, R., Kjelleberg, S., 2000. Inhibition of luminescence and virulence in the black tiger prawn (*Penaeus monodon*) pathogen *Vibrio harveyi* by intercellular signal antagonists. Appl. Environ. Microbiol. 66 (5), 2079–2084.

Manefield, M., Welch, M., Givskov, M., Salmond, G.P., Kjelleberg, S., 2001. Halogenated furanones from the red alga, *Delisea pulchra*, inhibit carbapenem antibiotic synthesis and exoenzyme virulence factor production in the phytopathogen *Erwinia carotovora*. FEMS Microbiol. Lett. 205 (1), 131–138.

Manefield, M., Rasmussen, T.B., Henzter, M., Andersen, J.B., Steinberg, P., Kjelleberg, S., Givskov, M., 2002. Halogenated furanones inhibit quorum sensing through accelerated LuxR turnover. Microbiology 148 (4), 1119–1127.

Martínez-Matamoros, D., Laiton Fonseca, M., Duque, C., Ramos, F.A., Castellanos, L., 2016. Screening of marine bacterial strains as source of quorum sensing inhibitors (QSI): first chemical study of *Oceanobacillus profundus* (RKHC-62B). Vitae 23 (1), 30–47.

Maximilien, R., de Nys, R., Holmstrom, C., Gram, L., Givskov, M., Crass, K., Kjelleberg, S., Steinberg, P.D., 1998. Chemical mediation of bacterial surface colonisation by secondary metabolites from the red alga *Delisea pulchra*. Aquat. Microb. Ecol. 15 (3), 233–246.

Mayer, C., Romero, M., Muras, A., Otero, A., 2015. Aii20J, a wide-spectrum thermostable N-acylhomoserine lactonase from the marine bacterium *Tenacibaculum* sp 20J, can quench AHL-mediated acid resistance in *Escherichia coli*. Appl. Microbiol. Biotechnol. 99 (22), 9523–9539.

McClean, K.H., Winson, M.K., Fish, L., Taylor, A., Chhabra, S.R., Camara, M., Daykin, M., Lamb, J.H., Swift, S., Bycroft, B.W., Stewart, G.S., Williams, P., 1997. Quorum sensing and *Chromobacterium violaceum*: exploitation of violacein production and inhibition for the detection of N-acylhomoserine lactones. Microbiology 143 (12), 3703–3711. Pt 12.

McLean, R.J., Whiteley, M., Stickler, D.J., Fuqua, W.C., 1997. Evidence of autoinducer activity in naturally occurring biofilms. FEMS Microbiol. Lett. 154 (2), 259–263.

Meyer, J.L., Gunasekera, S.P., Scott, R.M., Paul, V.J., Teplitski, M., 2016. Microbiome shifts and the inhibition of quorum sensing by black band disease cyanobacteria. ISME J. 10 (5), 1204–1216.

Miller, S.T., Xavier, K.B., Campagna, S.R., Taga, M.E., Semmelhack, M.F., Bassler, B.L., Hughson, F.M., 2004. *Salmonella typhimurium* recognizes a chemically distinct form of the bacterial quorum-sensing signal AI-2. Mol. Cell 15 (5), 677–687.

Miller, S.D., Haddock, S.H., Elvidge, C.D., Lee, T.F., 2005. Detection of a bioluminescent milky sea from space. Proc. Natl. Acad. Sci. U. S. A. 102 (40), 14181–14184.

Mohamed, N.M., Cicirelli, E.M., Kan, J., Chen, F., Fuqua, C., Hill, R.T., 2008. Diversity and quorum-sensing signal production of *Proteobacteria* associated with marine sponges. Environ. Microbiol. 10 (1), 75–86.

Mok, K.C., Wingreen, N.S., Bassler, B.L., 2003. *Vibrio harveyi* quorum sensing: a coincidence detector for two autoinducers controls gene expression. EMBO J. 22 (4), 870–881.

Montaser, R., Paul, V.J., Luesch, H., 2013. Modular strategies for structure and function employed by marine cyanobacteria: characterization and synthesis of pitinoic acids. Org. Lett. 15 (16), 4050–4053.

Muras, A., López-Pérez, M., Mayer, C., Parga, A., Amaro-Blanco, J., Otero, A., 2018a. High prevalence of quorum-sensing and quorum-quenching activity among cultivable bacteria and metagenomic sequences in the Mediterranean Sea. Genes 9 (2), 100.

Muras, A., Mayer, C., Romero, M., Camino, T., Ferrer, M.D., Mira, A., Otero, A., 2018b. Inhibition of *Steptococcus mutans* biofilm formation by extracts of *Tenacibaculum* sp. 20J, a bacterium with wide-spectrum quorum quenching activity. J. Oral Microbiol. 10(1), 1429788.

Nakashima, T., Miyazaki, Y., Matsuyama, Y., Muraoka, W., Yamaguchi, K., Oda, T., 2006. Producing mechanism of an algicidal compound against red tide phytoplankton in a marine bacterium gamma-proteobacterium. Appl. Microbiol. Biotechnol. 73 (3), 684–960.

Natrah, F.M., Kenmegne, M.M., Wiyoto, W., Sorgeloos, P., Bossier, P., Defoirdt, T., 2011a. Effects of micro-algae commonly used in aquaculture on acyl-homoserine lactone quorum sensing. Aquaculture 317 (1–4), 53–57.

Natrah, F.M., Ruwandeepika, H.A., Pawar, S., Karunasagar, I., Sorgeloos, P., Bossier, P., Defoirdt, T., 2011b. Regulation of virulence factors by quorum sensing in *Vibrio harveyi*. Vet. Microbiol. 154 (1–2), 124–129.

Nealson, K.H., Platt, T., Hastings, J.W., 1970. Cellular control of the synthesis and activity of the bacterial luminescent system. J. Bacteriol. 104 (1), 313–322.

Neumann, A., Patzelt, D., Wagner-Döbler, I., Schulz, S., 2013. Identification of new *N*-acylhomoserine lactone signalling compounds of *Dinoroseobacter shibae* DFL-12T by overexpression of *luxI* genes. ChemBioChem 14 (17), 2355–2361.

Newton, R.J., Griffin, L.E., Bowles, K.M., Meile, C., Gifford, S., Givens, C.E., Howard, E.C., King, E., Oakley, C.A., Reisch, C.R., Rinta-Kanto, J.M., Sharma, S., Sun, S., Varaljay, V., Vila-Costa, M., Westrich, J.R., Moran, M.A., 2010. Genome characteristics of a generalist marine bacterial lineage. ISME J. 4 (6), 784–798.

Ng, W.L., Perez, L.J., Wei, Y., Kraml, C., Semmelhack, M.F., Bassler, B.L., 2011. Signal production and detection specificity in *Vibrio* CqsA/CqsS quorum-sensing systems. Mol. Microbiol. 79 (6), 1407–1417.

Nhan, D.T., Cam, D.T., Wille, M., Defoirdt, T., Bossier, P., Sorgeloos, P., 2010. Quorum quenching bacteria protect *Macrobrachium rosenbergii* larvae from *Vibrio harveyi* infection. J. Appl. Microbiol. 109 (3), 1007–1016.

Paggi, R.A., Martone, C.B., Fuqua, C., De Castro, R.E., 2003. Detection of quorum sensing signals in the haloalkaliphilic archaeon *Natronococcus occultus*. FEMS Microbiol. Lett. 221 (1), 49–52.

Papenfort, K., Bassler, B.L., 2016. Quorum sensing signal-response systems in Gram-negative bacteria. Nat. Rev. Microbiol. 14 (9), 576–588.

Papenfort, K., Silpe, J.E., Schramma, K.R., Cong, J.-P., Seyedsayamdost, M.R., Bassler, B.L., 2017. A *Vibrio cholerae* autoinducer–receptor pair that controls biofilm formation. Nat. Chem. Biol. 13 (5), 551–557.

Patzelt, D., Wang, H., Buchholz, I., Rohde, M., Gröbe, L., Pradella, S., Neumann, A., Schulz, S., Heyber, S., Münch, K., 2013. You are what you talk: quorum sensing induces individual morphologies and cell division modes in *Dinoroseobacter shibae*. ISME J. 7 (12), 2274–2286.

Paul, C., Pohnert, G., 2011. Interactions of the algicidal bacterium *Kordia algicida* with diatoms: regulated protease excretion for specific algal lysis. PLoS One. 6(6), e21032.

Perez-Rodriguez, I., Bolognini, M., Ricci, J., Bini, E., Vetriani, C., 2015. From deep-sea volcanoes to human pathogens: a conserved quorum-sensing signal in *Epsilonproteobacteria*. ISME J. 9 (5), 1222–1234.

Peters, L., Konig, G.M., Wright, A.D., Pukall, R., Stackebrandt, E., Eberl, L., Riedel, K., 2003. Secondary metabolites of *Flustra foliacea* and their influence on bacteria. Appl. Environ. Microbiol. 69 (6), 3469–3475.

Porsby, C.H., Nielsen, K.F., Gram, L., 2008. *Phaeobacter* and *Ruegeria* species of the *Roseobacter* clade colonize separate niches in a Danish Turbot (*Scophthalmus maximus*)-rearing farm and antagonize *Vibrio anguillarum* under different growth conditions. Appl. Environ. Microbiol. 74 (23), 7356–7364.

Purohit, A.A., Johansen, J.A., Hansen, H., Leiros, H.K., Kashulin, A., Karlsen, C., Smalas, A., Haugen, P., Willassen, N.P., 2013. Presence of acyl-homoserine lactones in 57 members of the *Vibrionaceae* family. J. Appl. Microbiol. 115 (3), 835–847.

Ransome, E., Munn, C.B., Halliday, N., Cámara, M., Tait, K., 2014. Diverse profiles of *N*-acyl-homoserine lactone molecules found in cnidarians. FEMS Microbiol. Ecol. 87 (2), 315–329.

Rao, D., Webb, J.S., Holmstrom, C., Case, R., Low, A., Steinberg, P., Kjelleberg, S., 2007. Low densities of epiphytic bacteria from the marine alga *Ulva australis* inhibit settlement of fouling organisms. Appl. Environ. Microbiol. 73 (24), 7844–7852.

Rasch, M., Buch, C., Austin, B., Slierendrecht, W.J., Ekmann, K.S., Larsen, J.L., Johansen, C., Riedel, K., Eberl, L., Givskov, M., Gram, L., 2004. An inhibitor of bacterial quorum sensing reduces mortalities caused by Vibriosis in rainbow trout (*Oncorhynchus mykiss*, Walbaum). Syst. Appl. Microbiol. 27 (3), 350–359.

Rasmussen, T.B., Manefield, M., Andersen, J.B., Eberl, L., Anthoni, U., Christophersen, C., Steinberg, P., Kjelleberg, S., Givskov, M., 2000. How Delisea pulchra furanones affect quorum sensing and swarming motility in *Serratia liquefaciens* MG1. Microbiology 146 (12), 3237–3244.

Rasmussen, B.B., Nielsen, K.F., Machado, H., Melchiorsen, J., Gram, L., Sonnenschein, E.C., 2014. Global and phylogenetic distribution of quorum sensing signals, acyl homoserine lactones, in the family of *Vibrionaceae*. Mar. Drugs 12 (11), 5527–5546.

Reidl, J., Klose, K.E., 2002. *Vibrio cholerae* and cholera: out of the water and into the host. FEMS Microbiol. Rev. 26 (2), 125–139.

Rezzonico, F., Duffy, B., 2008. Lack of genomic evidence of AI-2 receptors suggests a non-quorum sensing role for *luxS* in most bacteria. BMC Microbiol. 8 (1), 154.

Rivas, M., Seeger, M., Jedlicki, E., Holmes, D.S., 2007. Second acyl homoserine lactone production system in the extreme acidophile *Acidithiobacillus ferrooxidans*. Appl. Environ. Microbiol. 73 (10), 3225–3231.

Rolland, J.L., Stien, D., Sanchez-Ferandin, S., Lami, R., 2016. Quorum sensing and quorum quenching in the phycosphere of phytoplankton: a case of chemical interactions in ecology. J. Chem. Ecol. 42 (12), 1–11.

Romero, M., Martin-Cuadrado, A.B., Roca-Rivada, A., Cabello, A.M., Otero, A., 2011. Quorum quenching in cultivable bacteria from dense marine coastal microbial communities. FEMS Microbiol. Ecol. 75 (2), 205–217.

Romero, M., Martin-Cuadrado, A.-B., Otero, A., 2012. Determination of whether quorum quenching is a common activity in marine bacteria by analysis of cultivable bacteria and metagenomic sequences. Appl. Environ. Microbiol. 78 (17), 6345–6348.

Romero, M., Muras, A., Mayer, C., Buján, N., Magariños, B., Otero, A., 2014. In vitro quenching of fish pathogen *Edwardsiella tarda* AHL production using marine bacterium *Tenacibaculum* sp. strain 20J cell extracts. Dis. Aquat. Organ. 108 (3), 217–225.

Romero, M., Mayer, C., Muras, A., Otero, A., 2015. Silencing bacterial communication through enzymatic quorum-sensing inhibition. In: Kalia, V.C. (Ed.), Quorum Sensing vs. Quorum Quenching: A Battle With No End in Sight. Springer, Delhi, pp. 219–236.

Rosenberg, E., Koren, O., Reshef, L., Efrony, R., Zilber-Rosenberg, I., 2007. The role of microorganisms in coral health, disease and evolution. Nat. Rev. Microbiol. 5 (5), 355.

Rui, H., Liu, Q., Ma, Y., Wang, Q., Zhang, Y., 2008. Roles of LuxR in regulating extracellular alkaline serine protease a, extracellular polysaccharide and mobility of *Vibrio alginolyticus*. FEMS Microbiol. Lett. 285 (2), 155–162.

Rutherford, S.T., van Kessel, J.C., Shao, Y., Bassler, B.L., 2011. AphA and LuxR/HapR reciprocally control quorum sensing in *Vibrios*. Genes Dev. 25 (4), 397–408.

Saurav, K., Bar-Shalom, R., Haber, M., Burgsdorf, I., Oliviero, G., Costantino, V., Morgenstern, D., Steindler, L., 2016a. In search of alternative antibiotic drugs: quorum-quenching activity in sponges and their bacterial isolates. Front. Microbiol. 7, 416.

Saurav, K., Burgsdorf, I., Teta, R., Esposito, G., Bar-Shalom, R., Costantino, V., Steindler, L., 2016b. Isolation of marine *Paracoccus* sp. Ss63 from the sponge *Sarcotragus* sp. and characterization of its quorum-sensing chemical-signaling molecules by LC-MS/MS analysis. Isr. J. Chem. 56 (5), 330–340.

Saurav, K., Costantino, V., Venturi, V., Steindler, L., 2017. Quorum sensing inhibitors from the sea discovered using bacterial *N*-acyl-homoserine lactone-based biosensors. Mar. Drugs 15 (3), 53.

Schaefer, A.L., Hanzelka, B.L., Parsek, M.R., Greenberg, E.P., 2000. Detection, purification, and structural elucidation of the acylhomoserine lactone inducer of *Vibrio fischeri* luminescence and other related molecules. Methods Enzymol. 305, 288–301.

Schaefer, A.L., Taylor, T.A., Beatty, J.T., Greenberg, E.P., 2002. Long-chain acyl-homoserine lactone quorum-sensing regulation of *Rhodobacter capsulatus* gene transfer agent production. J. Bacteriol. 184 (23), 6515–6521.

Schaefer, A.L., Greenberg, E.P., Oliver, C.M., Oda, Y., Huang, J.J., Bittan-Banin, G., Peres, C.M., Schmidt, S., Juhaszova, K., Sufrin, J.R., Harwood, C.S., 2008. A new class of homoserine lactone quorum-sensing signals. Nature 454 (7204), 595–599.

Shao, C.P., Hor, L.I., 2001. Regulation of metalloprotease gene expression in *Vibrio vulnificus* by a *Vibrio harveyi* LuxR homologue. J. Bacteriol. 183 (4), 1369–1375.

Sharif, D.I., Gallon, J., Smith, C.J., Dudley, E., 2008. Quorum sensing in *Cyanobacteria*: *N*-octanoyl-homoserine lactone release and response, by the epilithic colonial cyanobacterium *Gloeothece* PCC6909. ISME J. 2 (12), 1171–1182.

Simon, M., Grossart, H.-P., Schweitzer, B., Ploug, H., 2002. Microbial ecology of organic aggregates in aquatic ecosystems. Aquat. Microb. Ecol. 28 (2), 175–211.

Skerratt, J., Bowman, J., Hallegraeff, G., James, S., Nichols, P., 2002. Algicidal bacteria associated with blooms of a toxic dinoflagellate in a temperate Australian estuary. Mar. Ecol. Prog. Ser. 244, 1–15.

Slightom, R.N., Buchan, A., 2009. Surface colonization by marine roseobacters: integrating genotype and phenotype. Appl. Environ. Microbiol. 75 (19), 6027–6037.

Srivastava, D., Waters, C.M., 2012. A tangled web: regulatory connections between quorum sensing and cyclic di-GMP. J. Bacteriol. 194 (17), 4485–4493.

Srivastava, D., Harris, R.C., Waters, C.M., 2011. Integration of cyclic di-GMP and quorum sensing in the control of *vpsT* and *aphA* in *Vibrio cholerae*. J. Bacteriol. 193 (22), 6331–6341.

Steindler, L., Venturi, V., 2006. Detection of quorum-sensing *N*-acyl homoserine lactone signal molecules by bacterial biosensors. FEMS Microbiol. Lett. 266 (1), 1–9.

Surette, M.G., Miller, M.B., Bassler, B.L., 1999. Quorum sensing in *Escherichia coli*, *Salmonella typhimurium*, and *Vibrio harveyi*: a new family of genes responsible for autoinducer production. Proc. Natl. Acad. Sci. U. S. A. 96 (4), 1639–1644.

Svenningsen, S.L., Tu, K.C., Bassler, B.L., 2009. Gene dosage compensation calibrates four regulatory RNAs to control *Vibrio cholerae* quorum sensing. EMBO J. 28 (4), 429–439.

Tait, K., Havenhand, J., 2013. Investigating a possible role for the bacterial signal molecules *N*-acylhomoserine lactones in *Balanus improvisus* cyprid settlement. Mol. Ecol. 22 (9), 2588–2602.

Tait, K., Joint, I., Daykin, M., Milton, D.L., Williams, P., Camara, M., 2005. Disruption of quorum sensing in seawater abolishes attraction of zoospores of the green alga *Ulva* to bacterial biofilms. Environ. Microbiol. 7 (2), 229–240.

Tait, K., Hutchison, Z., Thompson, F.L., Munn, C.B., 2010. Quorum sensing signal production and inhibition by coral-associated vibrios. Environ. Microbiol. Rep. 2 (1), 145–150.

Tang, K., Su, Y., Brackman, G., Cui, F., Zhang, Y., Shi, X., Coenye, T., Zhang, X.H., 2015. MomL, a novel marine-derived *N*-acyl homoserine lactonase from *Muricauda olearia*. Appl. Environ. Microbiol. 81 (2), 774–782.

Taylor, M.W., Schupp, P.J., Baillie, H.J., Charlton, T.S., de Nys, R., Kjelleberg, S., Steinberg, P.D., 2004. Evidence for acyl homoserine lactone signal production in bacteria associated with marine sponges. Appl. Environ. Microbiol. 70 (7), 4387–4389.

Teasdale, M.E., Donovan, K.A., Forschner-Dancause, S.R., Rowley, D.C., 2011. Gram-positive marine bacteria as a potential resource for the discovery of quorum sensing inhibitors. Marine Biotechnol. 13 (4), 722–732.

Tello, E., Castellanos, L., Arevalo-Ferro, C., Duque, C., 2012. Disruption in quorum-sensing systems and bacterial biofilm inhibition by cembranoid diterpenes isolated from the octocoral *Eunicea knighti*. J. Nat. Prod. 75 (9), 1637–1642.

Teplitski, M., Krediet, C.J., Meyer, J.L., Ritchie, K.B., 2016. Microbial interactions on coral surfaces and within the coral holobiont. In: Goffredo, S., Dubinsky, Z. (Eds.), The Cnidaria, Past, Present and Future. 2016. Springer International Publishing, pp. 331–346.

Thiel, V., Brinkhoff, T., Dickschat, J.S., Wickel, S., Grunenberg, J., Wagner-Döbler, I., Simon, M., Schulz, S., 2010. Identification and biosynthesis of tropone derivatives and sulfur volatiles produced by bacteria of the marine *Roseobacter* clade. Org. Biomol. Chem. 8 (1), 234–246.

Thole, S., Kalhoefer, D., Voget, S., Berger, M., Engelhardt, T., Liesegang, H., Wollherr, A., Kjelleberg, S., Daniel, R., Simon, M., Thomas, T., Brinkhoff, T., 2012. *Phaeobacter gallaeciensis* genomes from globally opposite locations reveal high similarity of adaptation to surface life. ISME J. 6 (12), 2229–2244.

Thomas, K.V., Brooks, S., 2010. The environmental fate and effects of antifouling paint biocides. Biofouling 26 (1), 73–88.

Tinh, N.T.N., Yen, V.H.N., Dierckens, K., Sorgeloos, P., Bossier, P., 2008. An acyl homoserine lactone-degrading microbial community improves the survival of first-feeding turbot larvae (*Scophthalmus maximus* L.). Aquaculture 285 (1–4), 56–62.

Torres, M., Rubio-Portillo, E., Anton, J., Ramos-Espla, A.A., Quesada, E., Llamas, I., 2016. Selection of the *N*-acylhomoserine lactone-degrading bacterium *Alteromonas stellipolaris* PQQ-42 and of its potential for biocontrol in aquaculture. Front. Microbiol. 7, 646.

Tu, K.C., Bassler, B.L., 2007. Multiple small RNAs act additively to integrate sensory information and control quorum sensing in *Vibrio harveyi*. Genes Dev. 21 (2), 221–233.

Vacelet, J., Donadey, C., 1977. Electron microscope study of the association between some sponges and bacteria. J. Exp. Mar. Biol. Ecol. 30 (3), 301–314.

van Kessel, J.C., Rutherford, S.T., Shao, Y., Utria, A.F., Bassler, B.L., 2013. Individual and combined roles of the master regulators AphA and LuxR in control of the *Vibrio harveyi* quorum-sensing regulon. J. Bacteriol. 195 (3), 436–443.

Van Mooy, B.A., Hmelo, L.R., Sofen, L.E., Campagna, S.R., May, A.L., Dyhrman, S.T., Heithoff, A., Webb, E.A., Momper, L., Mincer, T.J., 2012. Quorum sensing control of phosphorus acquisition in *Trichodesmium consortia*. ISME J. 6 (2), 422–429.

Vinoj, G., Vaseeharan, B., Thomas, S., Spiers, A.J., Shanthi, S., 2014. Quorum-quenching activity of the AHL-lactonase from *Bacillus licheniformis* DAHB1 inhibits *Vibrio* biofilm formation in vitro and reduces shrimp intestinal colonisation and mortality. Marine Biotechnol. 16 (6), 707–715.

Wagner-Döbler, I., Thiel, V., Eberl, L., Allgaier, M., Bodor, A., Meyer, S., Ebner, S., Hennig, A., Pukall, R., Schulz, S., 2005. Discovery of complex mixtures of novel long-chain quorum sensing signals in free-living and host-associated marine *Alphaproteobacteria*. ChemBioChem 6 (12), 2195–2206.

Wang, L., Ling, Y., Jiang, H.W., Qiu, Y.F., Qiu, J.F., Chen, H.P., Yang, R.F., Zhou, D.S., 2013. AphA is required for biofilm formation, motility, and virulence in pandemic *Vibrio parahaemolyticus*. Int. J. Food Microbiol. 160 (3), 245–251.

Waters, C.M., Lu, W., Rabinowitz, J.D., Bassler, B.L., 2008. Quorum sensing controls biofilm formation in *Vibrio cholerae* through modulation of cyclic di-GMP levels and repression of *vpsT*. J. Bacteriol. 190 (7), 2527–2536.

Wei, Y., Perez, L.J., Ng, W.-L., Semmelhack, M.F., Bassler, B.L., 2011. Mechanism of *Vibrio cholerae* autoinducer-1 biosynthesis. ACS Chem. Biol. 6 (4), 356–365.

Whiteley, M., Diggle, S.P., Greenberg, E.P., 2017. Progress in and promise of bacterial quorum sensing research. Nature 551 (7680), 313.

Wieczorek, S.K., Todd, C.D., 1998. Inhibition and facilitation of settlement of epifaunal marine invertebrate larvae by microbial biofilm cues. Biofouling 12 (1–3), 81–118.

Wood, D.W., Pierson, L.S., 1996. The phzI gene of *Pseudomonas aureofaciens* 30-84 is responsible for the production of a diffusible signal required for phenazine antibiotic production. Gene 168 (1), 49–53.

Yang, Q., Defoirdt, T., 2015. Quorum sensing positively regulates flagellar motility in pathogenic *Vibrio harveyi*. Environ. Microbiol. 17 (4), 960–968.

Yang, M., Frey, E.M., Liu, Z., Bishar, R., Zhu, J., 2010. The virulence transcriptional activator AphA enhances biofilm formation by *Vibrio cholerae* by activating expression of the biofilm regulator VpsT. Infect. Immun. 78 (2), 697–703.

Yang, Q., Han, Y., Zhang, X.H., 2011. Detection of quorum sensing signal molecules in the family *Vibrionaceae*. J. Appl. Microbiol. 110 (6), 1438–1448.

Yates, E.A., Philipp, B., Buckley, C., Atkinson, S., Chhabra, S.R., Sockett, R.E., Goldner, M., Dessaux, Y., Camara, M., Smith, H., Williams, P., 2002. *N*-Acylhomoserine lactones undergo lactonolysis in a pH-, temperature-, and acyl chain length-dependent manner during growth of *Yersinia pseudotuberculosis* and *Pseudomonas aeruginosa*. Infect. Immun. 70 (10), 5635–5646.

Yildiz, F.H., Visick, K.L., 2009. *Vibrio* biofilms: so much the same yet so different. Trends Microbiol. 17 (3), 109–118.

Zan, J., Fricke, W.F., Fuqua, C., Ravel, J., Hill, R.T., 2011. Genome sequence of *Ruegeria* sp. strain KLH11, an *N*-acylhomoserine lactone-producing bacterium isolated from the marine sponge *Mycale laxissima*. J. Bacteriol. 193 (18), 5011–5012.

Zan, J., Cicirelli, E.M., Mohamed, N.M., Sibhatu, H., Kroll, S., Choi, O., Uhlson, C.L., Wysoczynski, C.L., Murphy, R.C., Churchill, M.E., Hill, R.T., Fuqua, C., 2012. A complex LuxR-LuxI type quorum sensing network in a roseobacterial marine sponge symbiont activates flagellar motility and inhibits biofilm formation. Mol. Microbiol. 85 (5), 916–933.

Zan, J.D., Heindl, J.E., Liu, Y., Fuqua, C., Hill, R.T., 2013. The CckA-ChpT-CtrA phosphorelay system is regulated by quorum sensing and controls flagellar motility in the marine sponge symbiont *Ruegeria* sp. KLH11. PLoS One 8 (6), 1932–6203.

Zan, J., Choi, O., Meharena, H., Uhlson, C.L., Churchill, M.E., Hill, R.T., Fuqua, C., 2015. A solo *luxI*-type gene directs acylhomoserine lactone synthesis and contributes to motility control in the marine sponge symbiont *Ruegeria* sp. KLH11. Microbiology 161 (Pt 1), 50–56.

Zhai, C., Zhang, P., Shen, F., Zhou, C., Liu, C., 2012. Does *Microcystis aeruginosa* have quorum sensing? FEMS Microbiol. Lett. 336 (1), 38–44.

Zhu, J., Mekalanos, J.J., 2003. Quorum sensing-dependent biofilms enhance colonization in *Vibrio cholerae*. Dev. Cell 5 (4), 647–656.

Zimmer, B.L., May, A.L., Bhedi, C.D., Dearth, S.P., Prevatte, C.W., Pratte, Z., Campagna, S.R., Richardson, L.L., 2014. Quorum sensing signal production and microbial interactions in a polymicrobial disease of corals and the coral surface mucopolysaccharide layer. PLoS One. 9(9), e108541.

CHAPTER 4

Quorum Sensing in Extremophiles

Gennaro Roberto Abbamondi*, Margarita Kambourova[†], Annarita Poli*,
Ilaria Finore*, Barbara Nicolaus*

**National Research Council of Italy—Institute of Biomolecular Chemistry, Pozzuoli, Italy*
[†]*Institute of Microbiology—Bulgarian Academy of Sciences, Sofia, Bulgaria*

1 Introduction to Extremophiles and Extreme Ecosystems

Extremophiles are organisms able to survive in prohibitive conditions from the physical and chemical point of view, where most of the living organisms on Earth would survive with great difficulty or perish. These microorganisms, which are able to perform any vital function in environments characterized by one or multiple chemical-physical stresses, are particularly interesting for the scientific world, for their potential application in different industrial fields and in the bioremediation processes (Antranikian et al., 2005; Canganella and Wiegel, 2014; Poli et al., 2017). Moreover, they are very useful for understanding the evolution because they probably were the first inhabitants of our planet at the origin of life. Extremophiles belong to the three domains, *Archaea*, *Bacteria* and *Eukarya*, and can be found in several environmental niches such as hydrothermal vents and springs, salty lakes, in halite crystals, in polar ice and lakes, in volcanic areas, in deserts, or under anaerobic conditions. The existence of life-forms beyond the Earth requires an extension of the classical limits of life: the resistance of extremophilic organisms to harsh conditions in terms of temperature, salinity, pH, pressure, dryness, and desiccation make these living organisms good candidates to assess the habitability of other planets (Di Donato et al., 2018; Mastascusa et al., 2014).

The growing availability of advanced genetic technologies allowed, and still allows, the identification of new metabolites and enzymes useful for applications in the medical, food, and industrial fields, so that microorganisms are now considered cellular factories (Dalmaso et al., 2015; van den Burg, 2003). The ability to survive and proliferate in extreme conditions (pH, temperature, pressure, salt, and nutrients) results in the production of a variety of biotechnologically useful molecules such as lipids, enzymes, polysaccharides, and compatible solutes that are employed in several industrial processes (Poli et al., 2017). There are many extremophilic enzymes and also endogenous compounds that are used with success in the food industry, in the preparation of detergents, in pharmacological applications, and also in genetic studies. In particular, enzymes that derive from thermophiles, for this reason called thermozymes, represent a good sources of novel catalysts of industrial interest (Finore et al., 2016; Poli et al., 2010). Examples of extremophiles include but are not limited to thermophiles (high temperature), psychrophiles (low temperature), acidophiles (low pH), alkaliphiles (high pH), piezophiles (high pressure, formerly called barophiles), halophiles (high salt concentration), osmophiles (high concentration of organic solutes), oligotrophs (low concentration of solutes and/or nutrients), and xerophiles (very dry environment). The term *extremophile* also includes microorganisms able to grow in the presence of high metal concentrations or high doses of radiations.

The most studied extreme environments are certainly those affected from high temperature, where the microorganisms are named thermophiles (optimal temperature $>45°C$) or hyperthermophiles ($>70°C$ and up to $113°C$). Hot places with extremely high temperatures are hot springs, deep oceanic vents, and deep subsurface areas. Hot springs are distributed all over the terrestrial crust, where the warm geothermal water emerges. Moreover, the groundwater transport causes the enrichment in mineral contents of water for the high temperature, giving therapeutic properties to the springs. Deep oceanic vents are hydrothermal water emissions on the deep ocean floor. The emerging water shows a temperature ranging between $60°C$ and $400°C$; the minerals transported by the groundwater in contact with the cold seawater form precipitates that are responsible to build a chimney structure that could be white or black smoker vents, according to the nature of minerals: barium, calcium, silicon, or sulfides, respectively. These niches are characterized by close cold and hot water, acid water (acidophiles), no solar energy, high pressure (barophiles), and no oxygen (anaerobe) (Reith, 2011).

In opposition to hot places, there are the cold and gelid sites, distributed all over the word, such as the alpine and arctic soils, high-latitude and deep ocean waters, polar ice, glaciers, and snowfields. The living microorganisms here are named *psychrophiles* or *cryophiles*. It has become widely accepted that bacteria isolated from deep-sea vents represent a valuable resource of unusual molecules for biotechnological purpose and at the same time provide insight into the role of the peculiar molecules produced. Most deep-sea environments are influenced by high pressure, low temperature, and low nutrient concentration. However, many kinds of deep-sea psychrotolerant bacteria live in the abyssal ecological community, and they have been largely investigated, as they may reveal important ecological characteristics enabling the morphological, physiological, and metabolic adaptation of bacteria to the continuously changing deep-sea ecosystem. Particularly, many psychrophiles are able to produce exopolysaccharides (EPSs) that protect cells from freezing; in fact, high concentrations of EPSs have been found in Antarctic and Arctic marine bacteria (Poli et al., 2010). It is very interesting to note that particulate aggregates are ubiquitous and abundant in the world's oceans. Marine bacteria benefit from living in aggregates, since their proximity to other cells and surfaces provides opportunities for interaction and nutrient uptake. EPSs excreted by bacteria offer a network to hold these structures together, providing a microhabitat suitable for the survival of microorganisms.

Another category of extreme habitat is that characterized by high salt concentrations, and the living microorganisms here are named *halophiles*. Salty environments are considered those containing a salt concentration five times higher than that observed in the sea; they can be both of natural and artificial origin. Some examples are the salt lakes, inland seas, evaporating ponds of seawater, curing brines, salted food products, and saline soils. All those ecosystems contain high amounts of salt tending to the saturation; usually the pH ranges between the neutrality and the alkalinity (halo-alkalophiles). Hypersaline environments are found in a wide variety of aquatic and terrestrial ecosystems. They are inhabited by halotolerant microorganisms as well as halophilic microorganisms (optimum salt in the range of 0.5 M and 2.5 M NaCl) and extreme halophiles (up above 2.5 M NaCl). Moderate and extreme halophiles were isolated not only from hypersaline ecosystems (salt lakes, marine salterns, and saline soils) but also from alkaline ecosystems (alkaline lakes). Halophilic microorganisms developed various biochemical strategies in order to survive in high saline conditions, including compatible solute synthesis to maintain cell structure and function. Their products such as ectoine, bacteriorhodopsins, exopolysaccharides, hydrolases, and biosurfactants are noticeably of industrial interest (Oren, 2010; Poli et al., 2017).

Moreover, there is the polyextremophile—that is, a microorganism needing at least three different stress parameters for growth, such as high salt concentration, high temperature, and alkaline pH value (Karan et al., 2013). In fact, many natural environments are characterized by two or more extreme parameters. To cite a few examples, many hot springs are acidic or alkaline at the same time, and usually rich in metal content; several hypersaline lakes are very alkaline; the deep ocean is generally cold, oligotrophic (very low nutrient content), and exposed to high pressure. Interestingly, an increasing number of species and strains isolated from these environments have been found to tolerate multiple extremes (Seckbach and Rampelotto, 2015).

The first studies on extremophiles from extreme habitats were based on strain isolation that employed classic culture-dependent approaches. With this method, only microorganisms whose metabolic and physiological requirements can be duplicated in the laboratory could be isolated. To overcome this limitation, metagenomic approaches have been recently developed to explore and access the uncultured microbial community (Finore et al., 2014; Hedlund et al., 2014; León et al., 2014).

The biodiversity of ecosystems became an object of intensive study, leading to a rich body of information regarding the distribution of microbial communities in the world. The isolation of new compounds from such natural sources remains the primary objective, together with studies regarding the compounds–biological systems relationship. In particular, enzymes from extreme marine microorganisms are widely used in the chemical industry, and they are also essential for other industries in which they are needed as biological catalysts. Examples range from the production of beer and biofuels to the biological detergents and paper industry. Marine organisms such as bacteria, fungi, sponges, and algae were identified as an unexplored source of enzymes, but they remain somewhat underused. Only a very small part of the marine enzymes in fact reached the commercialization stage. In addition, the search for high EPS-producing strains is an ongoing process, and the improvement of the fermentation conditions and the subsequent downstream steps for the recovery of the resulting EPS are still in progress. Genetic and metabolic engineering for the production of polymers with well-defined properties as well as the exploration of low-cost substrates for their production are necessary for the widespread use of biopolymers of microbial origins (Finore et al., 2014).

2 Quorum Sensing in Extremophiles

The term *quorum sensing* (QS) refers to the molecular communication mechanism through which several organisms regulate gene expression by means of a population-dependent signaling. Evidence of this collective behavior has been described within the three domains of life: *bacteria*, *archaea*, and *eukaryota*. Focusing on *bacteria* and *archaea*, several physiological activities are regulated by QS systems, including motility, biofilm development, growth inhibition, virulence expression, plasmid conjugation, and so on (Williams, 2007). At the basis of QS, there is the synthesis by the microorganisms of small diffusible compounds, named autoinducers (AIs), which are then released in the extracellular milieu. The most studied QS mechanism is based on the synthesis of N-acyl homoserine lactones (AHLs) (AI-1 QS), but other QS-systems with different signal molecules were found among the various microorganisms. In the AI-2 system, the microbial collective behavior is controlled by 4,5-Dihydroxy-2,3-pentanedione (DPD), (*2R,4S*)-2-Methyl-2,3,3,4-tetrahydroxytetrahydrofuran (R-THMF), and (*2S,4S*)-2-Methyl-2,3,3,4-tetrahydroxytetrahydrofuran-borate (S-THMF); the CAI-1 system is based on the synthesis of *Cholerae* autoinducer-1 (CAI-1) and (Z)-3-aminoundec-2-en-4-one (Ea-C8-CAI-1); also, peptides as diketopiperazines (DKPs) are involved in the regulation of gene expression (Tommonaro et al., 2015; Abbamondi et al., 2014).

These molecules accumulate in the surrounding environment, and when their concentration reaches a critical threshold, the strains start to coordinate gene expression and modulate specific phenotypic outcomes (Fuqua et al., 1994).

The molecular communication system in extreme environments is still unclear, but it is possible that QS plays a key role in the survival strategies of extremophilic microorganisms. Moreover, a better understanding of QS in these harsh habitats could contribute to the development of innovative strategies for the synthesis of "extreme biomolecules" and their employment in biotechnological processes. Here we present the recent advances in the study of QS in extremophilic microorganisms.

2.1 Halophiles

The halophiles are a group of organisms with the ability to prosper in hypersaline conditions. In this kind of extreme environments, the salt concentrations are in excess of seawater and can also reach 20%–30% NaCl. Actually, Halophiles require high salt concentrations to thrive in these unusual habitats. The lack of liquid water, which is essential for every living cell as metabolic reaction solvent, led this group of organisms to evolve different strategies to adapt to such extreme conditions. Bacteria mainly counterbalance the osmotic pressure of these harsh environments by means of the production of high concentrations of compatible solutes (small organic molecules), which are not involved in cell metabolism (amino, sugar, polyols, etc.). The accumulation of high concentration of salts (KCl) inside the cells is another strategy, which is mainly adopted by *Archaea* and some halophilic anaerobic bacteria (Saum and Müller, 2008).

Hypersaline environments are often characterized by alkaliphilic parameters (high pH) and/or high temperatures; thus the halophilic strains could also be thermophiles and/or alkaliphiles (Montgomery et al., 2013; Pikuta et al., 2007).

It is described in the literature that QS could be involved in the processes that regulate bacterial adaptation and growth in complex environmental niches—that is the case of hypersaline sites (Tommonaro et al., 2015). Among the halophiles, the main advances in the investigation of QS resulted from the study of the genus *Halomonas*. Llamas et al. (2005) analyzed the ability of different moderately halophilic bacteria belonging to the genus *Halomonas*, to produce signal molecules involved in QS mechanism. Eleven exopolysaccharide-producing strains were analyzed in total: *H. eurihalina* (3 strains), *H. maura* (4 strains), *H. ventosae* (1 strain), and *H. anticariensis* (3 strains). The presence of *N*-acyl-homoserine lactones (AHLs) was detected by means of the indicator strains *Agrobacterium tumefaciens* NTL4 (pZLR4) and *Chromobacterium violaceum* CV026 (Luo et al., 2001; McClean et al., 1997). AHL signal molecules were detected for all the tested *Halomonas* strains; this finding let the authors speculate that QS may play a key role in harsh habitats, in particular for biofilm formation and exopolysaccharide (EPS) production. EPSs have a protective function as a layer to prevent desiccation, and in addition, they facilitate cell-cell chemical communication (Decho, 2000). Some AHLs from *H. anticariens* FP35(T) were also identified by means of mass spectrometry (GC-MS and ESI tandem MS): *N*-butanoylhomoserine lactone (C4-HSL), *N*-hexanoylhomoserine lactone (C6-HSL), *N*-octanoylhomoserine lactone (C8-HSL), and *N*-dodecanoylhomoserine lactone (C12-HSL) (Llamas et al., 2005).

The same research group deeply investigated QS in *H. anticariens* FP35(T). They found that its QS system was composed of *luxR/luxI* homologues: *hanR* (the transcriptional regulator gene) and *hanI* (the AHL synthase gene) (Tahrioui et al., 2011). The draft genome sequence of this gammaproteobacterium was then reported on *Genome Announcements* (Tahrioui et al., 2013a). Tahrioui et al. (2013b) also performed a screening of AHLs production on 43 strains of the

Halomonadaceae family. Also in this case, all the examined bacteria activated the indicator strain (for medium-to-long acyl chains). PCR and DNA sequencing analysis demonstrated that *luxI* homolog was present in most of the studied species, and C6-HSL was detected as the most predominant AHL molecule (Tahrioui et al., 2013b). *Halomonas smyrnensis* AAD6 is a moderately halophilic, exopolysaccharide-producing bacterium isolated from a saltern area in the Aegean Region of Turkey (Poli et al., 2013). An AHL with an unsubstituted acyl side chain length of C16 was identified in the dichloromethane extracts from stationary phase cultures by means of LC-MS analysis. The authors hypothesized that the growth phase–dependent production of the EPS could be regulated by QS (Abbamondi et al., 2016).

Halobacillus halophilus is a moderately halophilic bacterium isolated from a salt marsh on the coast of Germany. It produces autoinducer-2 (AI-2) in a growth-phase dependent manner, highly dependent upon the presence of chloride ions. More specifically, the luxS homolog is upregulated by the anion concentration. It is possible that QS plays an important role in salt precipitation (chloride regulon), but also in cell motility (Averhoff and Müller, 2010; Sewald et al., 2007). The described examples of QS evidence in halophiles are reported in Table 1.

Table 1 Quorum sensing (QS) in halophiles

Halophiles				
Strain(s)	**QS Molecule(s)/ Genes Identified**	**Bioreporter Strain(s)**	**Modulated Activity/Activities**	**Reference(s)**
H. maura (4 strains), *H. ventosae* (1 strain) and *H. anticariensis* (3 strais)	N-acyl-homoserine lactones (AHLs)	*Agrobacterium tumefaciens* NTL4 (pZLR4) and *Chromobacteriumviolaceum* CV026	Putative regulation of biofilm formation and exopolysaccharide (EPS) production	Llamas et al. (2005)
H. anticariens FP35 (T)	C4-HSL, C6-HSL, C8-HSL and C12-HSL. hanR/hanI (luxR/ luxI homologues)	*A. tumefaciens* NTL4 (pZLR4), *C. violaceum* CV026		Llamas et al. (2005) and Tahrioui et al. (2011)
Halomonadaceae family (43 strains)	C_6-HSL(most predominant AHL) LuxI homolog	*A. tumefaciens* NTL4 (pZLR4)		Tahrioui et al. (2013b)
Halomonas smyrnensis AAD6	C16-HSL	*A. tumefaciens* NTL4 (pCF218; pCF372)	Putative regulation of EPS production	Abbamondi et al. (2016)
Halobacillus halophilus	autoinducer-2 (AI-2)		Modulation of salt precipitation (chloride regulon) and cell motility	Averhoff and Müller (2010) and Sewald et al. (2007)

2.2 Thermophiles

Thermophilic organisms have successfully colonized extremely hot terrestrial and marine sites, which are generally hostile for most living organisms on earth. It is possible to find thermophiles among the three domains of life (*Archaea, Bacteria*, and *Eukarya*), but mainly *Bacteria* and *Archaea* populate such harsh habitats. Thermophilic microorganisms are in fact able to grow and proliferate at unusually high temperatures of at least 60°C. According to the optimal growth temperature (OGT), it is possible to classify them in two categories: the thermophilic microorganisms which prefer temperatures between 60°C and 80°C, and the hyperthermophiles which are able to thrive at extremely hot parameters (80–110°C).

Over the past years, a growing number of researchers addressed their studies on thermophiles for different reasons. First, it is important to deeply investigate survival mechanisms of these organisms, in order to better understand how life can proliferate under extreme temperatures. In addition, this knowledge can give us information concerning the earliest known life-forms on Earth. Moreover, the unique proprieties of extremophilic metabolites make them highly suitable to biotechnological applications. Specifically, exopolysaccharides and extracellular enzymes from thermophilic strains can be effective even at extremely high temperature, which can be a great advantage for their application in biotechnological processes (Raddadi et al., 2015).

The discovery of the capability of thermophilic bacteria to communicate by means of the QS mechanism is quite recent. In fact, a limitation of natural *N*-Acyl homoserine lactones (AHLs) is their relative heat instability; thus it was not expected to find active signal molecules at extremely high temperatures (Schopf et al., 2008).

The first evidence of the presence of QS mechanism in hydrothermal environments was described by Johnson et al. (2005). The authors observed an increase in the cell density and in the exopolysaccharide production of *Thermotoga maritima* when it was cocultivated with *Methanococcus jannaschii*. In the coculture, they also detected the upregulation of a gene coding for a polypeptide (TM0504), which contains a motif common to peptides involved in QS mechanism. The exopolysaccharide production was also induced at low cell density when an synthetic exogenous peptide (based on TM0504) was added to the culture. This finding let the authors hypothesize that a peptide-based QS system could be implicated in the microbial ecological strategy within hydrothermal habitats, in particular in population density-dependent exopolysaccharide formation.

T. maritima, together with *Pyrococcus furiosus*, were chosen as hyperthermophilic model in order to investigate the role of autoinducer-2 (AI-2) based QS in hydrothermal environments, which is considered as a universal interspecies chemical communication system in mesophilic bacteria. Although the production of AI-2 was detected for both strains, the significance of AI-2 (and other furanones) in hydrothermal habitats needs further investigation. No detectable changes in QS phenotypes were in fact observed in the model strains (Nichols et al., 2009).

The presence of an AHL signaling system was detected in the *Thermus* sp. GH5, a strain isolated from a hot spring in North West province of Iran (Yousefi-Nejad et al., 2011). The authors studied the proteomes of the thermophilic bacterium after a decrease in temperature from 75°C (optimal growth temperature, OGT) to 45°C (for 2/5 h). The overexpression of acetyltransferase and S-adenosylmethioninsynthetase in early cold shock condition demonstrated a role for AHL signaling in the survival mechanisms of thermophilic microorganisms in response to cold shock. These proteins are in fact involved in the AHL synthesis cycle. In particular, the authors linked the synthesis of these AHL precursors to the accumulation and production of biofilms.

Geobacillus stearothermophilus T-6 is a thermophilic soil bacterium able of producing an extracellular xylanase (Xyn10A). It was observed that the activity of this enzyme was 50-fold higher during early exponential growth. Moreover, adding boiled-conditioned medium to low cell density cultures also increased the expression of Xyn10A. This finding suggested that a cell density regulation (QS) mechanism could control the xylanase production of *G. stearothermophilus* T-6 (Shulami et al., 2014).

The presence of QS activity was also detected in two Epsilonproteobacteria from deep-sea hydrothermal vents: *Sulfurovum lithotrophicum* and *Caminibacter mediatlanticus*. More specifically, the expression of the *luxS* gene and the production of a QS signal during the growth were detected. Moreover, *luxS* transcripts were identified in Epsilonproteobacteria-dominated biofilms from active deep-sea hydrothermal vents (Pérez-Rodríguez et al., 2015).

A small novel gene, Cthe_3383, was detected in the thermophilic anaerobe *Clostridium thermocellum*. This gene was highly expressed, and it codes for a peptide that is probably involved in QS mechanism (Wilson et al., 2013).

The group of processes that disrupt QS signaling, such as QS inhibitors (QSIs) or QQ enzymes, have been described with the term *quorum quenching* (QQ) (Grandclément et al., 2016). QQ enzymes, in particular, are really promising for biotechnological applications in medicine, agriculture, aquaculture, and biofouling as an innovative strategy to inhibit undesirable bacterial traits (Bzdrenga et al., 2017).

A strain belonging to *Geobacillus* genus, *Geobacillus caldoxylosilyticus*, was found to produce a metallo-β-lactamase involved in QQ mechanism; this enzyme, in fact, is able to hydrolyze different lactones, including the AHLs involved in QS system (Bergonzi et al., 2016).

QQ was also detected in the extremely thermophilic bacterium *Thermaerobacter marianensis*. This strain was isolated from the Mariana Trench, and its OGT is 75°C. An AHL-degrading gene homolog (*aiiT*) was in the genome of *T. marianensis*. AiiT resulted to function as a high thermostable AHL-lactonase, registering the highest activity at 60–80°C (Morohoshi et al., 2015).

Table 2 Quorum sensing (QS) and quorum quenching (QQ) in thermophiles

Thermophiles			
Strain(s)	**QS-QQ Molecule(s)/ Genes Identified**	**Modulated Activity/ Activities**	**Reference(s)**
Thermotoga maritima (cocultivated with *Methanococcus jannaschii*)	Gene coding for a polypeptide TM0504 (peptide-based QS system)	Putative role in exopolysaccharide (EPS) formation	Johnson et al. (2005)
T. maritima, Pyrococcus furiosus	AI-2		Nichols et al. (2009)
Thermus sp. GH5	AHL signaling system	Putative role in the accumulation and production of biofilms Implication in the survival mechanisms in response to cold shock	Yousefi-Nejad et al. (2011)
Geobacillus stearothermophilus T-6		Putative modulation of theextracellular xylanase (Xyn10A) production	Shulami et al. (2014)
Geobacillus caldoxylosilyticus	Metallo-β-lactamase (QQ)		Bergonzi et al. (2016)
Thermaerobacter marianensis	*aiiT* (AHL-degrading gene homolog) encoding a high thermostable AHL-lactonase (QQ)		Morohoshi et al. (2015)
Sulfurovum lithotrophicum, Caminibacter mediatlanticus	luxSgene		Pérez-Rodríguez et al. (2015)
Clostridium thermocellum	Cthe_3383 (small novel gene encoding a peptide probably involved in QS mechanism)		Wilson et al. (2013)

The more representative findings of QS/QQ in thermophiles are summarized in Table 2.

2.3 Acidophiles

Acidophile is the term adopted to describe the organism capable to inhabit strongly acidic habitats, which can reach really low pH values. It is possible to find acidic sites both in natural (e.g., acidic sulfur springs) and man-made (e.g., from mining metals and coal activities) environments (Oren, 2010). Acidic sites are often characterized by high levels of heavy metals, thus acidophilic microorganisms are often resistant to these potentially toxic elements. Moreover, there are examples of acidophiles which are also resistant to high temperatures. These microorganisms can thrive in such harsh environments, because they developed strategies to maintain their intracellular pH close to neutrality, their cytoplasmic membrane may in fact support extremely low pH gradients. As opposite to neutrophilic and alkaliphilic microorganisms, they often maintain a reversed (outside-negative) membrane potential.

There are only few reports about the evidence of QS mechanism in these acidophilic microorganisms. A Lux-like QS system was reported for the extremely acidophilic bacterium *Acidithiobacillus ferrooxidans* (Rivas et al., 2007). *A. ferrooxidans* is a chemolithoautotrophic strain, which can proliferate at extremely low pH values (pH 1–2) and at high concentrations of metals. The genomic analysis allowed to identify two genes (*afeI* and *afeR*), encoding for proteins significantly similar to the LuxI and LuxR families. Moreover, the authors reported the detection of an AHL with an unsubstituted acyl side chain length of C14 by means of GC/MS analysis. The genome of *A. ferrooxidans* was then deeper investigated by the same research group in 2007 (Rivas et al., 2007). They identified a gene encoding for an acyltransferase (*act*) which produces AHLs, mainly with C14 acyl chain according to the GC–MS results. The obtained results demonstrated that two QS systems respond to different external stimuli. Specifically, the Act-based QS system was highly expressed in medium containing iron, while in the Lux-like QS system, the divergently oriented genes encoding for the LuxI and LuxR-like proteins were more expressed in the sulfur medium. The link between QS and biofilm formation has been also investigated in *Acidithiobacillus ferrooxidans* ATCC 23270(T). The authors found that QS network represented at least 141 genes (4.5%) of its genome. Interestingly, 60 of these genes (42.5%) turned out to be related to biofilm formation (Mamani et al., 2016). A screening of five *A. ferrooxidans*, three *Acidithiobacillus thiooxidans*, and two *Leptospirillum ferrooxidans* strains was performed in order to assess their ability to produce AHLs. All the tested strains, except to the two *L. ferrooxidans*, were able to activate the bioreporter *Agrobacterium tumefaciens* NTL4 (pZLR4) (Ruiz et al., 2008). It was already reported for *A. ferrooxidans* an ability to produce nine different types of AHLs with acyl chains between 8 and 16 carbons and with oxo- and hydroxyl- substitutions in C-3 (Farah et al., 2005). Ruiz et al. (2008) also characterized two AHLs from the three *A. thiooxidans* strains. In detail, all the *A. thiooxidans* strains showed the ability to synthesize 3-oxo-C8-AHL, while only one strain (DSMZ 11478) was able to produce 3-oxo-C6-HSL. Another bacterium belonging to the genus *Leptospirillum*, *L. ferriphilum*[T] DSM 14647, was analyzed by means of a multi-omics approach. The analysis detected the presence of metabolic systems involved in the QS process (Christel et al., 2018). Finally, the screening of the genome sequence of the acidophilic iron-oxidizing bacterium *Ferrovum* sp. JA12 showed the presence of a thermophilic-like ene-reductase (FOYE-1). The authors suggest a putative involvement of FOYE-1 in the QS process (Scholtissek et al., 2017).

2.4 Archaea

Archaea are a separate group of prokaryotes which mainly inhabit harsh environments. One of the main characteristics used to distinguish these microorganisms from *Bacteria* is the difference in the composition of the cell wall. In fact, peptidoglycan is not a component of archaeal membranes. These microorganisms have various types of cell wall, but the most common wall structure is formed by a proteinaceous surface (S-) layer. In relatively limited cases, it is also possible to find polysaccharides like pseudomurein (pseudopeptidoglycan). These cell structures can be supported by an additional S-layer, but it is also possible to find archaeal species that completely lack a cell wall (Albers et al., 2017; Meyer and Albers, 2001).

Despite the growing scientific interest in microbial communication systems, there is still a lack of knowledge about QS mechanism in the archaea domain. The first indication about the presence of an AHL-based QS system in archaea was described by Paggi et al. (2003). The authors reported the detection of QS molecules in the haloalkaliphilic archaeon *Natronococcus occultus*, probably belonging to the AHL family. In more detail, the ethyl acetate extracts of the cell free medium were able to activate QS in the *Agrobacterium tumefaciens* NTL4 biosensor, which responds to a wide range of AHLs. It was also described that the adding of *N. occultus* late exponential conditioned medium to low density cultures induced the early production of a protease, which could be evidence of the presence of a cell density regulation mechanism controlling that enzymatic activity.

The first signal molecules that were chemically characterized in *Archaea* belong to the AHL family. Specifically, it was described that the methanogenic archaeon *Methanosaeta harundinacea* 6Ac regulates cell assembly and carbon metabolic flux by means of a carboxylated AHL-based QS system. Three molecules were purified and were determined to be *N*-carboxyl-decanoyl-homoserine, *N*-carboxyl-dodecanoyl-homoserine lactone, and *N*-carboxyl-tetradecanoylhomoserine lactone (Zhang et al., 2016). The presence of C4 and C6 acyl homoserine lactone-like signal molecules was also detected in the culture extract of the haloarchaeon *Haloferax volcanii* DS2; the extract could in fact induce a reaction it the QS Bioreporter*a. tumefaciens* ATCC BAA-2240 (Megaw and Gilmore, 2017).

QS was also investigated in a strain of the genus *Haloterrigena*, *Haloterrigena hispanica*, which was isolated from a salt lake in Fuente de Piedra (Spain). It was actually the first and only finding of diketopiperazines (DKPs) production from a halophilic archaeon. Tommonaro et al. (2012) purified and chemically characterized five DKPs: cyclo-(D-Pro-L-Tyr), cyclo-(L-Pro-L-Tyr), cyclo-(L-Pro-L-Val), cyclo-(L-Pro-L-Phe), and cyclo-(L-Pro-L-isoLeu). Among the isolated DKPs, only cyclo-(L-Pro-L-Val) stimulated QS activity in the two bioreporter strains *A. tumefaciens* NTL4 (pCF218; pCF372) and *Vibrio anguillarum* DM27.

QQ activity was detected in another strain belonging to the genus *Haloterrigena*. An archaeal strain with 99.6% phylogenetic similarity with *Haloterrigena saccharevitans* was isolated from a hypersaline cyanobacterial mat from a desert in Oman. Both the nonpolar (dichloromethane) and the polar (water) fractions inhibited violacein production in the *C. violeaceum* CV017 bioreporter (Abed et al., 2013). Biofilm formation can be under QS control; thus a halophilic strain with QQ ability could be a potentially antifouling agent in marine environments.

Finally, the inhibition of QS as lactonase activity (AHL hydrolysis) was detected in two hyperthermophilic archaea belonging to the genus *Sulfolobus*: *Sulfolobus solfataricus* and

Table 3 Quorum sensing (QS) and quorum quenching (QQ) in archaea

Archaea				
Strain(s)	QS-QQ Molecule(s)/Genes Identified	Bioreporter Strain(s)	Modulated Activity/Activities	Reference(s)
Natronococcus occultus	Putative AHL signal molecules	*Agrobacterium tumefaciens*NTL4 (pCF218)(pCF372)	Putative regulation of a protease production	Paggi et al. (2003)
Methanosaeta harundinacea 6Ac	*N*-carboxyl-C10-HSL, *N*-carboxyl-C12-HSL, *N*-carboxyl-C14-HSL	*A. tumefaciens* NTL4 (pZLR4)	Regulation of cell assembly and carbon metabolic flux	Zhang et al. (2012)
Haloferax volcanii DS2	C4 and C6 HSL-like signal molecules	*A. tumefaciens* ATCC BAA-2240		Megaw and Gilmore (2017)
Haloterrigena saccharevitans (99.6% phylogenetic similarity)	QQ activity	*Chromobacterium violaceum* CV017		Abed et al. (2013)
Haloterrigena hispanica	Cyclo-(L-Pro-L-Val)	*A. tumefaciens* NTL4 (pCF218; pCF372) and *Vibrio anguillarum* DM27		Tommonaro et al. (2012)
Sulfolobus solfataricus and *Sulfolobus islandicus*	QQ activity (lactonase - AHL hydrolysis)			Hiblot et al. (2012) and Ng et al. (2011)

Sulfolobus islandicus (Hiblot et al., 2012; Ng et al., 2011). Table 3 summarizes the main reports about QS/QQ findings in *Archaea*.

2.5 Other Extremophiles

There is a lack of knowledge about the presence of QS mechanisms in other groups of extremophiles.

Psychrophiles are organisms that have successfully colonized the environments characterized by exceptionally low temperatures. Actually, cold areas represent the majority of the biosphere on Earth (Margesin and Miteva, 2011). There are only a few articles describing the study of QS among psychrophiles. The screening of the genome sequence of the sea-ice bacterium *Psychromonas ingrahamii* 37 showed the presence of an orthologue of HapR (LuxR), which could be involved in biofilm formation (Riley et al., 2008). The production of bioluminescence in the psychrophilic bacterium *Aliivibrio logei* is controlled by QS. In more detail, the analysis of the structure of the *lux*-operons of 16 strains of *A. logei* showed the presence of a LuxI/LuxR QS-system, with two copies of the *luxR* gene (*luxR1* and *luxR2*). The experimental observations suggested that the transcription of the *lux*-operon in

A. logei is cooperatively regulated by LuxR1 and LuxR2 proteins (Khrulnova et al., 2016; Konopleva et al., 2016).

A piezophile (or barophile) is an organism that is adapted to live at relatively high pressure (up to 110 MPa), such as deep-sea sediments, subsurface rocks with high lithostatic pressures, or hydrothermal vents. These environments are also associated with exceptionally high (up to 375°C) or low (1–2°C) temperatures, and lack of light and nutrient availability (Kato, 1999). A large group of barophilic bacteria belongs to the genus *Shewanella*. The production of AI-2 in *Shewanella* species was described for the first time by Bodor et al. (2008). The presence of *luxS* gene and the production of AI-2 were assessed in all 10 species of *Shewanella* tested. Tait et al. (2009) also identified three major AHLs (OC4, OC10, and OC12-HSL) in the late exponential phase extracts of *Shewanella* sp. They also found an interesting link between the AHL production and the zoospore settlement of the macroalga *Ulva*, which is evidence of an interkingdom ecological effect of QS. The moderately piezophile *Photobacterium phosphoreum* ANT-2200 showed optimum growth conditions at 30°C and 10 MPa. It is described that it produced a higher bioluminescence intensity at 22 MPa compared with 0.1 MPa. *P. phosphoreum* responds to high pressure by forming cell aggregates, miming a higher cell density; therefore the authors supposed that a cell-density-dependent signaling system could modulate the bioluminescence production (Martini et al., 2013).

3 Biomolecule Synthesis by Extremophiles: QS Regulation

Extremophiles developed an ability to produce unusual biomolecules as one of the adaptive mechanisms to raw environmental conditions, among which thermostable enzymes and exopolysaccharides are essential for their function in extreme niches. Their production is QS regulated by small organic molecules or peptides defined as "autoinducers." QS mechanism gives a strategy for the evolving of microbial strains to adapt to extreme environments and is actively involved in intra- and interspecies interactions in extremophilic communities.

3.1 Extremozymes

Bioremediation is an environmentally friendly approach based on the exploration of microorganisms for destruction or transformation to harmless products of hazardous compounds like oil, heavy metals, and PAHs liberated in a growing number of industrial processes. The presence of multiple genes of autoinducer-like homoserine lactone (AHL) in bacterial hydrocarbon degraders and different structures of hydrocarbon degrading communities after severe pollution suggest QS regulation of the environmental detoxification process (Huang et al., 2013). Enzymes from the aminohydrolase family are able to detoxify organophosphates used as insecticides, herbicides, and chemical warfare agents. Extremely thermostable phosphotriesterase-like lactonases hydrolyzing organophosphates were isolated from hyperthermophilic archaeal species *Sulfolobus solfataricus* (Ng et al., 2011) and *Vulcanisaeta moutnovskia* (Kallnik et al., 2014). *S. solfataricus* enzyme expressed in *Pseudomonas aeruginosa* could attenuate QS signals for virulence factors production (Ng et al., 2011). Carboxylated AHLs alter carbon metabolisms that favor the conversion of acetate to methane as a result of an upregulation of initial enzymes in the methanogenesis pathway (Zhang et al., 2012). The cell density of many bacterial communities often is regulated by QS. QQ enzymes like lactonase, oxidoreductase, acylase, and paraoxonase interfere with the QS system in microbial community and could be used to limit cell growth by repression of the QS enzyme expression (Chen et al., 2013). The accumulation of autoinducer molecules stimulates the production of extracellular hydrolytic enzymes (including proteases) as the culture reaches a high cell density (Nakayama et al., 2001). Similar density-dependent induction of protease gene expression by cell-to-cell communication in bacteria via QS was observed for the archaeon *N. occultus* (Paggi et al., 2003). An accumulation of homoserine lactone molecules was observed in the late exponential growth when high cell density was reached.

3.2 Exopolysaccharides

Exopolysaccharides (EPSs) create an environment that allows cell adhesion in a biofilm with stability determined by EPS. Polymers keep bacteria in the biofilm together and protect them against severe environmental conditions in extreme niches. EPS matrix increases the rate of substance uptake by binding and accumulation of biodegradable compounds, a particularly important mechanism in oligotrophic environments (as many of extreme environments are), as well as ensuring a close contact between enzymes and their substrates. QS regulation of biofilm formation by acyl-homoserine lactones allows for the rapid switch from fast cell production to production of large EPS amounts as protection against environmental threats (Frederick et al., 2011).

A signaling system mediated by the two *N*-acylhomoserine lactone autoinducers *N*-butyryl-L-homoserine lactone (C4-HSL) and *N*-(3-oxododecanoyl)-L-homoserine lactone (OC12-HSL) was observed to control the genes essential for the formation of differentiated multicellular microcolonies of *P. aeruginosa* in bacterial biofilms (Whiteley et al., 1999). A different expression of the two identified QS genes lasI and rhlI coding for C4-HSL and OC12-HSL in biofilm-forming *P. aeruginosa* N6P6 was established when the strain grew on two polycyclic aromatic hydrocarbons and the formation of differentiated multicellular microcolonies in bacterial biofilms was observed (Mangwani et al., 2015).

4 Extreme Biomolecules: Potential Biotechnological Applications

The continuous scientific interest in extremophilic microorganisms is also due to their ability to produce biomolecules that exhibit high stability in the presence of diverse and simultaneous environmental stress factors, such as high or low temperature, pH, pressure, and so on. Indeed, it is exactly thanks to these biomolecules, together with the whole adapted cellular mechanism, that extremophiles optimally proliferate in peculiar niches in which any life-form is imaginable (Poli et al., 2017). In fact, the properties of the extreme molecules are attractive for the possible biotechnological applications that they can have in many fields, from pharmacology to ecology, from molecular biology to medicine. Therefore, molecules from extremophilic microbes, such as enzymes, lipids, and exopolysaccharides, are already useful in numerous applicative sectors, with an outcome to increase the efficiency and/or the sustainability of the corresponding industrial production process (Finore et al., 2015).

In order to improve the industrial procedures based on biocatalysis, new stable enzymes are necessary (Woodley, 2013).

Extremozymes, which are synthesized by microorganisms that prosper in extremes of temperature (from 122°C to −12°C), pH (from 0 to 12), pressure (up to 1000 atm), and salinity (up to saturating concentration) work under the extreme conditions of many manufacturing processes (Coker, 2016). Up to now, only a small amount of extremozymes found large-scale biotechnologically applications.

Among them, it is indispensable to cite the DNA polymerases, of which the essential activity relies on the synthesis of the deoxyribonucleotide chain. They are involved in the physiologic genome reparation and maintenance cellular mechanisms during replication. The thermostable DNA polymerases purified from the thermophilic microorganisms *Thermusaquaticus*, *Pyrococcusfuriosus*, and *Thermococcus litoralis*, commonly recognized as Taq22, Pfu23, and Vent24, respectively (Ishino and Ishino, 2014), found large space in the polymerase chain reaction (PCR) techniques in molecular biology. The development and the efficiency of this practice generated many benefits not only limited to research field, but also in the general human society; PCR-based technologies are crucial in all circumstances wherein the amount of biological samples are exiguous, like in forensic science (Gunn et al., 2014).

The glycosyl hydrolases are responsible for glycosidic bond hydrolysis in complex sugars. They are very common in nature and are classified into well over 100 families. Over time, these enzymes found numerous biotechnological applications.

The inability to digest lactose is widespread in the world's population and can produce embarrassing affects. It is due to the lactose intolerance and the lack of the β-galactosidase enzyme (Coker, 2016); therefore, to resolve this inconvenience, the consumption of lactose-free dietary products is preferred in many cases, obtained by pretreating products with β-galactosidase from *Kluyveromyces lactis* (Messia et al., 2007). The reaction temperature of the enzyme-catalyzed hydrolysis is higher than the food maintenance conditions (5–25°C), which implies great contamination and deterioration risk. A solution could be the utilization of β-galactosidases from psychrophilic microorganisms, which are active at lower temperatures. At present, many psychrophilic enzymes were studied to produce effects that are similar to the presently used mesophilic enzymes with an optimal working temperature ranging from 15°C to 37°C (Coker and Brenchley, 2006).

Analogously, the cellulose, hemicellulose, and starch hydrolysis are processes that involve the use of enzymatic pools, all belonging to the glycosyl hydrolases. The enzymatic degradation of the previously mentioned polymers is favorite at high temperature, in which the contamination risk, due to the gradual enrichment in oligo and monosaccharides concentration, is reduced. Then, the use of thermophilic enzymes is preferred. Therefore, the search for thermostable cellulase, α-glucosidase, xilanase, β-xilosidase, pullulanase, glucoamylase, and β-amylase is ongoing. The corresponding hydrolysis products are used in many industrial sectors, such as laundry and porcelain detergents, antistaling agents in baking, and in the production of bioethanol and value-added organic compounds from renewable agricultural residues (Dornez et al., 2011; Finore et al., 2011; Finore et al., 2015; Finore et al., 2016; Mohammad et al., 2017; van der Maarel et al., 2002).

In addition, the extremophiles represent a surprising source of hydrogen, as metabolic product of anaerobic fermentation. *Aeropyrum, Caldicellulosiruptor, Thermoanaerobacterium* (Ren et al., 2008), and *Pyrococcus* (Baker et al., 2009) genera have shown enormous potential.

A different sector that exploits the catalytic properties of enzymes from extremophilic microbes is biomining. It is a biotechnological and eco-friendly process for metal recovering by employing microorganisms. Indeed, instead of the conventional heap leaching, that requires many chemicals for binding and separating metals, the utilization of acidophilic and thermophilic microorganisms able to oxidize iron (II) or sulfur compounds is very attractive (Coker, 2016). The genera *Acidithiobacillus, Leptospirillum, Ferroplasma, Sulfolobus, Metallosphaera, Acidianus,* and *Sulfurisphaera* were employed in the biomining of copper, gold, cobalt, nickel, zinc, and uranium (Podar and Reysenbach, 2006; Vera et al., 2013). Clearly, this microbial application can be extended to the leaching of processing residues or mine waste dumps (mine tailings) (Chen et al., 2016), as well as for the extraction of metals from industrial waste residues (Mishra and Rhee, 2014). In addition, advances in the DNA sequencing of the bioleaching microorganism community are desirable, because this could furnish unprecedented opportunities to acquire information about the genomes, allowing predictive models of metabolic potential and ecosystem-level interactions to be constructed (Cárdenas et al., 2016).

Along with these commercially valuable enzymes is the protease that carries out the hydrolysis of peptide bonds. Among them, the alkaline proteases show high stability when used in detergents; they found applications in many bioindustries as washing powders, leather processing, the food industry, and pharmaceuticals (Mohammad et al., 2017; Sarmiento et al., 2015; Sellami-Kamoun et al., 2008). Normally, the commercially used proteases are from mesophilic microbes, essentially from *Bacillus* species, and are sold by companies such as Novozymes and Genencor. However, the investigation of psychrophilic proteases to improve cold water washing has been initialized. Most of the tested psychrophilic enzymes proved to be unsuitable for their low stability at room temperature. Anyway, an engineered psychrophilic/mesophilic protease was generated that improved performance during cold water washing (Tindbaek et al., 2004).

Lipases are enzymes that catalyze the hydrolysis of fats. They are very attractive for industrial purposes thanks to their wide range of substrates, high specificity, and stability (Coker, 2016; Hasan et al., 2006; Sarmiento et al., 2015). Many mesophilic lipases, which typically come from organisms like *Bacillus* and *Aspergillus* species, are active at high temperatures. As a result, extremophilic lipases are often ignored; however, lipases from thermophilic *Bacillus* species have been showed to be more efficient than currently used enzymes (Imamura and Kitaura, 2000). Recent and attractive studies concerning lipases from psychrophilic microorganisms displayed a high catalytic efficiency with low thermal stability, which is a differentiating mark with respect to their mesophilic and thermophilic counterparts (Maiangwa et al., 2015).

The exopolysaccharides are polymers with a high molecular weight that are composed of sugar residues and are secreted by microorganisms into the surrounding environment. Actually, they can be distinguished in intracellular, structural, and extracellular polysaccharides according to their store position and/or function. They are in general composed of monosaccharides and some substituents such as acetate, pyruvate, succinate, and phosphate. The vast diversity in term of EPS composition implies a different chemical structure and physical properties, and therefore different potential applications in diverse sectors (Nicolaus et al., 2010).

There are many studies showing the possibility to induce and improve the EPS production, by acting on growth conditions, by means of temperature, pressure, salt concentration, and nutrient composition. In addition, genomic and statistical studies can support the selection of a suitable substrate for the polymer production (Finore et al., 2014). It was proven that waste material can be used as a microbial carbon source for EPS production (Kucukasik et al., 2011).

The extremophilic microorganisms, as reported previously, produce these kinds of molecules, often in response to environmental stress (Poli et al., 2017). An example of the industrial use of EPSs is the application of dextran in the bakery industry (Kothari et al., 2014). In general, the EPSs with an industrial use are produced by mesophilic microorganisms; the exploitation of polysaccharides produced by extremophilic prokaryotes is limited, but their atypical properties are attractive for defining possible new biotechnological applications (Nicolaus et al., 2010). In fact, there are many case studies aiming to demonstrate the great potential of EPSs from extreme microbes; for instance, the halophilic *Halomonas smyrnensis* strain AAD6[T] produces high levels of levan (Poli et al., 2009; Poli et al., 2013). The deep investigation of this polymer revealed that it did not influence cellular viability and proliferation in two different cellular systems tested, osteoblasts and murine macrophages, showing its high biocompatibility. Besides, it displayed a protective effect against the toxic activity of avarol, a rearranged sesquiterpenoid hydroquinone isolated from the marine sponge *Dysidea avara* which showed different biological activities (e.g., antiviral, antimicrobial, antioxidant, cytotoxic, antiinflammatory, and antipsoriatic) (Minale et al., 1974; Crispino et al., 1989). The cytoprotective effect of levan isolated from the halophilic *H. smyrnensis* strain AAD6^{T} implied its additional use as an anticytotoxic agent. The potential applications of levan as a blood plasma extender, an industrial gum, a stabilizer, a thickener, an emulsifier, a sweetener, a formulation aid, a surface-finishing agent, an encapsulating agent, and a carrier for flavor and fragrances are known (Beine et al., 2008; Shih et al., 2005). Then, *Halomonas* sp. AAD6 could be considered a different source of levan when grown on defined media, hypothesizing its larger employment in industrial application being a nonpathogenic microorganism (Sam et al., 2011; Sezer et al., 2011).

The lipid composition of cellular membrane belonging to extreme *Archaea* and *Bacteria* was deeply studied; it is one of the most distinctive taxonomic factors and results to be essentially affected by the temperature, pressure, and salt concentration to which the microorganisms is exposed (Poli et al., 2017; Siliakus et al., 2017). Archaeal lipids are promising devices for biological, medical, and biotechnological applications. Tetraether lipids are able to self-arrange in order to generate a supramolecular structure such as lipid films or liposomes that could be use as drug and gene delivery systems and vaccines (Jacquemet et al., 2009).

In addition, Müller et al. (2006) used ultrathin films of natural tetraether derivatives to cover nanoporous aluminium oxide membranes to change the corresponding filtration properties for biological applications. As result, the new membranes showed less permeability and could be easily sterilized. It makes tetraether lipid a suitable systems for bionanotechnology and material science, by means of the design of peculiar membrane nanosystems (Chugunov et al., 2014). However, further study is still required to increase the knowledge concerning these lipids and to expand the range of their biotechnological applications (Jacquemet et al., 2009).

A different, interesting application coming from extremophiles is related to the production of trehalose, useful as stabilizers for vaccines and antibodies (Guo et al., 2000). The trehalose production from the extremely thermophilic archaeon *Sulfolobus solfataricus* was set, and it could substitute without difficulty the presently utilized mesophilic enzymes from *Arthrobacter* sp. Q36 (Schiraldi et al., 2002).

Another microbial product is ectoine, which belongs to the osmolyte category and works as an osmoregulator to balance the osmotic pressure to which the cells are subjected. Ectoine found an application as a skin protectant against the damage induced from UVA. RonaCare Ectoin is produced by Merck KGaA (Darmstadt, Germany), and it is utilized as a hydrating agent and is recovered from halophilic microorganisms (DasSarma et al., 2009).

Nowadays, the roles and mechanisms of microbial communication in extreme environments are still complex and unclear. Therefore, it seems that the production of all those biomolecules and the QS could be related and influenced by each other (Tommonaro et al., 2015). Indeed, communication mechanisms can affect the production yield of these biomolecules and/or inhibit them. As the production of biomolecules in extreme areas can be interpreted as a response to the physico-chemical environmental stress factors, the QS is also involved in the adaptation mechanisms that allow these extraordinary life-forms to proliferate in inhospitable niches.

5 Conclusions

The finding of organisms able to proliferate in "extreme" environments demonstrated that life could exist at atypical chemical and physical conditions. Extremophiles can in fact thrive at high or low temperatures, exceptionally high or low pH values, high salinity, pressure or toxicity, and even high radiation levels. These extreme environmental parameters unexpectedly do not represent a limiting factor, but ensure them the optimal growth. The study of life in extreme environments, with special attention to the mechanisms that extremophiles developed to counteract hostile conditions, represents an area of particular interest for researchers. The particular attention paid to these organisms is not limited to the investigation of the fundamental characteristics of their biological processes, but it is also focused on the potential use of "extreme biomolecules" in industrial biotechnology. Extremophilic metabolites, in particular the extracellular enzymes, are in fact stable and active at the harsh parameters of extreme environments. The unique resistance properties of extreme biomolecules represent a clear advantage for their application in biotechnological processes, compared to labile mesophilic molecules. Quorum sensing (QS) is a well-characterized cell-to-cell communication strategy that has been described in a wide variety of microorganisms. It regulates diverse social behaviors such as production of luminescence, biofilm formation, exopolysaccharide (EPS) production, motility, and so on. Despite the ubiquity of QS systems among microorganisms, the cell-to-cell communication mechanisms in extreme environments need further investigation. The advent of new technologies, such as bioinformatics and genomic sequencing, led to the identification of the genes involved in QS mechanism in a number of extremophilic microorganisms. Different signal molecules, mainly N-Acyl homoserine lactones (AHLs), were detected in the cell free media of different strains living in harsh habitats by means of plasmid-based biosensors. Nevertheless, a comprehensive knowledge of the role of QS in extremophiles requires further research. The study of the genetic basis of autoinducer production, and a deeper investigation of the mechanisms by which these signal molecules modulate the different phenotypes in extremophilic microorganisms, could play a pivotal role in the understanding of how these organisms survive in extreme conditions. Moreover, as it has been described in different microbial models, QS could play a role in the synthesis of some "extreme molecules" (in particular exopolisaccharides and exoenzymes) that can find interesting industrial applications. Therefore it is clear that investigating QS in harsh environments could be an important contribution to the improvement of biotechnological processes.

Glossary

Archaea Domain of life containing prokaryotic microorganisms, genetically distinct from bacteria and often living in extreme environments.

Autoinducer Signaling molecule involved in quorum sensing (QS) that is synthesized when cell population reaches a threshold density.

Bacteriorhodopsin Photosynthetic protein used by archaea.

Biocatalysis An enzyme or enzyme complex consisting of, or derived from, an organism or cell culture (in cell-free or whole-cell forms) that catalyzes metabolic reactions in living organisms and/or substrate conversions in various chemical reactions.

Biofilm Aggregate of microorganisms held together by a self-produced matrix mostly composed of exopolysaccharides and proteins.

Bioremediation The use of living organisms to treat contaminants or remediate contaminated soil, water, or air.

Biosurfactant Amphiphilic compounds produced on living surfaces.

Catalyst A substance that increases the rate of a chemical reaction without itself undergoing any permanent chemical change.

Conjugation Mechanism by which a bacterium (donor) transfer genetic material, most often a plasmid or transposon, to another cell (recipient).

Exopolysaccharide High molecular weight polymers composed of sugar residues, secreted by a microorganism into the surrounding environment.

Extremophile Organism that thrives in physically or geochemically extreme conditions.

Halite The mineral (natural) form of sodium chloride (NaCl).

Mesophile Any of various organisms, especially certain bacteria and fungi, that thrive at moderate temperatures between about 25°C and 40°C.

Metagenomics The genomic analysis of microorganisms by means of DNA shearing and sequencing, and data processing using bioinformatics tools.

Peptidoglycan A biopolymer consisting of amino acids and carbohydrates, forming the cell wall of most bacteria.

Polyol A generic name for low molecular weight, water-soluble polymers, and oligomers containing a large number of hydroxyl groups.

Polysaccharide High molecular weight polymer composed of sugar residues.

Quorum quenching The group of processes which disrupt quorum sensing (QS) signaling.

Quorum sensing Molecular communication mechanism through which several organisms regulate gene expression by means of a population-dependent signaling.

Regulon A group of genes that is regulated by the same regulatory molecule.

Violacein A violet pigment produced by certain bacteria.

Zoospore A tiny asexual reproductive cell of certain algae and fungi that is capable of independent motion.

Abbreviations

R-THMF	$(2R,4S)$-2-Methyl-2,3,3,4-tetrahy droxytetrahydrofuran
S-THMF	$(2S,4S)$-2-Methyl-2,3,3,4-tetrahydroxytetrahydrofuran-borate
DPD	(Z)-3-aminoundec-2-en-4-one (Ea-C8-CAI-1); 4,5-Dihydroxy-2, 3-pentanedione
AI	Autoinducer
AI-2	Autoinducer-2
CAI-1	*Cholerae* autoinducer-1
DKP	Diketopiperazine
ESI tandem MS	Electrospray ionization mass spectrometry
EPS	Exopolysaccharide
GC–MS	Gas chromatography–mass spectrometry
LC–MS	Liquid chromatography–mass spectrometry
OC4-HSL	N-(3-oxobutyryl)-L-homoserine
OC10-HSL	N-(3-oxodecanoyl)-L-homoserine lactone
OC12-HSL	N-(3-oxododecanoyl)-L-homoserine lactone
AHL	N-acyl homoserine lactone
C4-HSL	N-butanoylhomoserine lactone
C12-HSL	N-dodecanoylhomoserine lactone
C6-HSL	N-hexanoylhomoserine lactone
C8-HSL	N-octanoylhomoserine lactone
OGT	Optimal growth temperature
PAHs	Polycyclic aromatic hydrocarbons
QQ	Quorum quenching
QS	Quorum sensing
FOYE-1	Thermophilic-like ene-reductase

Acknowledgments

The authors thank the BAS/CNR Joint Projects 2016–18 for supporting this work.

References

Abbamondi, G.R., De Rosa, S., Iodice, C., Tommonaro, G., 2014. Cyclic dipeptides produced by marine sponge-associated bacteria as quorum sensing signals. Nat. Prod. Commun. 9 (2), 229–232.

Abbamondi, G.R., Suner, S., Cutignano, A., Grauso, L., Nicolaus, B., Toksoy Oner, E., Tommonaro, G., 2016. Identification of N-Hexadecanoyl-L-homoserine lactone (C16-AHL) as signal molecule in halophilic bacterium *Halomonas smyrnensis* AAD6. Ann. Microbiol. 66 (3), 1329–1333.

Abed, R.M.M., Dobretsov, S., Al-Fori, M., Gunasekera, S.P., Sudesh, K., Paul, V.J., 2013. Quorum-sensing inhibitory compounds from extremophilic microorganisms isolated from a hypersaline cyanobacterial mat. J. Ind. Microbiol. Biotechnol. 40 (7), 759–772.

Albers, S., Eichler, J., Aebi, M., 2017. Archaea. In: Varki, A., Cummings, R.D., Esko, J.D., Stanley, P., Hart, G.W., Aebi, M., Darvill, A.G., Kinoshita, T., Packer, N.H., Prestegard, J.H., Schnaar, R.L., Seeberger, P.H. (Eds.), Essentials of Glycobiology, third ed. Cold Spring Harbor, NY.

Antranikian, G., Vorgias, C.E., Bertoldo, C., 2005. Extreme environments as a resource for microorganisms and novel biocatalysts. In: Ulber, R., Le Gal, Y. (Eds.), Marine Biotechnology I. Springer Berlin Heidelberg, Berlin, Heidelberg, pp. 219–262.

Averhoff, B., Müller, V., 2010. Exploring research frontiers in microbiology: recent advances in halophilic and thermophilic extremophiles. Res. Microbiol. 161 (6), 506–514.

Baker, S.E., Hopkins, R.C., Blanchette, C.D., Walsworth, V.L., Sumbad, R., Fischer, N.O., Kuhn, E.A., Coleman, M., Chromy, B.A., Létant, S.E., Hoeprich, P.D., Adams, M.W.W., Henderson, P.T., 2009. Hydrogen production by a hyperthermophilic membrane-bound hydrogenase in water-soluble nanolipoprotein particles. J. Am. Chem. Soc. 131 (22), 7508–7509.

Beine, R., Moraru, R., Nimtz, M., Na'amnieh, S., Pawlowski, A., Buchholz, K., Seibel, J., 2008. Synthesis of novel fructooligosaccharides by substrate and enzyme engineering. J. Biotechnol. 138 (1), 33–41.

Bergonzi, C., Schwab, M., Elias, M., 2016. The quorum-quenching lactonase from *Geobacillus caldoxylosilyticus*: Purification, characterization, crystallization and crystallographic analysis. Acta Crystallogr. Sect. F 72 (9), 681–686.

Bodor, A., Elxnat, B., Thiel, V., Schulz, S., Wagner-Döbler, I., 2008. Potential for luxS related signalling in marine bacteria and production of autoinducer-2 in the genus *Shewanella*. BMC Microbiol. 8, 13.

Bzdrenga, J., Daudé, D., Rémy, B., Jacquet, P., Plener, L., Elias, M., Chabrière, E., 2017. Biotechnological applications of quorum quenching enzymes. Chem. Biol. Interact. 267, 104–115.

Canganella, F., Wiegel, J., 2014. Anaerobic thermophiles. Life 4 (1), 77–104.

Cárdenas, J.P., Quatrini, R., Holmes, D.S., 2016. Genomic and metagenomic challenges and opportunities for bioleaching: A mini-review. Res. Microbiol. 167 (7), 529–538.

Chen, F., Gao, Y., Chen, X., Yu, Z., Li, X., 2013. Quorum quenching enzymes and their application in degrading signal molecules to block quorum sensing-dependent infection. Int. J. Mol. Sci. 14 (9), 17477–17500.

Chen, L.-x., Huang, L.-n., Méndez-García, C., Kuang, J.-l., Hua, Z.-s., Liu, J., Shu, W.-s., 2016. Microbial communities, processes and functions in acid mine drainage ecosystems. Curr. Opin. Biotechnol. 38, 150–158.

Christel, S., Herold, M., Bellenberg, S., El Hajjami, M., Buetti-Dinh, A., Pivkin, I.V., Sand, W., Wilmes, P., Poetsch, A., Dopson, M., 2018. Multi-omics reveals the lifestyle of the acidophilic, mineral-oxidizing model species *Leptospirillum ferriphilum*^{T}. Appl. Environ. Microbiol 84 (3), e02091–17.

Chugunov, A.O., Volynsky, P.E., Krylov, N.A., Boldyrev, I.A., Efremov, R.G., 2014. Liquid but durable: Molecular dynamics simulations explain the unique properties of archaeal-like membranes. Sci. Rep. 4, 7462.

Coker, J.A., 2016. Extremophiles and biotechnology: current uses and prospects. F1000Research 5, F1000 Faculty Rev-1396.

Coker, J.A., Brenchley, J.E., 2006. Protein engineering of a cold-active β-galactosidase from Arthrobacter sp. SB to increase lactose hydrolysis reveals new sites affecting low temperature activity. Extremophiles 10 (6), 515–524.

Crispino, A., de Giulio, A., de Rosa, S., Strazzullo, G., 1989. A new bioactive derivative of avarol from the marine sponge Dysidea avara. J. Nat. Prod. 52 (3), 646–648.

Dalmaso, Z.G., Ferreira, D., Vermelho, B.A., 2015. Marine extremophiles: a source of hydrolases for biotechnological applications. Mar. Drugs 13(4).

DasSarma, P., Coker, J.A., Huse, V., DasSarma, S., Flickinger, M.C., 2009. Halophiles, Industrial Applications, Encyclopedia of Industrial Biotechnology. John Wiley & Sons, Inc.

Decho, A.W., 2000. Microbial biofilms in intertidal systems: An overview. Cont. Shelf Res. 20 (10), 1257–1273.

Di Donato, P., Romano, I., Mastascusa, V., Poli, A., Orlando, P., Pugliese, M., Nicolaus, B., 2018. Survival and adaptation of the thermophilic species *Geobacillus thermantarcticus* in simulated spatial conditions. Origins Life Evol. Biospheres 48 (1), 141–158.

Dornez, E., Cuyvers, S., Holopainen, U., Nordlund, E., Poutanen, K., Delcour, J.A., Courtin, C.M., 2011. Inactive fluorescently labeled xylanase as a novel probe for microscopic analysis of arabinoxylan containing cereal cell walls. J. Agric. Food Chem. 59 (12), 6369–6375.

Farah, C., Vera, M., Morin, D., Haras, D., Jerez, C.A., Guiliani, N., 2005. Evidence for a functional quorum-sensing type ai-1 system in the extremophilic bacterium *Acidithiobacillus ferrooxidans*. Appl. Environ. Microbiol. 71 (11), 7033–7040.

Finore, I., Di Donato, P., Mastascusa, V., Nicolaus, B., Poli, A., 2014. Fermentation technologies for the optimization of marine microbial exopolysaccharide production. Mar. Drugs 12 (5), 3005–3024.

Finore, I., Kasavi, C., Poli, A., Romano, I., Oner, E.T., Kirdar, B., Dipasquale, L., Nicolaus, B., Lama, L., 2011. Purification, biochemical characterization and gene sequencing of a thermostable raw starch digesting alpha-amylase from *Geobacillus thermoleovorans* subsp *stromboliensis* subsp nov. World J. Microbiol. Biotechnol. 27 (10), 2425–2433.

Finore, I., Lama, L., Poli, A., Di Donato, P., Nicolaus, B., 2015. Biotechnology implications of extremophiles as life pioneers and wellspring of valuable biomolecules. In: Kalia, V.C. (Ed.), Microbial Factories: Biodiversity, Biopolymers, Bioactive Molecules: Volume 2. Springer India, New Delhi, pp. 193–216.

Finore, I., Poli, A., Di Donato, P., Lama, L., Trincone, A., Fagnano, M., Mori, M., Nicolaus, B., Tramice, A., 2016. The hemicellulose extract from *Cynara cardunculus*: a source of value-added biomolecules produced by xylanolytic thermozymes. Green Chem. 18 (8), 2460–2472.

Frederick, M.R., Kuttler, C., Hense, B.A., Eberl, H.J., 2011. A mathematical model of quorum sensing regulated EPS production in biofilm communities. Theor. Biol. Med. Modell. 8, 8.

Fuqua, W.C., Winans, S.C., Greenberg, E.P., 1994. Quorum sensing in bacteria: the LuxR-LuxI family of cell density-responsive transcriptional regulators. J. Bacteriol. 176 (2), 269–275.

Grandclément, C., Tannières, M., Moréra, S., Dessaux, Y., Faure, D., 2016. Quorum quenching: role in nature and applied developments. FEMS Microbiol. Rev. 40 (1), 86–116.

Gunn, P., Walsh, S., Roux, C., 2014. The nucleic acid revolution continues—will forensic biology become forensic molecular biology? Front. Genet. 5, 44.

Guo, N., Puhlev, I., Brown, D.R., Mansbridge, J., Levine, F., 2000. Trehalose expression confers desiccation tolerance on human cells. Nat. Biotechnol. 18, 168.

Hasan, F., Shah, A.A., Hameed, A., 2006. Industrial applications of microbial lipases. Enzym. Microb. Technol. 39 (2), 235–251.

Hedlund, B.P., Dodsworth, J.A., Murugapiran, S.K., Rinke, C., Woyke, T., 2014. Impact of single-cell genomics and metagenomics on the emerging view of extremophile "microbial dark matter". Extremophiles 18 (5), 865–875.

Hiblot, J., Gotthard, G., Chabriere, E., Elias, M., 2012. Structural and enzymatic characterization of the lactonase SisLac from *Sulfolobus islandicus*. PLoS One 7(10), e47028.

Huang, Y.L., Zeng, Y., Yu, Z., Zhang, J., Feng, H., Lin, X., 2013. In silico and experimental methods revealed highly diverse bacteria with quorum sensing and aromatics biodegradation systems—a potential broad application on bioremediation. Bioresour. Technol. 148, 311–316.

Imamura, S., Kitaura, S., 2000. Purification and characterization of a monoacylglycerol lipase from the moderately thermophilic *Bacillus* sp. H-257. J. Biochem. 127 (3), 419–425.

Ishino, S., Ishino, Y., 2014. DNA polymerases as useful reagents for biotechnology—the history of developmental research in the field. Front. Microbiol. 5, 465.

Jacquemet, A., Barbeau, J., Lemiègre, L., Benvegnu, T., 2009. Archaeal tetraether bipolar lipids: structures, functions and applications. Biochimie 91 (6), 711–717.

Johnson, M.R., Montero, C.I., Conners, S.B., Shockley, K.R., Bridger, S.L., Kelly, R.M., 2005. Population density-dependent regulation of exopolysaccharide formation in the hyperthermophilic bacterium *Thermotoga maritima*. Mol. Microbiol. 55 (3), 664–674.

Kallnik, V., Bunescu, A., Sayer, C., Bräsen, C., Wohlgemuth, R., Littlechild, J., Siebers, B., 2014. Characterization of a phosphotriesterase-like lactonase from the hyperthermoacidophilic crenarchaeon *Vulcanisaeta moutnovskia*. J. Biotechnol. 190, 11–17.

Karan, R., Capes, M.D., DasSarma, P., DasSarma, S., 2013. Cloning, overexpression, purification, and characterization of a polyextremophilic β-galactosidase from the Antarctic haloarchaeon *Halorubrum lacusprofundi*. BMC Biotechnol. 13, 3.

Kato, C., 1999. Barophiles (Piezophiles). In: Horikoshi, K., Tsujii, K. (Eds.), Extremophiles in Deep-Sea Environments. Springer Japan, Tokyo, pp. 91–111.

Khrulnova, S.A., Baranova, A., Bazhenov, S.V., Goryanin, I.I., Konopleva, M.N., Maryshev, I.V., Salykhova, A.I., Vasilyeva, A.V., Manukhov, I.V., Zavilgelsky, G.B., 2016. Lux-operon of the marine psychrophilic bacterium *Aliivibrio logei*: A comparative analysis of the LuxR1/LuxR2 regulatory activity in *Escherichia coli* cells. Microbiology 162 (4), 717–724.

Konopleva, M.N., Khrulnova, S.A., Baranova, A., Ekimov, L.V., Bazhenov, S.V., Goryanin, I.I., Manukhov, I.V., 2016. A combination of luxR1 and luxR2 genes activates Pr-promoters of psychrophilic *Aliivibrio logei* lux-operon independently of chaperonin GroEL/ES and protease Lon at high concentrations of autoinducer. Biochem. Biophys. Res. Commun. 473 (4), 1158–1162.

Kothari, D., Das, D., Patel, S., Goyal, A., 2014. Dextran and food application. In: Ramawat, K.G., Mérillon, J.-M. (Eds.), Polysaccharides: Bioactivity and Biotechnology. Springer International Publishing, Cham, pp. 1–16.

Kucukasik, F., Kazak, H., Guney, D., Finore, I., Poli, A., Yenigun, O., Nicolaus, B., Oner, E.T., 2011. Molasses as fermentation substrate for Levan production by *Halomonas* sp. Appl. Microbiol. Biotechnol. 89 (6), 1729–1740.

León, M.J., Fernández, A.B., Ghai, R., Sánchez-Porro, C., Rodriguez-Valera, F., Ventosa, A., 2014. From metagenomics to pure culture: Isolation and characterization of the moderately halophilic bacterium *Spiribacter salinus* gen. Nov., sp. nov. Appl. Environ. Microbiol. 80 (13), 3850–3857.

Llamas, I., Quesada, E., Martínez-Cánovas, M.J., Gronquist, M., Eberhard, A., González, J.E., 2005. Quorum sensing in halophilic bacteria: detection of *N*-acyl-homoserine lactones in the exopolysaccharide-producing species of *Halomonas*. Extremophiles 9 (4), 333–341.

Luo, Z.Q., Clemente, T.E., Farrand, S.K., 2001. Construction of a derivative of agrobacterium tumefaciens C58 that does not mutate to tetracycline resistance. Molecular plant-microbe interactions. Journal 14 (1), 98–103.

Maiangwa, J., Ali, M.S.M., Salleh, A.B., Rahman, R.N.Z.R.A., Shariff, F.M., Leow, T.C., 2015. Adaptational properties and applications of cold-active lipases from psychrophilic bacteria. Extremophiles 19 (2), 235–247.

Mamani, S., Moinier, D., Denis, Y., Soulère, L., Queneau, Y., Talla, E., Bonnefoy, V., Guiliani, N., 2016. Insights into the quorum sensing regulon of the acidophilic *Acidithiobacillus ferrooxidans* revealed by transcriptomic in the presence of an acyl homoserine lactone superagonist analog. Front. Microbiol. 7(1365).

Mangwani, N., Kumari, S., Das, S., 2015. Involvement of quorum sensing genes in biofilm development and degradation of polycyclic aromatic hydrocarbons by a marine bacterium *Pseudomonas aeruginosa* N6P6. Appl. Microbiol. Biotechnol. 99 (23), 10283–10297.

Margesin, R., Miteva, V., 2011. Diversity and ecology of psychrophilic microorganisms. Res. Microbiol. 162 (3), 346–361.

Martini, S., Al Ali, B., Garel, M., Nerini, D., Grossi, V., Pacton, M., Casalot, L., Cuny, P., Tamburini, C., 2013. Effects of hydrostatic pressure on growth and luminescence of a moderately-piezophilic luminous bacteria *Photobacterium phosphoreum* ANT-2200. PLoS One 8(6), e66580.

Mastascusa, V., Romano, I., Di Donato, P., Poli, A., Della Corte, V., Rotundi, A., Bussoletti, E., Quarto, M., Pugliese, M., Nicolaus, B., 2014. Extremophiles survival to simulated space conditions: an astrobiology model study. Orig. Life Evol. Biosph. 44 (3), 231–237.

McClean, K.H., Winson, M.K., Fish, L., Taylor, A., Chhabra, S.R., Camara, M., Daykin, M., Lamb, J.H., Swift, S., Bycroft, B.W., Stewart, G.S., Williams, P., 1997. Quorum sensing and *Chromobacterium violaceum*: Exploitation of violacein production and inhibition for the detection of N-acyl homoserine lactones. Microbiology 143, 3703–3711.

Megaw, J., Gilmore, B.F., 2017. Archaeal persisters: Persister cell formation as a stress response in *Haloferax volcanii*. Front. Microbiol. 8, 1589.

Messia, M.C., Candigliota, T., Marconi, E., 2007. Assessment of quality and technological characterization of lactose-hydrolyzed milk. Food Chem. 104 (3), 910–917.

Meyer, B.H., Albers, S.-V., 2001. Archaeal Cell Walls, eLS. John Wiley & Sons, Ltd.

Minale, L., Riccio, R., Sodano, G., 1974. Avarol a novel sesquiterpenoid hydroquinone with a rearranged drimane skeleton from the sponge *Dysidea avara*. Tetrahedron Lett. 38 (15), 3401–3404.

Mishra, D., Rhee, Y.H., 2014. Microbial leaching of metals from solid industrial wastes. J. Microbiol. 52 (1), 1–7.

Mohammad, B.T., Al Daghistani, H.I., Jaouani, A., Abdel-Latif, S., Kennes, C., 2017. Isolation and characterization of thermophilic bacteria from jordanian hot springs: *Bacillus licheniformis* and *Thermomonas hydrothermalis* isolates as potential producers of thermostable enzymes. Int. J. Microbiol. 2017, 12.

Montgomery, K., Charlesworth, J.C., LeBard, R., Visscher, P.T., Burns, B.P., 2013. Quorum sensing in extreme environments. Life 3 (1), 131–148.

Morohoshi, T., Tominaga, Y., Someya, N., Ikeda, T., 2015. Characterization of a novel thermostable N-acylhomoserine lactonase from the thermophilic bacterium *Thermaerobacter marianensis*. J. Biosci. Bioeng. 120 (1), 1–5.

Müller, S., Pfannmöller, M., Teuscher, N., Heilmann, A., Rothe, U., 2006. New method for surface modification of nanoporous aluminum oxide membranes using tetraether lipid. J. Biomed. Nanotechnol. 2 (1), 16–22.

Nakayama, J., Cao, Y., Horii, T., Sakuda, S., Akkermans, A.D.L., de Vos, W.M., Nagasawa, H., 2001. Gelatinase biosynthesis-activating pheromone: A peptide lactone that mediates a quorum sensing in *Enterococcus faecalis*. Mol. Microbiol. 41 (1), 145–154.

Ng, F.S.W., Wright, D.M., Seah, S.Y.K., 2011. Characterization of a phosphotriesterase-like lactonase from *Sulfolobus solfataricus* and its immobilization for disruption of quorum sensing. Appl. Environ. Microbiol. 77 (4), 1181–1186.

Nichols, J.D., Johnson, M.R., Chou, C.-J., Kelly, R.M., 2009. Temperature, not LuxS, mediates AI-2 formation in hydrothermal habitats. FEMS Microbiol. Ecol. 68 (2), 173–181.

Nicolaus, B., Kambourova, M., Oner, E.T., 2010. Exopolysaccharides from extremophiles: From fundamentals to biotechnology. Environ. Technol. 31 (10), 1145–1158.

Oren, A., 2010. Industrial and environmental applications of halophilic microorganisms. Environ. Technol. 31 (8–9), 825–834.

Paggi, R.A., Martone, C.B., Fuqua, C., De Castro, R.E., 2003. Detection of quorum sensing signals in the haloalkaliphilic archaeon *Natronococcus occultus*. FEMS Microbiol. Lett. 221 (1), 49–52.

Pérez-Rodríguez, I., Bolognini, M., Ricci, J., Bini, E., Vetriani, C., 2015. From deep-sea volcanoes to human pathogens: a conserved quorum-sensing signal in Epsilonproteobacteria. ISME J. 9 (5), 1222–1234.

Pikuta, E.V., Hoover, R.B., Tang, J., 2007. Microbial extremophiles at the limits of life. Crit. Rev. Microbiol. 33 (3), 183–209.

Podar, M., Reysenbach, A.-L., 2006. New opportunities revealed by biotechnological explorations of extremophiles. Curr. Opin. Biotechnol. 17 (3), 250–255.

Poli, A., Anzelmo, G., Nicolaus, B., 2010. Bacterial exopolysaccharides from extreme marine habitats: Production, characterization and biological activities. Mar. Drugs 8 (6), 1779.

Poli, A., Finore, I., Romano, I., Gioiello, A., Lama, L., Nicolaus, B., 2017. Microbial diversity in extreme marine habitats and their biomolecules. Microorganisms 5 (2), 25.

Poli, A., Kazak, H., Gürleyendağ, B., Tommonaro, G., Pieretti, G., Öner, E.T., Nicolaus, B., 2009. High level synthesis of Levan by a novel *Halomonas* species growing on defined media. Carbohydr. Polym. 78 (4), 651–657.

Poli, A., Nicolaus, B., Denizci, A.A., Yavuzturk, B., Kazan, D., 2013. *Halomonas smyrnensis* sp. nov., a moderately halophilic, exopolysaccharide-producing bacterium. Int. J. Syst. Evol. Microbiol. 63 (1), 10–18.

Raddadi, N., Cherif, A., Daffonchio, D., Neifar, M., Fava, F., 2015. Biotechnological applications of extremophiles, extremozymes and extremolytes. Appl. Microbiol. Biotechnol. 99 (19), 7907–7913.

Reith, F., 2011. Life in the deep subsurface. Geology 39 (3), 287–288.

Ren, N., Cao, G., Wang, A., Lee, D.-J., Guo, W., Zhu, Y., 2008. Dark fermentation of xylose and glucose mix using isolated *Thermoanaerobacterium thermosaccharolyticum* W16. Int. J. Hydrog. Energy 33 (21), 6124–6132.

Riley, M., Staley, J.T., Danchin, A., Wang, T.Z., Brettin, T.S., Hauser, L.J., Land, M.L., Thompson, L.S., 2008. Genomics of an extreme psychrophile, *Psychromonas ingrahamii*. BMC Genomics 9, 210.

Rivas, M., Seeger, M., Jedlicki, E., Holmes, D.S., 2007. Second acyl homoserine lactone production system in the extreme acidophile *Acidithiobacillus ferrooxidans*. Appl. Environ. Microbiol. 73 (10), 3225–3231.

Ruiz, L.M., Valenzuela, S., Castro, M., Gonzalez, A., Frezza, M., Soulère, L., Rohwerder, T., Queneau, Y., Doutheau, A., Sand, W., Jerez, C.A., Guiliani, N., 2008. AHL communication is a widespread phenomenon in biomining bacteria and seems to be involved in mineral-adhesion efficiency. Hydrometallurgy 94 (1), 133–137.

Sam, S., Kucukasik, F., Yenigun, O., Nicolaus, B., Oner, E.T., Yukselen, M.A., 2011. Flocculating performances of exopolysaccharides produced by a halophilic bacterial strain cultivated on agro-industrial waste. Bioresour. Technol. 102 (2), 1788–1794.

Sarmiento, F., Peralta, R., Blamey, J.M., 2015. Cold and hot extremozymes: Industrial relevance and current trends. Front Bioeng Biotechnol 3, 148.

Saum, S.H., Müller, V., 2008. Regulation of osmoadaptation in the moderate halophile *Halobacillus halophilus*: chloride, glutamate and switching osmolyte strategies. Saline Systems 4 (1), 4.

Schiraldi, C., Giuliano, M., De Rosa, M., 2002. Perspectives on biotechnological applications of archaea. Archaea 1 (2), 75–86.

Scholtissek, A., Ullrich, S.R., Mühling, M., Schlömann, M., Paul, C.E., Tischler, D., 2017. A thermophilic-like ene-reductase originating from an acidophilic iron oxidizer. Appl. Microbiol. Biotechnol. 101 (2), 609–619.

Schopf, S., Wanner, G., Rachel, R., Wirth, R., 2008. An archaeal bi-species biofilm formed by *Pyrococcus furiosus* and *Methanopyrus kandleri*. Arch. Microbiol. 190 (3), 371–377.

Seckbach, J., Rampelotto, P.H., 2015. Chapter 8: Polyextremophiles. In: Bakermans, C. (Ed.), Microbial Evolution under Extreme Conditions. De Gruyter.

Sellami-Kamoun, A., Haddar, A., Ali, N.E.-H., Ghorbel-Frikha, B., Kanoun, S., Nasri, M., 2008. Stability of thermostable alkaline protease from *Bacillus licheniformis* RP1 in commercial solid laundry detergent formulations. Microbiol. Res. 163 (3), 299–306.

Sewald, X., Saum, S.H., Palm, P., Pfeiffer, F., Oesterhelt, D., Müller, V., 2007. Autoinducer-2-producing protein LuxS, a novel salt- and chloride-induced protein in the moderately halophilic bacterium *Halobacillus halophilus*. Appl. Environ. Microbiol. 73 (2), 371–379.

Sezer, A.D., Kazak, H., Öner, E.T., Akbuğa, J., 2011. Levan-based nanocarrier system for peptide and protein drug delivery: Optimization and influence of experimental parameters on the nanoparticle characteristics. Carbohydr. Polym. 84 (1), 358–363.

Shih, I.-L., Yu, Y.-T., Shieh, C.-J., Hsieh, C.-Y., 2005. Selective production and characterization of Levan by *Bacillus subtilis* (natto) Takahashi. J. Agric. Food Chem. 53 (21), 8211–8215.

Shulami, S., Shenker, O., Langut, Y., Lavid, N., Gat, O., Zaide, G., Zehavi, A., Sonenshein, A.L., Shoham, Y., 2014. Multiple regulatory mechanisms control the expression of the *Geobacillus stearothermophilus* gene for extracellular xylanase. J. Biol. Chem. 289 (37), 25957–25975.

Siliakus, M.F., van der Oost, J., Kengen, S.W.M., 2017. Adaptations of archaeal and bacterial membranes to variations in temperature, pH and pressure. Extremophiles 21 (4), 651–670.

Tahrioui, A., Quesada, E., Llamas, I., 2011. The hanR/hanI quorum-sensing system of *Halomonas anticariensis*, a moderately halophilic bacterium. Microbiology 157 (12), 3378–3387.

Tahrioui, A., Quesada, E., Llamas, I., 2013a. Draft genome sequence of the moderately halophilic gammaproteobacterium *Halomonas anticariensis* FP35(T). Genome Announc. 1 (4), e00497–00413.

Tahrioui, A., Schwab, M., Quesada, E., Llamas, I., 2013b. Quorum sensing in some representative species of *Halomonadaceae*. Life 3 (1), 260–275.

Tait, K., Williamson, H., Atkinson, S., Williams, P., Cámara, M., Joint, I., 2009. Turnover of quorum sensing signal molecules modulates cross-kingdom signalling. Environ. Microbiol. 11 (7), 1792–1802.

Tindbaek, N., Svendsen, A., Oestergaard, P.R., Draborg, H., 2004. Engineering a substrate-specific cold-adapted subtilisin. Protein Eng. Des. Sel. 17 (2), 149–156.

Tommonaro, G., Abbamondi, G.R., Iodice, C., Tait, K., De Rosa, S., 2012. Diketopiperazines produced by the halophilic archaeon, *Haloterrigena hispanica*, activate AHL bioreporters. Microb. Ecol. 63 (3), 490–495.

Tommonaro, G., Abbamondi, G.r., Toksoy Oner, E., Nicolaus, B., 2015. Investigating the quorum sensing system in halophilic bacteria. In: Maheshwari, D.K., Saraf, M. (Eds.), Halophiles: Biodiversity and Sustainable Exploitation. Springer International Publishing, Cham, pp. 189–207.

van den Burg, B., 2003. Extremophiles as a source for novel enzymes. Curr. Opin. Microbiol. 6 (3), 213–218.

van der Maarel, M.J.E.C., van der Veen, B., Uitdehaag, J.C.M., Leemhuis, H., Dijkhuizen, L., 2002. Properties and applications of starch-converting enzymes of the α-amylase family. J. Biotechnol. 94 (2), 137–155.

Vera, M., Schippers, A., Sand, W., 2013. Progress in bioleaching: fundamentals and mechanisms of bacterial metal sulfide oxidation—Part A. Appl. Microbiol. Biotechnol. 97 (17), 7529–7541.

Whiteley, M., Lee, K.M., Greenberg, E.P., 1999. Identification of genes controlled by quorum sensing in *Pseudomonas aeruginosa*. Proc. Natl. Acad. Sci. U. S. A. 96 (24), 13904–13909.

Williams, P., 2007. Quorum sensing, communication and cross-kingdom signalling in the bacterial world. Microbiology 153 (12), 3923–3938.

Wilson, C.M., Rodriguez, M., Johnson, C.M., Martin, S.L., Chu, T.M., Wolfinger, R.D., Hauser, L.J., Land, M.L., Klingeman, D.M., Syed, M.H., Ragauskas, A.J., Tschaplinski, T.J., Mielenz, J.R., Brown, S.D., 2013. Global transcriptome analysis of *Clostridium thermocellum* ATCC 27405 during growth on dilute acid pretreated Populus and switchgrass. Biotechnol. Biofuels 6, 179.

Woodley, J.M., 2013. Protein engineering of enzymes for process applications. Curr. Opin. Chem. Biol. 17 (2), 310–316.

Yousefi-Nejad, M., Manesh, H.N., Khajeh, K., 2011. Proteomics of early and late cold shock stress on thermophilic bacterium, *Thermus* sp. GH5. J. Proteome 74 (10), 2100–2111.

Zhang, G., Zhang, F., Ding, G., Li, J., Guo, X., Zhu, J., Zhou, L., Cai, S., Liu, X., Luo, Y., Zhang, G., Shi, W., Dong, X., 2012. Acyl homoserine lactone-based quorum sensing in a methanogenic archaeon. ISME J. 6 (7), 1336–1344.

Zhang, X., Yu, S., Gong, Y., Li, Y., 2016. Optimization design for turbodrill blades based on response surface method. Adv. Mech. Eng. 8 (2), 1–12.

PART 2

Inter-Kingdom Communication

CHAPTER 5

Quorum Sensing in Phytopathogenesis

Onur Kırtel[*,a], Maxime Versluys[†,a], Wim Van den Ende[†], Ebru Toksoy Öner[*]

[*]Industrial Biotechnology and Systems Biology Research Group, Marmara University, Bioengineering Department, Istanbul, Turkey, [†]Laboratory of Molecular Plant Biology, KU Leuven, Leuven, Belgium

1 Introduction

Microorganisms are everywhere. Although miniscule in size, they take part in innumerable interactions with their environment or their hosts, thus causing significant changes. The human microbiome, for instance, has been a popular research area in recent years, since strong interplays between the profile and actions of the microbiome and human health have been demonstrated (Althani et al., 2016 and references therein). Just like animals, plants are also hosts to significantly diverse groups of symbiotic microorganisms. Whether they act beneficially (mutualistic relationship) or detrimentally (parasitic relationship; pathogens) to the plants, these symbionts are constantly involved in complex inter- and intraspecies signaling events. The term *symbiosis* encompasses all pathogenic, commensalistic, or mutualistic relationships between the microorganisms and their hosts (Martin and Schwab, 2013). Thus, in this review, plant-pathogenesis is evaluated as a part of those symbiotic interactions. As seen in Fig. 1, this symbiosis is consolidated by a complex interplay between plants and microbes through the exchange of numerous signals. In this chain of events, rhizosphere microbes, which reside in the soil enveloping the plant roots, play a central role not only by initiating the crosstalk by root colonization but also through modulating the host immunity (Mendes et al., 2013). Colonization of the roots is achieved through the attraction of microorganisms with root exudates that encompass diverse molecules from simple sugars, polysaccharides and amino acids, proteins to phenolic compounds, vitamins, and hormones (Lareen et al., 2016). The chemotaxic response is stimulated via receptors with unknown specificity. Also, the transport of ions and protons at the root surface creates electrical currents that are able to attract motile zoospores. However, the relative significance of chemotaxis and electrotaxis in shaping the plant microbiome is still not clear (Mendes et al., 2013). Besides the soil-associated microbiome in the rhizosphere, phyllosphere microbes are also attracted by the exudates of the stems and leaves which include volatile organic compounds or hormones (Vorholt, 2012).

---

[a] These authors have contributed equally to this work.

Quorum Sensing. https://doi.org/10.1016/B978-0-12-814905-8.00005-8

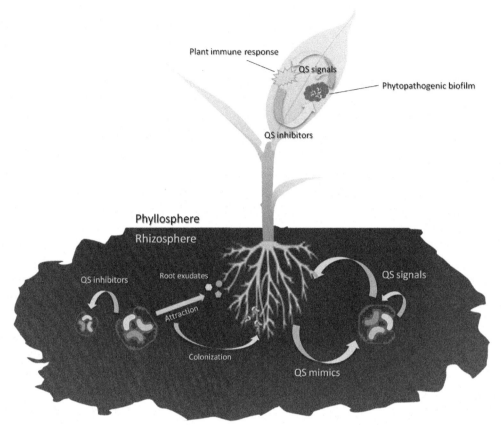

Fig. 1

Mutualistic and parasitic relationship between plants and microorganisms as mediated by quorum sensing (QS) systems. Root exudates as well as QS mimics by the plant attract mutualistic microorganisms to the rhizosphere for their colonization. Both mutualistic and phytopathogenic microorganisms rely on QS systems for various actions such as exopolysaccharide (EPS) synthesis, secretion of virulence factors, evading recognition by the plant, and so on. The presence of pathogens may create both pathogen-associated molecular patterns (PAMPs) or damage-associated molecular patterns (DAMPs), which trigger immune response pathways of the plant, resulting in the secretion of QS inhibitors. Various enzymes that show QS-inhibiting effects are also secreted by bacteria while competing for resources with other species.

Though not as abundant as in rhizospheres, bacteria are also known to be present in the plant organs like fruits, seeds, flowers, and nectar (Compant et al., 2011).

According to their lifestyle, phytopathogens are divided into three groups. The first group, biotrophs, acquire necessary metabolites from living cells; thus they do not kill their hosts rapidly. The second group, necrotrophs, are saprotrophs that rapidly kill their hosts and then derive the nutrients from dead host cells. The last group, hemibiotrophs, show biotrophic characteristics at the initial stages of infection, then turn to necrotrophs at later stages

(Moore et al., 2011). Phytopathogens are found across different groups of life, like bacteria, fungi, viruses, and nematodes (Chagas et al., 2018). Although some phytopathogenic species such as *Pseudomonas syringae* (Baltrus et al., 2017), *Erwinia amylovora* (McNally et al., 2015), or *Magnaporthe oryzae* (Martin-Urdiroz et al., 2016) are considered to be "model organisms" since they have been very well-studied over the years, -omics approaches and the identification of signaling events between these symbionts and their hosts in recent years reveal that there is much more to the complex mechanisms of phytopathogenesis than what is known today.

Plant pathogens devastate crops worldwide, costing tremendous amounts of money in the agricultural and food industries. In the United States alone, the value of crops lost due to phytopathogens is estimated to be around $33 billion every year (Pimentel, 2011). The rice blast fungus *Magnaporthe grisea* is responsible for the loss of 10%–30% of all harvested rice worldwide, which is estimated to be more than enough for feeding approximately 60 million people (Talbot, 2003). It is clear that humankind is still unable to combat these pests in an effective manner. The usage of -cidal chemicals may be effective on a short-term basis, but in addition to their detrimental effects on the environment and human health, these agents create significant selection pressure on the pathogens, paving the way for the emergence of resistant strains or their fast replacement by other pathogens. Understanding the underlying modes of action in phytopathogenesis is crucial to overcoming these losses. Over the recent years, the scientific world gained a better understanding of the signaling events between these pathogenic communities with the identification of diverse signaling molecules such as *N*-acylated homoserine lactones (AHLs), mainly in Gram-negative bacteria; several oligopeptides in Gram-positive bacteria; tyrosol and farnesol in yeasts; and the interspecies signaling molecule autoinducer furanosyl borate diester (AI-2) in many bacteria (Chagas et al., 2018). Identification and characterization of these signaling molecules make it possible to develop novel strategies to combat phytopathogenesis.

2 Quorum Sensing Phenomenon

Quorum sensing (QS) is an important phenomenon within the crosstalk between plants and microbes. By enabling the signaling with the host plant and inter- and intramicrobial communications, QS molecules play important roles in microbial colonization and symbiotic associations. The main driving forces for the pathogenesis of microorganisms mostly rely upon population density–dependent chemical signaling molecules, which make up the quorum sensing system as a whole. QS system was first identified in the marine bacterium *Vibrio fischeri* (Nealson and Hastings, 1979), which colonizes the light organs of its marine eukaryotic hosts such as the Hawaiian bobtail squid, *Euprymna scolopes*. After reaching a certain cell density in the light organ, *V. fischeri* colonies initiate bioluminescence via the luciferase enzyme. This mutualistic relationship enables *V. fischeri* to colonize a stationary surface in the nutrient-poor seawater, while providing the squid an effective camouflage by getting rid of its shadow on the water column at night. Basically, the synthesis of luciferase starts with the extracellular concentration of an AHL molecule (3-oxo-hexanoyl-homoserine lactone; 3-oxo-C6-HSL) reaching a certain threshold concentration (100–200 nM), after which it binds to the LuxR protein. The AHL-LuxR complex then binds to the luciferase promoter region, starting the transcription of both luciferase and the LuxI protein, creating a feed-forward loop that results in an exponential increase in AHL synthesis. The luciferase oxidizes a reduced flavin mononucleotide and a long-chain fatty acid to flavin mononucleotide and an aliphatic acid, respectively, also creating bioluminescence in the process (Verma and Miyashiro, 2013). Since the concentration of AHL (or any other autoinducer) in the community is directly proportional to the concentration of the cells that synthesize it, this QS system can be considered as a cell density–dependent chemical signaling cascade that leads to differential gene expression, in turn creating various phenotypic characteristics such as bioluminescence, biofilm formation, or the secretion of virulence factors.

2.1 Quorum Sensing in Plant-Associated Microorganisms

Plants have close relationships with a wide variety of microorganisms. The phyllosphere, comprising the aboveground microbiome of the plant, and the rhizosphere, the microbiome in the narrow zone around plant roots, harbor a large number of microorganisms—either epiphytes (attached to the plant) or endophytes (residing within the plant). The species composition is extremely dependent on plant species as well as abiotic and biotic environmental factors. Rhizosphere microbiota influence plant growth and development, and belowground microbial species richness can even be used to predict plant productivity. Apart from soil pathogens, beneficial rhizosphere microorganisms, including mycorrhizal fungi, plant growth–promoting rhizobacteria, and nitrogen-fixing bacteria, are topics of extensive research, clearly improving plant-growth dynamics (Mendes et al., 2013). Phyllosphere microbiota also manipulate many plant processes, such as immune responses, nutrient acquisition, and overall growth and development. The production and secretion of phytohormones is a common strategy to manipulate the plant host in favor of microbial survival. Exopolysaccharide (EPS) production is one of the most important adaptations for plant-associated bacteria to tolerate stresses such as drought (Chagas et al., 2018; and references therein).

The high abundance of microorganisms in close relationship with plants is closely linked to QS mechanisms. Approximately 2 decades ago, it was first shown that the production of AHL QS molecules, including 3-oxo- and 3-hydroxy AHLs with different acyl chain lengths, is widespread among plant-associated bacteria, including members of the genera *Agrobacterium*, *Erwinia*, *Pantoea*, *Pseudomonas*, and *Rhizobium* (Cha et al., 1998). *Xanthomonas* and *Burkholderia* species, on the other hand, mainly produce a different class of QS molecule—namely, DSFs (diffusible signal factors). These QS molecules often play a very important role in the regulation of biofilm formation and intra- and interspecies communication. Studies have revealed that AHL QS is common in both the rhizo- and phyllosphere, and across beneficial as well as pathogenic microorganisms. In plant pathogenic bacteria, QS also influences virulence factor expression. In *Pectobacterium carotovorum* subsp. *carotovorum* (formerly known as *Erwinia carotovora* subsp. *carotovora*), for example, QS influences the expression of extracellular enzymes and antibiotic production (von Bodman et al., 2003). Interestingly, species of the genus *Rhizobium* produce the greatest diversity in AHL structures, which clearly indicates the importance of QS in these nitrogen-fixing bacteria and the well-established interaction with their host plants (Cha et al., 1998; Sanchez-Contreras et al., 2007).

2.1.1 Quorum sensing in phytopathogenic bacteria

After the discovery of the AHL-dependent signaling system in *V. fischeri,* homologous systems in many other bacteria were found. A significant part of these belong to Gram-negative plant pathogens such as *Agrobacterium tumefaciens* (Lang and Faure, 2014; Subramoni et al., 2014), *Pectobacterium carotovorum* (Pirhonen et al., 1993), *Pantoea stewartii* (von Bodman and Farrand, 1995), and *Pseudomonas syringae* (Quiñones et al., 2005). Although the QS systems in these bacteria show the general pattern of the classical population density–dependent signaling, the molecules and the genes involved show great variability, thus demonstrating strikingly different pathogenicity strategies.

Agrobacterium is a genus of Gram-negative bacteria that causes crown gall disease in many plant species. Their virulence is based on the transfer and integration of transferred DNA (T-DNA) from a tumor inducing (Ti) plasmid into the nucleus of the plant cell, where it induces the expression of phytohormones auxin and cytokinin, as well as opine production. In the *Agrobacterium tumefaciens* QS system, the *TraR* gene, a homologue of *LuxR* in *V. fischeri*, was characterized as the transcriptional activator, which is located on the Ti plasmid and thus controlled by opines at the level of gene expression. Opines are produced by plants with crown gall disease by genes in the T-DNA and upregulate *TraR* expression. The coinducer of TraR is 3-oxo-C8-HSL, although other 3-oxo-AHLs also induce TraR to a certain extent. This AHL is produced by TraI from 3-oxo-C8 acyl carrier protein and S-adenosylmethionine. Active TraR-3-oxo-C8-HSL dimers regulate QS-dependent genes through a tra box in the promotor region. These QS-regulated genes are also located on the Ti plasmid. Upregulation of *tra*, *rep*, and *traM* genes induces Ti plasmid conjugation. TraM suppresses TraR activity by direct interaction, inhibiting TraR binding to DNA. Active TraR also upregulates *traM* to regulate basal TraR levels. QS signaling is also negatively regulated by lactonases, enzymes degrading the AHLs, which are induced by signals from plant origin (discussed in "Plant QS inhibitors"; Lang and Faure, 2014; Subramoni et al., 2014).

Erwinia amylovora, the pathogenic agent of the fire blight disease that devastates rosaceous plants like apples and pears, invades its hosts via a pilus-like structure, a part of the type III secretion system (T3SS; Piqué et al., 2015). *E. amylovora* also harbors a type VI secretion system (T6SS) that consists of 15 to 25 genes. Although the exact function of the T6SS is still undiscovered, some researchers reported that T6SS-deficient mutants showed impaired motility (Kamber et al., 2017) or EPS production (Tian et al., 2017), which are both crucial factors for the survival and virulence of the bacteria. *E. amylovora* produces both amylovoran, a heteropolysaccharide composed of glucose, galactose, and glucuronic acid, and levan, a homopolysaccharide of fructose, as EPSs. It was shown by Koczan et al. (2009) that amylovoran-deficient strains were nonpathogenic and died rapidly after inoculation to apple, while levan-deficient strains exhibited reduced motility and could not infect xylem vessels. *Pectobacterium* spp. were reported to harbor the LuxI homologue ExpI, which represses the transcription of various virulence factors such as T3SS, motility, adhesion, and plant cell wall degrading enzymes below a certain AHL concentration (Moleleki et al., 2017), thus ensuring the success of the pathogenic attack to the plant only after the community reaches a certain cell density. *Pantoea stewartii* is the causal agent of Stewart's wilt disease in corn, which mainly relies on the blockage of water transport in the xylem by the capsular polysaccharide (CPS) called *stewartan*, which is a branched, acidic heteropolysaccharide composed of glucose, galactose, and glucuronic acid residues (Nimtz et al., 1996). When the AHL concentration is below the threshold value, the LuxR homologue EsaR in *P. stewartii* represses *rcsA,* the transcription factor for CPS synthesis, and activates *lrhA*, which controls the surface motility of the bacteria (Duong and Stevens, 2017). *Pseudomonas syringae*, which heavily colonizes leaf surfaces in large aggregates before causing the brown spot disease on bean, also relies on an AHL-mediated QS system that regulates oxidative stress tolerance, EPS production, and motility (Quiñones et al., 2005).

Although most Gram-negative bacteria rely on AHL-mediated QS systems as seen previously, members of the genera *Burkholderia* and *Xanthomonas* use DSFs such as *cis*-2-dodecenoic acid and *cis*-11-methyl-2-dodecenoic acid, respectively (Carlier et al., 2014). In *Xanthomonas* species, DSFs are synthesized by RpfF homologues, an enzyme from the crotonase superfamily. After reaching a certain concentration in the extracellular space, DSFs are sensed by a sensor kinase on the membrane, RpfC, which in turn degrades bis-(3′,5′)-cyclic dimeric guanosine monophosphate (cyclic-di-GMP) through activating the cytoplasmic protein RpfG. Cyclic-di-GMP acts as a repressor of virulence factors in both *Xanthomonas* and *Burkholderia* species. Differently from *Xanthomonas* spp., in *Burkholderia* spp. DSFs are directly sensed by the cytoplasmic RpfR, which also degrades cyclic-di-GMP (Dow, 2017). It should be noted that *Burkholderia* spp. also harbor the classical AHL-mediated QS system, and there is evidence for an interplay between these two systems (Udine et al., 2013).

There are few Gram-positive phytopathogenic bacteria as compared with Gram-negative ones. The best characterized species are *Clavibacter michiganensis* and *Rhodococcus fascians,* which cause bacterial canker disease and leafy gall, respectively. Some *Streptomyces* spp., such as *S. acidiscabies, S. scabies* and *S. turgidiscabies,* are known to cause potato scab (Kers et al., 2005). In Gram-positive QS systems, no AHL-mediated signaling pathways were found thus far. Instead, Gram-positive bacteria use some posttranslationally modified oligopeptides as signaling molecules. Secreted oligopeptides are either detected by two-component systems on the membrane surface or taken into the cytoplasm by an oligopeptide transport system, after which they interact with different regulators to alter the expression of genes related to virulence, competence, or conjugation (Monnet and Gardan, 2015).

2.1.2 Quorum sensing in phytopathogenic fungi

Phytopathogenic fungi and oomycetes are responsible for 70%–80% of the worldwide annual crop loss caused by microbial agents (Oerke, 2006). Top phytopathogenic fungi/oomycetes and the diseases they cause include *Armillaria* spp. (honey fungus disease), *Blumeria graminis* (powdery mildew), *Botrytis cinerea* (gray mold), *Cercospora* spp. (leaf spots), *Fusarium oxysporum* (vascular wilt), *Magnaporthe oryzae* (rice blast disease), *Melampsora lini* (flax rust), *Ophiostoma novo-ulmi* (Dutch elm disease), *Phytophthora infestans* (potato late blight, the disease that caused the Irish famine in years 1845–46, resulting in the death of one-in-eight in the Irish population), *Puccinia graminis tritici* (stem rust), and *Ustilago* spp. (smut fungus disease; Moore et al., 2011). Fungi can be extremely aggressive pathogens, using diverse strategies to invade their plant hosts. *Armillaria* spp. use highly-protected subterrestrial multihyphal structures called *rhizomorphs* to transport water and nutrients across the forest floor. One clone of *Armillaria ostoyae* living in a forest in Oregon killed 30% of the ponderosa pines and covers a $9.65\,km^2$ area of the forest floor, easily classifying this as the largest organism alive today (Ferguson et al., 2003). A similar one covering an area of $450\,m^2$ was also found in a coniferous forest in the Black Sea region of Turkey (Lehtijärvi et al., 2017). Fungi like *M. limi* and *B. graminis* produce invasive structures called *haustoria* to penetrate plant cells to acquire nutrients and repress the defense systems of the host. Similarly, *Ustilago* spp. form intracellular hyphae for the same purposes (Moore et al., 2011).

As summarized by Oliveira-Garcia and Valent (2015; and references therein), parasitic fungi have developed highly specialized strategies to avoid being detected by the plant immune system. The first one establishes a gel–matrix interface between the haustoria/intracellular hyphae and the plant cell. This matrix contains high amounts of carbohydrates and proteins originating from both the fungus and the plant, and also sugar and amino acid transporters expressed by the plant. These transporters efficiently transfer the carbon sources to the invading fungus, thus minimizing the risk of triggering the formation of apoplastic signals that take part in plant defense mechanisms. The second strategy is the synthesis of specialized effectors that manipulate the host metabolism. These effectors are small proteins that may act via inhibiting host lytic enzyme activities or host immune response pathways, preventing the release of chitin oligomers from the fungal cell wall or sequestering already released ones, inhibiting peroxidases from the plant to protect the invading hyphae from reactive oxygen species (ROS), or directly manipulating the host metabolism to synthesize virulence factors that aid pathogenicity. The last and the third strategy includes structural modifications on the fungal cell wall to evade recognition by the host. For this, invading fungi may deacylate cell wall chitins to chitosans, produce α-1,3 glucans that protect the hyphae from plant lytic enzymes, or modulate β-1,3/1,6 glucan contents to evade pathogen-associated molecular pattern (PAMP) triggered immunity.

QS signaling molecules utilized by fungi and oomycetes are completely different from the ones in bacteria, with the best-known ones being farnesol, farnesoic acid, and tyrosol from *Candida albicans*; dimethoxycinnamate from *Uromyces phaseoli*; trisporic acid from zygomycetes; and a-factor from *Saccharomyces cerevisiae* (Hogan, 2006). Compared to bacteria, QS systems in fungi were only recently discovered, and are known to affect morphogenesis between different life stages in polymorphic fungi, which indirectly affects their virulence capacity. Similar to bacteria, signal molecules are produced intracellularly and secreted to the extracellular space, which then bind to signal receptor proteins on the membrane after reaching a threshold concentration, altering various gene expressions as a result. In *U. maydis*, short farnesylated peptides act as mating pheromones on the cells of the opposite mating type, inducing morphogenesis from budding to filamentous dikaryon (Spellig et al., 1994). Cyclic adenosine monophosphate (cAMP) has the same effect in this fungus (Gold et al., 1994). *Phytophthora parasitica* forms biofilms via the action of cell-density dependent extracellular signaling molecules (Galiana et al., 2008). In *B. graminis,* short-chain peptide signaling molecules similar to ones in Gram-positive bacteria were detected (Rajput et al., 2015). Since fungi are always in close contact with bacteria in nature, bacterial QS signaling molecules like AHLs affect fungi as well, inducing morphological changes or triggering enzymes involved in AHL degradation (Hartmann and Schikora, 2012). Most research in the literature that encompasses QS and fungi is limited to the context of inducing plant systemic resistance against pathogenic fungi via bacterial QS molecules, while the characterization of the QS systems in phytopathogenic fungi themselves is largely missing.

3 Effect of Bacterial Quorum Sensing on Plant Immunity and Physiology

3.1 Plant Innate Immunity

Plants, much like animals, have an innate immune system to maintain homeostasis under stressful environmental conditions, including biotic interactions. They respond to the presence of microorganisms via the recognition of microbe-associated molecular patterns (MAMPs) as part of their immune response. Originally, the plant immune system was visualized in two phases, PTI (pattern-triggered immunity) and ETI (effector-triggered immunity), in a zigzag model suggested by Jones and Dangl (2006), which has been adjusted and improved in recent years to better explain the immune responses *in planta*. ETI depends on specific interactions between Avr proteins from the pathogen and R proteins from the host, leading to programmed cell death pathways. PTI, on the other hand, is an evolutionary, more ancient mechanism based on the recognition of MAMPs by host receptors.

Perception of MAMPs is carried out by plant PRRs (pattern recognition receptors), located at the plasma membrane, and receptor activation induces downstream signaling cascades involving ROS dynamics, MAPK cascades, and hormone signals. LRR (leucine rich repeat) or LysM (lysine motif) are the most common domains found in PRRs, with the latter involved in the recognition of glycan-based molecules. One of the best-studied MAMPs is flg22, a conserved epitope of bacterial flagellin, which is recognized by the FLS2 receptor in plants. In the presence of flg22, the receptor heterodimerizes with the BAK1 kinase to induce downstream signaling (Bigeard et al., 2015). Among fungal MAMPs, chitin is the best characterized, recognized by plant LysM receptors based on *N*-acetylglucosamine oligomer fragments. Other MAMPs include EF-Tu, peptidoglycans, fructans, and so on. In case of pathogenic microorganisms, the recognition of MAMPs can be accompanied by DAMP (damage-associated molecular pattern) perception, as infection may lead to cellular leakage and/or cell wall degradation (Versluys et al., 2016).

Downstream of the MAMP receptor, a signaling cascade is activated upon perception. The earliest signaling responses involve Ca^{2+} and ROS signals. Ca^{2+}-dependent signals involve Ca^{2+}-binding proteins such as calcium-dependent protein kinases, which lead to the activation of transcription factors involved in biotic stress responses. The main source of ROS in biotic stress are the NADPH oxidase RBOHs, starting ROS waves for long distance signaling (Choi et al., 2017 and references therein). MAPKs (mitogen-activated protein kinases) are central players in the phosphorylation cascades downstream of MAMP receptors. They phosphorylate serine/threonine residues in the Thr-X-Tyr activation motifs of other MAPKs. While plants contain a large array of MAPKs, only some are involved in immune response activation, particularly MPK3 and MPK6. It has been suggested that MPK3 is more important in PTI-based signaling (Meng and Zhang, 2013). These pathways lead to a downstream induction of appropriate plant defenses against microbial attacks.

Plants possess several forms of constitutive defenses, which include physical barriers such as the cell wall and the cuticle, but also antimicrobial compounds such as saponins and glucosinolates. The latter is an inactive storage compound degraded during cellular damage to release the toxic isothiocyanates, a mechanism common among members of the Brassicales (Bigeard et al., 2015 and references therein). If these defenses prove insufficient to halt the pathogen, MAMP perception can trigger a wide variety of defenses, including lignin and callose deposition, programmed cell death, and the production of antimicrobial secondary metabolites and enzymes (Taiz and Zeiger, 2012). Phenylalanine ammonia lyase is an important enzyme involved in the production of lignin and flavonoids, among others. Moreover, it is also involved in the synthesis of salicylic acid (SA), a phytohormone involved in plant immune responses (Gao et al., 2015).

Besides an innate immunity, plants also have an induced systemic immune system. During pathogen attack, resistance mechanisms are induced in distal plant parts, leading to a systemic acquired resistance (SAR), conferring lower susceptibility to future infections. A similar mechanism occurs due to interaction with plant beneficial bacteria and is called induced systemic resistance (ISR; Gao et al., 2015). These systemic pathways require mobile signals, such as azelaic acid and pipecolic acid, that can be transported throughout the plant (Shah and Zeier, 2013). The phytohormone involved in SAR is SA, which increases at the site of infection and stimulates long-distance signaling for systemic resistance. ISR mechanisms are regulated by the phytohormones jasmonic acid (JA) and ethylene. SA and JA are generally antagonizing signals used for the induction of different defense responses (Gimenez-Ibanez and Solano, 2013).

3.2 Quorum Sensing Molecules and Plant Immunity

Since QS molecules are widespread among bacteria, it can be predicted that they may be recognized by PRRs in the plant as MAMPs. However, to date only the MAMP Ax21 peptide has been proposed as a QS molecule. This sulfated peptide binds to the XA21 PRR of rice, a receptor-like kinase that, when activated, translocates the intracellular kinase domain to the nucleus for downstream immune responses (Park and Ronald, 2012). Recent studies suggest that this Ax21 peptide is found in both plant and animal pathogens and is secreted through a type I secretion system. While some studies hinted at a role for Ax21 in QS, until now confirmation of this is still missing. It can be argued that due to the large diversity of QS molecules, they may not be appropriate for recognition as MAMPs. Nevertheless, several studies are discussed in the following paragraphs that highlight the effects of QS molecules on plant immunity.

The detection of AHLs by animal systems has been studied before, indicating effects on immunomodulatory mechanisms such as macrophages, interleukin production, and inflammatory responses. In plants, Mathesius et al. (2003) showed how *Medicago truncatula* roots are able to detect nanomolar concentrations of bacterial AHL, altering the accumulation of more than 150 proteins, among which are stress responsive factors such as PR proteins, peroxidases and superoxide dismutases, enzymes involved in flavonoid metabolism, and proteins involved in phytohormone signaling (Hartmann and Schikora, 2012). Most studies use the root application of QS molecules, which may be because foliar application causes less efficient detection by the plant, as shown in a study by You et al. (2006). Foliar spray of *Arabidopsis* with a 3-oxo-C8-HSL from *Agrobacterium tumefaciens* led to very limited changes in gene expression.

In tomato, Schuhegger et al. (2006) reported the involvement of a C6-HSL from *Serratia liquefaciens* in the induction of tomato disease resistance against *Alternaria alternata*. von Rad et al. (2008) studied the effects of a C6-HSL in *Arabidopsis* and found that this molecule causes only limited changes in defense-related transcription, thus indicating differences in detection among plant species. In a later study, Schikora et al. (2011) evidenced that the specific response of the host plant depends on the length of acyl moieties as well as the functional group present at the γ position of the AHL. Treatment with a 3-oxo-C14-HSL increased resistance of *Arabidopsis* to both the biotrophic fungus *Golovinomyces orontii* and the hemibiotrophic bacterium *Pseudomonas syringae* pv *tomato* DC3000. Intriguingly, by studying *atmpk6* mutants, the authors showed the requirement for AtMPK6, a central player in plant immune signaling pathways, in AHL-induced immunity.

In addition, Schikora et al. (2011) studied the effects of 3-oxo-C14-HSL on barley roots and found a resistance induction against *Blumeria graminis* f. sp. *hordei* similar to the responses found in *Arabidopsis*. In a later study, they showed a higher induction of immune responses, and thus resistance to infection, for longer acyl chain AHLs in the *Arabidopsis–P. syringae* pv. *tomato* DC3000 system (Schenk et al., 2012). In follow-up experiments, these researchers further characterized the specific defenses and signaling pathways upregulated by AHLs, using the 3-oxo-C14-HSL in *Arabidopsis*. Primed plants showed strengthening of the cell wall by lignification and callose depositions, as well as the production of phenolic compounds, providing resistance against *P. syringae* infection (Schenk and Schikora, 2015 and references therein).

AHL detection not only induces local defenses, but also seems to induce a systemic resistance pathway in the host plant. SA levels were elevated in tomato leaves after rhizosphere interaction with a C6-HSL from *S. liquefaciens*. Thus the QS molecules from this species can induce a systemic resistance response, as evidenced by a systemic induction of SA- and ethylene-dependent genes (Schuhegger et al., 2006). In the same line, Pang et al. (2009) showed an AHL-based ISR response in bean and tomato against *B. cinerea* infection, while Schenk and Schikora (2015) evidenced systemic disease resistance in *Arabidopsis* after C14-HSL root treatment. These authors later observed changes in oxylipins, such as JA and related metabolites, and SA in *Arabidopsis* treated with 3-oxo-C14-HSL. While SA and JA are known to behave antagonistically in systemic resistance pathways, in this AHL-priming system, they seem to have synergistic effects, indicating a new and AHL-specific signaling pathway (Schenk and Schikora, 2015 and references therein).

3.3 Quorum Sensing Molecules Affect Plant Physiology

In order to be detected by plants in a systemic way, QS molecules have to be taken up by the host plant. The uptake and transport of QS molecules through the roots depends on the length of AHLs, as well as the plant species. Götz et al. (2007) show how barley roots take up both C6- and C8-HSLs, while *Pachyrhizus erosus* roots only transport detectable levels of C6-HSL. In *Arabidopsis*, von Rad et al. (2008) show how transportation depends on the length of the acyl chain for AHLs. While a C6-HSL was readily transported from roots to shoots, C10-HSL was not transported, arguably due to higher hydrophobicity. As a result, C6-HSL treatment led to responses in all plant parts, while C10-HSL accumulated in the roots, leading to toxic effects. In addition to transport in plants, Ortíz-Castro et al. (2008) suggested the possibility of *Arabidopsis* to metabolize AHLs by studying fatty acid amide hydrolases, while this was also proposed earlier in *Lotus corniculatus* for AHLs with different acyl chain length (Delalande et al., 2005).

While there was no real induction of plant defenses after C6-HSL treatment in *Arabidopsis* in the study by von Rad et al. (2008), several genes related to growth hormones were significantly up- or downregulated and root elongation was apparent. Main changes were found in auxin to cytokinin ratios. Schenk and Schikora (2015) also found a positive effect on growth in *Arabidopsis* due to alterations in the ratio of these hormones. Shoot growth was most prominent with C6-HSL, while longer acyl chain AHLs showed decreasing growth levels, and the same effects were observed for root growth, indicating how shorter AHLs have a more profound effect on the plant. This is in sharp contrast with the effects on the plant immune system, where longer AHLs led to higher resistance. Nevertheless, Ortíz-Castro et al. (2008) showed more prominent effects on *Arabidopsis* root architecture for C10-HSL as compared with shorter or longer AHLs, this without the induction of the classic auxin-signaling pathway.

Root growth in *Arabidopsis* was also confirmed by Liu et al. (2012) after treatment of roots with C6- or C8-HSLs. In their study, *gcr1* mutants were insensitive to these treatments, indicating a role for GPCR1 in AHL signaling. In addition, they showed an induction in *GPCR1* gene expression after AHL treatment, confirming GPCR's role in AHL-based root elongation. Another study by Jin et al. (2012) also investigated the role of two putative GPCRs in Arabidopsis and further evidenced a role for these receptors in root elongation responses to AHLs. In mung bean, AHL treatment induced adventitious root formation, but only for the C3 substituted 3-oxo-C10-HSL. This was mediated by auxin-dependent signaling under control of H_2O_2- and NO-dependent cGMP signaling in the plants (Bai et al., 2012). A later study further confirmed the role of AHL detection in ROS dynamics in barley and *Pachyrhizus erosus*. The authors found an increase in antioxidative enzymes after treatment with AHLs of different acyl chain length (Götz-Rösch et al., 2015). These studies reveal the first players in plant AHL detection and signaling pathways, but they also point out the complexity of these interactions due to high structural differences between AHLs and potentially different plant responses to such signals.

4 Quorum Quenching in Phytopathogenesis

4.1 Plant Quorum Sensing Inhibitors

While QS compounds from plant-associated bacteria can have positive effects on plant health and growth, it is known that plants can produce QS inhibitors, which are especially important against pathogenic microorganisms. As QS is a way of communication between microbes, disrupting these signals will have no direct toxic effect on the bacterium, thus not leading to selection pressures for the development of resistant strains. This QS inhibitory mechanism is also known as *quorum quenching* (QQ; Rasmussen and Givskov, 2006). Plants can target the QS system at different levels, including inhibition of QS compound biosynthesis, inhibition of receptor binding, and QS compound degradation (Asfour, 2017).

A general strategy against AHLs can be alkalinization, leading to opening of the lactone ring. This strategy has been observed in plants interacting with the pathogen *Pectobacterium carotovorum*. Oxidized AHLs, such as 3-oxo-C14-HSL, can also be targeted, as *Laminaria digitata* is known to secrete oxidized halogen compounds that interfere with these oxidized AHLs. AHL degradation has also been proven for plant extracts from several legume species, although only degradation of short acyl chain AHLs was observed (Delalande et al., 2005; Rasmussen and Givskov, 2006).

A key study for QQ at the level of the receptor was performed in the red alga *Delisea pulchra*, which interferes with AHL-based QS using halogenated furanones (Manefield et al., 1999). In recent years, quorum sensing inhibitors targeting the bacterial signal receptors have been discovered in extracts of several plant species, including carrot, tomato, and garlic, although the identity of such compounds in often still unknown (Truchado et al., 2015). In garlic extract, at least three different inhibitors have been discovered, including dimethyl disulfide, ρ-coumaric acid, and ajoene. ρ-Coumaric acid is ubiquitous in plants, and the discovery of QS inhibitory effects in garlic indicates a similar role in other plant species, although earlier studies pointed in the direction of a stimulatory role for this compound in relation to other bacterial species (Kalia, 2013 and references therein).

Since then, many studies investigated the QQ potential in different plant systems, including seeds, seedlings, and whole plant extracts. In rice, seed extracts showed the presence of AHL mimics lacking the lactone ring. The bulk of these QQ studies used *Chromobacterium violaceum* as a model QS system to test QQ effects, but other bacterial species have also been assayed. QS biofilm formation was specifically inhibited in some bacterial strains due to these AHL mimics. The isothiocyanates sulforaphane and erucin, both found in *Brassica oleracea*, significantly inhibited virulence of the pathogen *Pseudomonas aeruginosa* through QQ mechanisms (Kalia, 2013).

Several other plant secondary metabolites have been identified as QS inhibitors in plant extracts. Monoterpenes binene, pulegone, limonene, and the sesquiterpene α-zingiberene from Colombian plant species worked as inhibitors in the *E. coli* short chain AHL QS system. The terpenoid citral also worked against the long chain QS system of *Pseudomonas putida*. The monoterpenoid carvacrol from oregano oils inhibited QS-regulated biofilm formation in *C. violaceum*. The phenolic compounds curcumin and ε-viniferin were shown to be involved in QQ in *Curcuma longa* and *Carex pumila*, respectively, against *P. aeruginosa*. From extracts of *Citrus* species, several flavonoids such as naringenin inhibited QS systems of *Vibrio harveyi*, *E. coli*, and *Yersinia enterocolitica* (Helman and Chernin, 2015, and references therein).

Interestingly, naringenin appears to inhibit multiple levels in the QS pathways, both at the level of AHL biosynthesis and AHL recognition (Asfour, 2017; Truchado et al., 2015).

It is known that *A. tumefaciens* QS is inhibited by plant GABA (γ-aminobutyric acid) and SA. GABA, which accumulates after wounding, and SA, an important player in plant defense signaling as indicated earlier, cause the induction of AHL-degrading γ-butyrolactonase expression. Interestingly, this mechanism is also beneficial for *A. tumefaciens*, since high levels of QS compounds might induce strong defense responses in the plant. In addition, this AHL-degrading enzyme can degrade other plant compounds, which the pathogen can further metabolize and may lead to a competitive advantage over other bacterial species, whose AHLs can also be degraded by this enzyme (Subramoni et al., 2014). Acetosyringone, a phenolic compound produced by plants, is also known to induce *TraM* expression, thus inhibiting TraR, providing another level of QS fine tuning (Lang and Faure, 2014).

In recent years, researchers also found QS inhibiting effects in honey from several plant species. Mainly the phenolic compounds in the honey were found to reduce AHL concentrations in *Erwinia amylovora* and *Y. enterocolitica*. They tested 29 different unifloral honeys and found QS inhibitor activity for most of them, mainly in chestnut and linden honey. In addition, many essential oils show QQ activity, as well as extracts of medicinal plants, fruits, and vegetables (Asfour, 2017). This indicates the omnipresence of QQ systems in plants to control bacterial populations in the rhizosphere and phyllosphere, and also provides sources for QS inhibitors, which can be used in numerous applications to reduce phytopathogen-based crop losses.

4.2 Bacterial Quorum Sensing Inhibitors against Plant Pathogenic Bacteria

Pathogenic or nonpathogenic, many bacteria can also inhibit QS systems of other species (including pathogenic ones). Considering the heavy competition for nutrients and space among different bacteria, it is only natural that they developed various strategies to inhibit each other's signaling systems. Several quorum sensing inhibitors (QSIs) have been characterized in five phyla of bacteria: Actinobacteria, Bacteroidetes, Cyanobacteria, Firmicutes, and Proteobacteria, all harboring a large number of species (Romero et al., 2015). Most studies on QSIs are focused on the inhibitors of AHL-dependent QS systems. AHL-QSIs are enzymes that degrade AHLs in different ways, and they can be classified into three groups: AHL lactonases, which cleave the ester bond on the homoserine lactone ring (Torres et al., 2013); AHL acylases, which hydrolyze the amide bond between the fatty acid and the homoserine lactone ring (Terwagne et al., 2013); and AHL oxidoreductases, which oxidize the acyl chain, thus altering the structure of the signal molecule (Schipper et al., 2009). Among plant-associated bacteria, *A. tumefaciens* (Carlier et al., 2003), *Bacillus* sp. (Dong et al., 2001), *Mesorhizobium loti* (Funami et al., 2005), *Microbacterium testaceum* (Wang et al., 2010), and *Rhizobium* sp. (Krysciak et al., 2011) have been reported to produce AHL lactonases, while *Pseudomonas syringae* has been reported to use two AHL acylases (Shepherd and Lindow, 2009). *Bacillus megaterium* can oxidize AHLs (Chowdhary et al., 2007), as well as farnesol from *Candida albicans* (Ramage et al., 2002; Murataliev et al., 2004). Employing bacterial QSI enzymes to fight phytopathogenesis appears to be an effective strategy with enormous potential in view of agronomical applications (Romero et al., 2015).

5 Sugars and Sugar Signaling in Plant–Microbe Interactions

It is clear that the interaction between plants and the microbiome are of utmost importance and is tightly regulated from both sides. Among the players involved, it is becoming more apparent that sugars are an important class of biomolecules in this context. In the very first stage, the establishment of the interaction, the plant attracts microbes through root exudates. The composition of these exudates is very diverse and means a significant cost for the plant in terms of carbon. Among the compounds, sugars and polysaccharides are abundantly present (Chagas et al., 2018). Interaction with beneficial microorganisms is dependent on the flux of sugars from the host to the bacterium. For example, in nodules the expression of sugar transporters mediates efficient efflux toward rhizobia. In the case of pathogenic bacteria, the release of sugars from the host has to be tightly regulated in order to block pathogen growth. In this sense, two hypotheses are proposed. When apoplasmic sugar levels are very low, the plant limits the sugar availability to the pathogen. Nevertheless, the leakage of sugars in the infection zone by sugar transporters has been described. Proposedly, these sugars can also be used to fuel plant defenses. Based on sugar signaling and sweet immunity principles, changing levels of small metabolic sugars can trigger the activation of defense pathways (Versluys et al., 2018; Bezrutczyk et al., 2018).

While the role of sugars in plant–microbe interactions is very clear, a specific role in QS mechanisms is less apparent. QS often regulates EPS production, ultimately controlling biofilm formation. Of specific interest are glucans and fructans, glucose- and fructose-based polysaccharides, respectively, which are produced from sucrose (Gangoiti et al., 2018; Versluys et al., 2018). In *Sinorhizobium meliloti*, the *bgsBA* operon, which is involved in β-glucan synthesis, is regulated by a QS system through c-di-GMP-dependent mechanisms (Pérez-Mendoza et al., 2015). It was previously reported in *P. aeruginosa* that the biosynthesis of this messenger is regulated by an AHL-based QS (Ueda and Wood, 2009). In Gram-negative bacteria, an outer membrane is found consisting of lipopolysaccharides (LPS), comprising a hydrophilic lipid A and a hydrophilic sugar chain with a core of heptose residues, originating from a unique ADP-heptose biosynthetic pathway. One intermediate, heptose-1,7-biphosphate (HBP), is known to elicit immune responses in animals (Gaudet and Gray-Owen, 2016). Whether HBP also functions as MAMP in plants is unknown. Earlier studies report the control of LPS biosynthesis by an AI-2 dependent QS pathway, thus indicating that AI-2 signaling may influence the amount of HBP available (De Araujo et al., 2010). Interestingly, AI-2 is a borate furanosyl, so it contains a sugar-like component.

Small metabolic sugars and enzymes from sugar metabolic pathways are also connected to QS, as the bacterium uses its metabolic status to regulate QS. Two recent studies point out the role of glucose-6-phosphate dehydrogenase and isomerase enzymes in *Xanthomonas oryzae* pv. *oryzicola*. Strains mutated in these enzymes show reduced virulence in rice and lower EPS production. In addition, these mutations led to changes in transcription of several genes involved in DSF signaling, thus indicating how sugar-related metabolism influences QS in this species (Guo et al., 2015, 2017).

A direct signaling effect of sugars on QS in plant interactions is, however, still unclear, although some indications may be present in the literature. It is known that plant cellobiose influences the production of thaxtomins in *Streptomyces scabies*. These are secondary metabolites used as a toxin by the bacterium. Lerat et al. (2010) showed that both cellobiose and suberin positively influence thaxtomin A production. Although the specific mechanism of cellobiose-based activation is unknown, it is possible that cellobiose influences QS, which is known to regulate secondary metabolite production in *Streptomyces* (Du et al., 2011). In the animal pathogen *Escherichia coli* O157:H7, Lee et al. (2011) proposed that glucose and fructose from acacia honey inhibit biofilm formation and virulence. Intriguingly, at the basis of these inhibitory effects was a downregulation of AI-2 expression and import, even by very low concentrations of these sugars. In the commensal *E. coli* K-12, however, these concentrations had no negative impact on biofilm formation, indicating how such QQ mechanisms can specifically target the harmful bacterial strains. This study links sugar signaling with QS, and while it is so far only demonstrated for animal pathogens, likely similar effects remain to be discovered for plant pathogens as well.

6 Concluding Remarks and Perspectives

Population density-dependent inter- and intraspecies signaling events between pathogenic microorganisms make up the driving force of phytopathogenesis. The mechanisms of their QS systems are diverse, and these pathogens not only employ their QS systems to initiate coordinated attacks on their hosts, but also manage to evade recognition by the plant via different ways. QS signaling molecules employed by phytopathogenic bacteria are relatively well-characterized, such as AHLs from Gram-negative bacteria or oligopeptides from Gram-positive ones. However, the knowledge on QS systems in fungi is rather limited to the morphological changes induced by signaling molecules, with research mostly focusing on only a few species like *Candida albicans* or *Ustilago maydis*. Considering the fact that phytopathogenic fungi (and oomycetes) are responsible for 70%–80% of crop losses worldwide, it is of utmost importance to understand the effects of their QS systems on plants during infection. Targeting QS signaling pathways to block the communication between pathogenic microorganisms, namely quorum quenching, is a promising alternative to chemical pest control agents, but requires extensive knowledge on the plant microbiome and the QS systems that take part in phytopathogenesis to develop combat methods that will be effective on a long-term basis.

Glossary

Biofilm A multifunctional, complex aggregate of various extracellular polymeric substances such as polysaccharides, proteins, and DNA, which are produced by the community of microorganisms it contains.

Bioluminescence Capacity of an organism to emit light via various biochemical events.

Damage-associated molecular patterns (DAMP) An array of host biomolecules that are released as a result of damage to the plant and recognized by specific plant receptors, which initiate an immune response.

Homeostasis Autoregulated ability of an organism to maintain stability while adapting to environmental changes.

Pathogen-associated molecular patterns (PAMP) An array of common and conserved biomolecules belonging to pathogens, such as flg22 or chitin, which are sensed by plant receptors to initiate immune response.

Phyllosphere Aboveground parts of a plant as a habitat for microorganisms.

Phytopathogenesis Parasitic ability of an organism to infect plants, which in turn results in damage to the plant.

Polymorphic fungi Fungi that exhibit various morphological forms during different stages of their life-cycle.

Quorum quenching (QQ) Disruption of quorum-sensing systems of organisms via inhibiting the actions of chemical signals or receptors involved.

Quorum sensing (QS) A microbial population density–dependent phenomenon that involves differential expression of various genes according to the presence of several signaling molecules at certain threshold concentrations in the environment, which in turn results in drastic changes in different phenotypes (i.e. virulence, bioluminescence).

Rhizosphere The soil area around the roots of a plant as a habitat for microorganisms.

Sweet immunity The connection between immune processes and carbohydrate dynamics in plants.

Abbreviations

AHL	acylated homoserine lactone
AI	autoinducer
CPS	capsular polysaccharide
cyclic-di-GMP	bis-(3′,5′)-cyclic dimeric guanosine monophosphate
DAMP	damage-associated molecular pattern
DSF	diffusible signal factor
EF-Tu	elongation factor Tu
EPS	exopolysaccharide
ETI	effector-triggered immunity

GABA	γ-aminobutyric acid
HBP	heptose-1,7-biphosphate
HSL	homoserine lactone
ISR	induced systemic resistance
JA	jasmonic acid
LPS	Lipopolysaccharide
MAMP	microbe-associated molecular pattern
MAPK	mitogen-activated protein kinase
PAMP	pathogen-associated molecular pattern
PR protein	pathogenesis related protein
PRR	pattern recognition receptor
PTI	pattern-triggered immunity
QQ	quorum quenching
QS	quorum sensing
QSI	quorum sensing inhibitor
ROS	reactive oxygen species
SA	salicylic acid
SAR	systemic acquired resistance
T3SS	type III secretion system
T6SS	type VI secretion system
T-DNA	transferred deoxyribonucleic acid
Ti	tumor-inducing

References

Althani, A.A., Marei, H.E., Hamdi, W.S., Nasrallah, G.K., El Zowalaty, M.E., Al Khodor, S., Al-Asmakh, M., Abdel-Aziz, H., Cenciarelli, C., 2016. Human microbiome and its association with health and diseases. J. Cell. Physiol. 231 (8), 1688–1694.

Asfour, H.Z., 2017. Antiquorum sensing natural compounds. J. Microsc. Ultrastruc. 6, 2–10.

Bai, X., Todd, C.D., Desikan, R., Yang, Y., Hu, X., 2012. *N*-3-oxo-decanoyl-L-homoserine-lactone activates auxin-induced adventitious root formation via hydrogen peroxide- and nitric oxide-dependent cyclic GMP signaling in mung bean. Plant Physiol. 158 (2), 725–736.

Baltrus, D.A., McCann, H.C., Guttman, D.S., 2017. Evolution, genomics and epidemiology of *Pseudomonas syringae*. Mol. Plant Pathol. 18 (1), 152–168.

Bezrutczyk, M., Yang, J., Eom, J., Prior, M., Sosso, D., Hartwig, T., Szurek, B., Oliva, R., Vera-Cruz, C., White, F.F., Yang, B., Frommer, W.B., 2018. Sugar flux and signaling in plant-microbe interactions. Plant J. 93 (4), 675–685.

Bigeard, J., Colcombet, J., Hirt, H., 2015. Signaling mechanisms in pattern-triggered immunity (PTI). Mol. Plant 8 (4), 521–539.

Carlier, A., Uroz, S., Smadja, B., Fray, R., Latour, X., Dessaux, Y., Faure, D., 2003. The Ti plasmid of *Agrobacterium tumefaciens* harbors an *attM*-paralogous gene, *aiiB*, also encoding *N*-acyl homoserine lactonase activity. Appl. Environ. Microbiol. 69 (8), 4989–4993.

Carlier, A., Pessi, G., Eberl, L., 2014. Microbial biofilms and quorum sensing. In: Lugtenberg, B. (Ed.), Principles of Plant-Microbe Interactions. Springer, Cham, pp. 45–52.

Cha, C., Gao, P., Chen, Y., Shaw, P.D., Farrand, S.K., 1998. Production of acyl-homoserine lactone quorum-sensing signals by gram-negative plant-associated bacteria. Mol. Plant Microbe Interact. 11 (11), 1119–1129.

Chagas, F.O., Pessotti, R.C., Carabello-Rodríguez, A.M., Pupo, M.T., 2018. Chemical signaling involved in plant-microbe interactions. Chem. Soc. Rev. 47 (5), 1652–1704.

Choi, W., Miller, G., Wallace, I., Harper, J., Mittler, R., Gilroy, S., 2017. Orchestrating rapid long-distance signaling in plants with Ca^{2+}, ROS and electrical signals. Plant J. 90 (4), 698–707.

Chowdhary, P.K., Keshavan, N., Nguyen, H.Q., Peterson, J.A., González, J.E., Haines, D.C., 2007. *Bacillus megaterium* CYP102A1 oxidation of acyl homoserine lactones and acyl homoserines. Biochemistry 46 (50), 14429–14437.

Compant, S., Mitter, B., Colli-Mull, J.G., Gangl, H., Sessitsch, A., 2011. Endophytes of grapevine flowers, berries, and seeds: Identification of cultivable bacteria, comparison with other plant parts, and visualization of niches of colonization. Microb. Ecol. 62 (1), 188–197.

De Araujo, C., Balestrino, D., Roth, L., Charbonnel, N., Forestier, C., 2010. Quorum sensing affects biofilm formation through lipopolysaccharide synthesis in *Klebsiella pneumoniae*. Res. Microbiol. 161 (7), 595–603.

Delalande, L., Faure, F., Raffoux, A., Uroz, S., D'Angelo-Picard, C., Elasri, M., Carlier, A., Berruyer, R., Petit, A., Williams, P., Dessaux, Y., 2005. *N*-hexanoyl-L-homoserine lactone, a mediator of bacterial quorum-sensing regulation, exhibits plant-dependent stability and may be inactivated by germinating *Lotus corniculatus* seedlings. FEMS Microbiol. Ecol. 52 (1), 13–20.

Dong, Y.H., Wang, L.H., Xu, J.L., Zhang, H.B., Zhang, X.F., Zhang, L.H., 2001. Quenching quorum-sensing-dependent bacterial infection by an *N*-acyl homoserine lactonase. Nature 411 (6839), 813–817.

Dow, J.M., 2017. Diffusible signal factor-dependent quorum sensing in pathogenic bacteria and its exploitation for disease control. J. Appl. Microbiol. 122 (1), 2–11.

Du, Y., Shen, X., Yu, P., Bai, L., Li, Y., 2011. Gamma-butyrolactone regulatory system of *Streptomyces chattanoogensis* links nutrient utilization, metabolism, and development. Appl. Environ. Microbiol. 77 (23), 8415–8426.

Duong, D.A., Stevens, A.M., 2017. Integrated downstream regulation by the quorum-sensing controlled transcription factors LrhA and RcsA impacts phenotypic outputs associated with virulence in the phytopathogen *Pantoea stewartii* subsp. *stewartii*. PeerJ. 5e4145.

Ferguson, B.A., Dreisbach, T.A., Parks, C.G., Filip, G.M., Schmitt, C.L., 2003. Coarse-scale population structure of pathogenic *Armillaria* species in a mixed-conifer forest in the Blue Mountains of northeast Oregon. Can. J. For. Res. 33 (4), 612–623.

Funami, J., Yoshikane, Y., Kobayashi, H., Yokochi, N., Yuan, B., Iwasaki, K., Ohnishi, K., Yagi, T., 2005. 4-Pyridoxolactonase from a symbiotic nitrogen-fixing bacterium *Mesorhizobium loti*: cloning, expression, and characterization. Biochim. Biophys. Acta 1753 (2), 234–239.

Galiana, E., Fourré, S., Engler, G., 2008. *Phytophthora parasitica* biofilm formation: installation and organization of microcolonies on the surface of a host plant. Environ. Microbiol. 10 (8), 2164–2171.

Gangoiti, J., Pijning, T., Dijkhuizen, L., 2018. Biotechnological potential of novel glycoside hydrolase family 70 enzymes synthesizing α-glucans from starch and sucrose. Biotechnol. Adv. 36 (1), 196–207.

Gao, Q., Zhu, S., Kachroo, P., Kachroo, A., 2015. Signal regulators of systemic acquired resistance. Front. Plant Sci. 6, 228.

Gaudet, R.G., Gray-Owen, S.D., 2016. Heptose sounds the alarm: innate sensing of a bacterial sugar stimulates immunity. PLoS Pathog. 12(9), e1005807.

Gimenez-Ibanez, S., Solano, R., 2013. Nuclear jasmonate and salicylate signaling and crosstalk in defense against pathogens. Front. Plant Sci. 4, 72.

Gold, S., Duncan, G., Barrett, K., Kronstad, J., 1994. cAMP regulates morphogenesis in the fungal pathogen *Ustilago maydis*. Genes Dev. 8 (23), 2805–2816.

Götz, C., Fekete, A., Gebefuegi, I., Forczek, S.T., Fuksová, K., Li, X., Englmann, M., Gryndler, M., Hartmann, A., Matucha, M., Schmitt-Kopplin, P., Schröder, P., 2007. Uptake, degradation and chiral discrimination of *N*-acyl-D/L-homoserine lactones by barley (*Hordeum vulgare*) and yam bean (*Pachyrhizus erosus*) plants. Anal. Bioanal. Chem. 389 (5), 1447–1457.

Götz-Rösch, C., Sieper, T., Fekete, A., Schmitt-Kopplin, P., Hartmann, A., Schröder, P., 2015. Influence of bacterial *N*-acyl-homoserine lactones on growth parameters, pigments, antioxidative capacities and the xenobiotic phase II detoxification enzymes in barley and yam bean. Front. Plant Sci. 6, 205.

Guo, W., Zou, L., Cai, L., Chen, G., 2015. Glucose-6-phosphate dehydrogenase is required for extracellular polysaccharide production, cell motility and the full virulence of *Xanthomonas oryzae* pv. *oryzicola*. Microb. Pathog. 78, 87–94.

Guo, W., Zou, L., Ji, Z., Cai, L., Chen, G., 2017. Glucose 6-phosphate isomerase (Pgi) is required for extracellular polysaccharide biosynthesis, DSF signals production and full virulence of *Xanthomonas oryzae* pv. *oryzicola* in rice. Physiol. Mol. Plant Pathol. 100, 209–219.

Hartmann, A., Schikora, A., 2012. Quorum sensing of bacteria and trans-kingdom interactions of *N*-acyl homoserine lactones with eukaryotes. J. Chem. Ecol. 38 (6), 704–713.

Helman, Y., Chernin, L., 2015. Silencing the mob: disrupting quorum sensing as a means to fight plant disease. Mol. Plant Pathol. 16 (3), 316–329.

Hogan, D.A., 2006. Talking to themselves: autoregulation and quorum sensing in fungi. Eukaryot. Cell 5 (4), 613–619.

Jin, G., Liu, F., Ma, H., Hao, S., Zhao, Q., Bian, Z., Jia, Z., Song, S., 2012. Two G-protein-coupled-receptor candidates, Cand2 and Cand7, are involved in Arabidopsis root growth mediated by the bacterial quorum-sensing signals *N*-acyl-homoserine lactones. Biochem. Biophys. Res. Commun. 417 (3), 991–995.

Jones, J. D. G., Dangl, J. L., 2006. The plant immune system. Nature 444(7117), 323–329.

Kalia, V.C., 2013. Quorum sensing inhibitors: an overview. Biotechnol. Adv. 31 (2), 224–245.

Kamber, T., Pothier, J.F., Pelludat, C., Rezzonico, F., Duffy, B., Smits, T.H., 2017. Role of the type VI secretion systems during disease interactions of *Erwinia amylovora* with its plant host. BMC Genomics 18 (1), 628.

Kers, J.A., Cameron, K.D., Joshi, M.V., Bukhalid, R.A., Morello, J.E., Wach, M.J., Gibson, D.M., Loria, R., 2005. A large, mobile pathogenicity island confers plant pathogenicity on *Streptomyces* species. Mol. Microbiol. 55 (4), 1025–1033.

Koczan, J.M., McGrath, M.J., Zhao, Y., Sundin, G.W., 2009. Contribution of *Erwinia amylovora* exopolysaccharides amylovoran and levan to biofilm formation: implications in pathogenicity. Phytopathology 99 (11), 1237–1244.

Krysciak, D., Schmeisser, C., Preuss, S., Riethausen, J., Quitschau, M., Grond, S., Streit, W.R., 2011. Involvement of multiple loci in quorum quenching of autoinducer I molecules in the nitrogen-fixing symbiont *Rhizobium (Sinorhizobium)* sp. strain NGR234. Appl. Environ. Microbiol. 77 (15), 5089–5099.

Lang, J., Faure, D., 2014. Functions and regulation of quorum sensing in *Agrobacterium tumefaciens*. Front. Plant Sci. 5, 14.

Lareen, A., Burton, F., Schäfer, P., 2016. Plant root-microbe communication in shaping root microbiomes. Plant Mol. Biol. 90 (6), 575–587.

Lee, J., Park, J., Kim, J., Neupane, G.P., Cho, M.H., Lee, C., Lee, J., 2011. Low concentrations of honey reduce biofilm formation, quorum sensing, and virulence in *Escherichia coli* O157:H7. Biofouling 27 (10), 1095–1104.

Lehtijärvi, A., Doğmuş-Lehtijärvi, H.T., Aday Kaya, A.G., Ünal, S., Woodward, S., 2017. *Armillaria ostoyae* in managed coniferous forests in Kastamonu in Turkey. Forest Pathol. 47(6), e12364.

Lerat, S., Simao-Beaunoir, A.-M., Wu, R., Beaudoin, N., Beaulieu, C., 2010. Involvement of the plant polymer suberin and the disaccharide cellobiose in triggering thaxtomin A biosynthesis, a phytotoxin produced by the pathogenic agent *Streptomyces scabies*. Phytopathology 100 (1), 91–96.

Liu, F., Bian, Z., Jia, Z., Zhao, Q., Song, S., 2012. The GCR1 and GPA1 participate in promotion of *Arabidopsis* primary root elongation induced by *N*-acyl-homoserine lactones, the bacterial quorum-sensing signals. Mol. Plant Microbe Interact. 25 (5), 677–683.

Manefield, M., de Nys, R., Kumar, N., Read, R., Givskov, M., Steinberg, P., Kjelleberg, S., 1999. Evidence that halogenated furanones from *Delisea pulchra* inhibit acylated homoserine lactone (AHL)-mediated gene expression by displacing the AHL signal from its receptor protein. Microbiology 145 (2), 283–291.

Martin, B.D., Schwab, E., 2013. Current usage of symbiosis and associated terminology. Int. J. Biol. 5 (1), 32–45.

Martin-Urdiroz, M., Oses-Ruiz, M., Ryder, L.S., Talbot, N.J., 2016. Investigating the biology of plant infection by the rice blast fungus *Magnaporthe oryzae*. Fungal Genet. Biol. 90, 61–68.

Mathesius, U., Mulders, S., Gao, M., Teplitski, M., Caetano-Anollés, G., Rolfe, B.G., Bauer, W.D., 2003. Extensive and specific responses of a eukaryote to bacterial quorum-sensing signals. PNAS 100 (3), 1444–1449.

McNally, R.R., Zhao, Y., Sundin, G.W., 2015. Towards understanding fire blight: virulence mechanisms and their regulation in *Erwinia amylovora*. In: Murillo, J., Vinatzer, B.A., Jackson, R.W., Arnold, D.L. (Eds.), Bacteria-Plant Interactions. Caister Academic Press, pp. 61–82.

Mendes, R., Garbeva, P., Raaijmakers, J.M., 2013. The rhizosphere microbiome: significance of plant beneficial, plant pathogenic, and human pathogenic microorganisms. FEMS Microbiol. Rev. 37 (5), 634–663.

Meng, X., Zhang, S., 2013. MAPK cascades in plant disease resistance signaling. Annu. Rev. Phytopathol. 51, 245–266.

Moleleki, L.N., Pretorius, R.G., Tanui, C.K., Mosina, G., Theron, J., 2017. A quorum sensing-defective mutant of *Pectobacterium carotovorum* ssp. *brasiliense* 1692 is attenuated in virulence and unable to occlude xylem tissue of susceptible potato plant stems. Mol. Plant Pathol. 18 (1), 32–44.

Monnet, V., Gardan, R., 2015. Quorum-sensing regulators in Gram-positive bacteria:'cherchez le peptide. Mol. Microbiol. 97 (2), 181–184.

Moore, D., Robson, G. D., & Trinci, A. P., 2011. 21st Century Guidebook to Fungi with CD. Cambridge, UK: Cambridge University Press.

Murataliev, M.B., Trinh, L.N., Moser, L.V., Bates, R.B., Feyereisen, R., Walker, F.A., 2004. Chimeragenesis of the fatty acid binding site of cytochrome P450BM3. Replacement of residues 73–84 with the homologous residues from the insect cytochrome P450 CYP4C7. Biochemistry 43 (7), 1771–1780.

Nealson, K.H., Hastings, J.W., 1979. Bacterial bioluminescence: its control and ecological significance. Microbiol. Rev. 43 (4), 496–518.

Nimtz, M., Mort, A., Wray, V., Domke, T., Zhang, Y., Coplin, D.L., Geider, K., 1996. Structure of stewartan, the capsular exopolysaccharide from the corn pathogen *Erwinia stewartii*. Carbohydr. Res. 288, 189–201.

Oerke, E.C., 2006. Crop losses to pests. J. Agric. Sci. 144 (1), 31–43.

Oliveira-Garcia, E., Valent, B., 2015. How eukaryotic filamentous pathogens evade plant recognition. Curr. Opin. Microbiol. 26, 92–101.

Ortíz-Castro, R., Martìnez-Trujillo, M., Lòpez-Bucio, J., 2008. *N*-acyl-L-homoserine lactones: a class of bacterial quorum-sensing signals alter post-embryonic root development in *Arabidopsis thaliana*. Plant Cell Environ. 31 (10), 1497–1509.

Pang, Y., Liu, X., Ma, Y., Chernin, L., Berg, G., Gao, K., 2009. Induction of systemic resistance, root colonisation and biocontrol activities of the rhizospheric strain of *Serratia plymuthica* are dependent on N-acyl homoserine lactones. Eur. J. Plant Pathol. 124 (2), 261–268.

Park, C., Ronald, P.C., 2012. Cleavage and nuclear localization of the rice XA21 immune receptor. Nat. Commun. 3, 920.

Pérez-Mendoza, D., Rodríguez-Carvajal, M.Á., Romero-Jiménez, L., Farias, G.A., Lloret, J., Gallegos, M.T., Sanjuán, J., 2015. Novel mixed-linkage β-glucan activated by c-di-GMP in *Sinorhizobium meliloti*. PNAS E757–E765.

Pimentel, D. (Ed.), 2011. Biological Invasions: Economic and Environmental Costs of Alien Plant, Animal, and Microbe Species. CRC Press.

Piqué, N., Miñana-Galbis, D., Merino, S., Tomás, J.M., 2015. Virulence factors of *Erwinia amylovora*: a review. Int. J. Mol. Sci. 16 (12), 12836–12854.

Pirhonen, M., Flego, D., Heikinheimo, R., Palva, E.T., 1993. A small diffusible signal molecule is responsible for the global control of virulence and exoenzyme production in the plant pathogen *Erwinia carotovora*. EMBO J. 12 (6), 2467–2476.

Quiñones, B., Dulla, G., Lindow, S.E., 2005. Quorum sensing regulates exopolysaccharide production, motility, and virulence in *Pseudomonas syringae*. Mol. Plant Microbe Interact. 18 (7), 682–693.

Rajput, A., Gupta, A.K., Kumar, M., 2015. Prediction and analysis of quorum sensing peptides based on sequence features. PLoS One. 10(3), e0120066.

Ramage, G., Saville, S.P., Wickes, B.L., López-Ribot, J.L., 2002. Inhibition of *Candida albicans* biofilm formation by farnesol, a quorum-sensing molecule. Appl. Environ. Microbiol. 68 (11), 5459–5463.

Rasmussen, T.B., Givskov, M., 2006. Quorum sensing inhibitors: a bargain of effects. Microbiology 152 (4), 895–904.

Romero, M., Mayer, C., Muras, A., Otero, A., 2015. Silencing bacterial communication through enzymatic quorum-sensing inhibition. In: Kalia, V.C. (Ed.), Quorum Sensing vs Quorum Quenching: A Battle with No End in Sight. Springer, India, pp. 219–236.

Sanchez-Contreras, M., Bauer, W.D., Gao, M., Robinson, J.B., Downie, J.A., 2007. Quorum-sensing regulation in rhizobia and its role in symbiotic interactions with legumes. Philos. Trans. R. Soc. B 362, 1149–1163.

Schenk, S.T., Schikora, A., 2015. AHL-priming functions via oxylipin and salicylic acid. Front. Plant Sci. 5, 784.

Schenk, S.T., Stein, E., Kogel, K., Schikora, A., 2012. Arabidopsis growth and defense are modulated by bacterial quorum sensing molecules. Plant Signal. Behav. 7 (2), 178–181.

Schikora, A., Schenk, S.T., Stein, E., Molitor, A., Zuccaro, A., Kogel, K., 2011. *N*-acyl-homoserine lactone confers resistance toward biotrophic and hemibiotropic pathogens via altered activation of AtMPK6. Plant Physiol. 157 (3), 1407–1418.

Schipper, C., Hornung, C., Bijtenhoorn, P., Quitschau, M., Grond, S., Streit, W.R., 2009. Metagenome-derived clones encoding two novel lactonase family proteins involved in biofilm inhibition in *Pseudomonas aeruginosa*. Appl. Environ. Microbiol. 75 (1), 224–233.

Schuhegger, R., Ihring, A., Gantner, S., Bahnweg, G., Knappe, C., Vogg, G., Hutzler, P., Schmid, M., Van Breusegem, F., Eberl, L., Hartmann, A., Langebartels, C., 2006. Induction of systemic resistance in tomato by *N*-acyl-L-homoserine lactone-producing rhizosphere bacteria. Plant Cell Environ. 29 (5), 909–918.

Shah, J., Zeier, J., 2013. Long-distance communication and signal amplification in systemic acquired resistance. Front. Plant Sci. 4, 30.

Shepherd, R.W., Lindow, S.E., 2009. Two dissimilar *N*-acyl-homoserine lactone acylases of *Pseudomonas syringae* influence colony and biofilm morphology. Appl. Environ. Microbiol. 75 (1), 45–53.

Spellig, T., Bölker, M., Lottspeich, F., Frank, R.W., Kahmann, R., 1994. Pheromones trigger filamentous growth in *Ustilago maydis*. EMBO J. 13 (7), 1620–1627.

Subramoni, S., Nathoo, N., Klimov, E., Yuan, Z., 2014. *Agrobacterium tumefaciens* responses to plant-derived signalling molecules. Front. Plant Sci. 5, 322.

Taiz, L., Zeiger, E., 2012. Chapter 13: Secondary metabolites and plant defense. In: Taiz, L., Zeiger, E. (Eds.), Plant Physiology. fifth ed. Sinauer Associated Inc., Sunderland, MA, pp. 369–400

Talbot, N.J., 2003. On the trail of a cereal killer: exploring the biology of *Magnaporthe grisea*. Annu. Rev. Microbiol. 57 (1), 177–202.

Terwagne, M., Mirabella, A., Lemaire, J., Deschamps, C., De Bolle, X., Letesson, J.J., 2013. Quorum sensing and self-quorum quenching in the intracellular pathogen *Brucella melitensis*. PLoS One. 8(12), e82514.

Tian, Y., Zhao, Y., Shi, L., Cui, Z., Hu, B., Zhao, Y., 2017. Type VI secretion systems of *Erwinia amylovora* contribute to bacterial competition, virulence, and exopolysaccharide production. Phytopathology 107 (6), 654–661.

Torres, M., Romero, M., Prado, S., Dubert, J., Tahrioui, A., Otero, A., Llamas, I., 2013. *N*-acylhomoserine lactone-degrading bacteria isolated from hatchery bivalve larval cultures. Microbiol. Res. 168 (9), 547–554.

Truchado, P., Larrosa, M., Castro-Ibáñez, I., Allende, A., 2015. Plant food extracts and phytochemicals: their role as quorum sensing inhibitors. Trends Food Sci. Technol. 43 (2), 189–204.

Udine, C., Brackman, G., Bazzini, S., Buroni, S., Van Acker, H., Pasca, M.R., Ricardi, G., Coenye, T., 2013. Phenotypic and genotypic characterisation of *Burkholderia cenocepacia* J2315 mutants affected in homoserine lactone and diffusible signal factor-based quorum sensing systems suggests interplay between both types of systems. PLoS One. 8(1), e55112.

Ueda, A., Wood, T.K., 2009. Connecting quorum sensing, c-di-GMP, Pel polysaccharide, and biofilm formation in *Pseudomonas aeruginosa* through tyrosine phosphatase TpbA (PA3885). PLoS Pathog. 5(6), e1000483.

Verma, S.C., Miyashiro, T., 2013. Quorum sensing in the squid-*Vibrio* symbiosis. Int. J. Mol. Sci. 14 (8), 16386–16401.

Versluys, M., Tarkowski, Ł.P., Van den Ende, W., 2016. Fructans as DAMPs or MAMPs: evolutionary prospects, cross-tolerance, and multistress resistance potential. Front. Plant Sci. 7, 2061.

Versluys, M., Kirtel, O., Öner, E.T., Van den Ende, W., 2018. The Fructan syndrome: evolutionary aspects and common themes among plants and microbes. Plant Cell Environ. 41 (1), 16–38.

von Bodman, S.B., Farrand, S.K., 1995. Capsular polysaccharide biosynthesis and pathogenicity in *Erwinia stewartii* require induction by an *N*-acylhomoserine lactone autoinducer. J. Bacteriol. 177 (17), 5000–5008.

von Bodman, S.B., Bauer, W.D., Coplin, D.L., 2003. Quorum sensing in plant-pathogenic bacteria. Annu. Rev. Phytopathol. 41 (1), 455–482.

von Rad, U., Klein, I., Dobrev, P.I., Kottova, J., Zazimalova, E., Fekete, A., Hartmann, A., Schmitt-Kopplin, P., Durner, J., 2008. Response of *Arabidopsis thaliana* to *N*-hexanoyl-DL-homoserine-lactone, a bacterial quorum sensing molecule produced in the rhizosphere. Planta 229 (1), 73–85.

Vorholt, J.A., 2012. Microbial life in the phyllosphere. Nat. Rev. Microbiol. 10 (12), 828.

Wang, W.Z., Morohoshi, T., Ikenoya, M., Someya, N., Ikeda, T., 2010. AiiM, a novel class of N-acylhomoserine lactonase from the leaf-associated bacterium *Microbacterium testaceum*. Appl. Environ. Microbiol. 76 (8), 2524–2530.

You, Y., Marella, H., Zentella, R., Zhou, Y., Ulmasov, T., Ho, T.D., Quatrano, R.S., 2006. Use of bacterial quorum-sensing components to regulate gene expression in plants. Plant Physiol. 140 (4), 1205–1212.

Further Reading

Ronald, P.C., 2011. Small protein-mediated quorum sensing in a gram-negative bacterium: novel targets for control of infectious disease. Discov. Med. 12 (67), 461–470.

CHAPTER 6

Quorum Sensing and the Gut Microbiome

Angel G. Jimenez*, Vanessa Sperandio[†]

Department of Microbiology, University of Texas Southwestern Medical Center, Dallas, TX, United States, [†]Department of Biochemistry, University of Texas Southwestern Medical Center, Dallas, TX, United States

1 Introduction

Humans are populated by an extensive community of microorganisms, primarily in organs such as the skin, mucosal membranes in the mouth, reproductive organs, and the gut. This complex community, termed the *microbiota*, is primarily composed of bacteria, many of which have an intimate association with their host to promote a healthy status. These organisms, commonly termed *commensal microbes*, have a rich and long history with their human host, and have evolved mechanisms to communicate and gauge host physiology to effectively accomplish important functions such as providing the host with nutrients, developing the immune system, and preventing colonization by pathogenic organisms (Gordon and Klaenhammer, 2011). These functions are especially apparent within the gastrointestinal (GI) tract, which contains the richest and most densely populated community in the body. Healthy gut function relies on proper structure and balance of this microbial community. Disruption of the community or dysbiosis has been associated with a plethora of diseases such as neurological disorders, altered inflammation, and cancer progression (Grenham et al., 2011). Proper structure of the microbiota relies on cell-to-cell communication through the exchange of chemical messages. The most studied form of bacterial cell-to-cell communication is known as *quorum sensing*. Quorum sensing enables bacterial populations to efficiently synchronize bacterial behaviors that are costly when undertaken by a small percentage of the population. Deployment of virulence factors, toxin production and secretion, the formation of secretion systems, and biofilm formation become more efficient when undertaken by the community. Quorum sensing triggers changes in global gene expression by responding to small molecules called *autoinducers*. These molecules are thought to increase in concentration as the bacteria replicate and increase in number.

Quorum sensing has been extensively studied in simple systems like those found in monocultures that are well shaken and steadily oxygenated. These systems fail to account for the complexity found in multispecies consortiums like the gut. In this environment, bacteria must respond to a milieu of chemical signals and effectively integrate them into a reliable message. In this context, autoinducers and other small molecules serve to communicate not only with cells of their species or close relatives but with those of even different kingdoms like the host. This broadens the definition of quorum sensing to include multidirectional communication pathways like interspecies communication and interkingdom signaling. This chapter will focus on several small molecules that are important signaling molecules in the gut, the responses to these molecules by the host, the microbiota, and incoming pathogens.

2 Autoinducers

The gut is a complex environment with a diverse consortium of microbes. These microbes produce a vast array of chemicals and therefore rely on this complex chemistry to regulate their gene expression. One important class of signaling molecules are known as *autoinducers*, which are the major signaling molecules of quorum sensing. The most common class of autoinducers are acyl homoserine lactones (AHLs). These molecules have an N-acylated homoserine-lactone ring and a 4–18 carbon acyl chain (Galloway et al., 2011). The acyl chain may contain modifications on the third carbon, which are sometimes unique to certain microbes and close relatives. These modifications on the acyl chain can alter the stability of the molecule, as well as provide specificity for their sensors (von Bodman et al., 2008).

AHLs are produced through the activity of LuxI enzymes, which employ S-adenosylmethionine (SAM) and intermediates from fatty acid metabolism as precursors to synthesize AHLs (Case et al., 2008). AHLs are sensed through their cognate receptor LuxR, a DNA-binding transcriptional regulator (Zhang et al., 2002). LuxI/LuxR often constitute functional pairs that coevolve for increased sensitivity and specificity for specific AHLs. LuxR-type receptors are cytoplasmic and detect a cognate AHL from a partner LuxI synthase. The receptor has an N-terminal AHL binding domain and a C-terminal DNA-binding domain. LuxR-type receptors are usually unstable in the absence of an AHL, as they fail to fold properly and are quickly degraded. Upon binding to their cognate AHL, the LuxR-type transcription factor dimerize and bind to short sequences termed *lux boxes* that are upstream of their target genes (Engebrecht et al., 1983; Engebrecht and Silverman, 1984; Stevens et al., 1994; Zhang et al., 2002; Zhu and Winans, 1999).

There is a great deal of understanding of the signal transduction mechanisms related to sensing of AHLs, and the importance of these small molecules in conducting functions associated with host–microbial interactions. One of the most studied systems of quorum sensing through AHLs comes from the interaction between *Vibrio fischeri* and its host the bobtail squid, *Euprymna scolopes*. *V. fisheri* and the bobtail squid have a natural symbiosis where the production and accumulation of AHL by the bacterium at high cell density triggers the production of bioluminescence (Engebrecht et al., 1983; Engebrecht and Silverman, 1984). This bioluminescence prevents nighttime predation of the squid. This symbiosis illustrates one example where quorum sensing mechanisms are utilized by bacteria for host-microbial interactions.

In mammals, the gut is the organ with the highest microbial colonization in the body, and as such, it has the highest potential for host-microbial interactions. Members of the *Gammaproteobacteria* such as *Salmonella enterica, Escherichia coli, Klebsiella pneumoniae,* and *Enterobacter* encode a homolog of the AHL sensor LuxR called SdiA, yet do not encode a partner LuxI homolog, and do not synthesize their own AHLs (Hudaiberdiev et al., 2015). It has been shown, however, that these bacteria are in fact capable of sensing and responding to a variety of AHLs (Dyszel et al., 2010; Michael et al., 2001; Nguyen et al., 2015; Sheng et al., 2013; Smith and Ahmer, 2003; Smith et al., 2008; Sperandio, 2010). This suggests that these bacteria might respond to AHLs produced by other members of the microbiota. However, the evidence for production of AHLs by common members of the microbiota such as *Bacteroides* is lacking. Also, studies have endeavored to extract AHLs from samples from the mammalian intestine and have failed to detect these compounds (Hughes et al., 2010). Efforts to detect AHL genetically have also come out short (Swearingen et al., 2013). *Salmonella* reporter systems using SdiA as the sensor were constructed and were used to inoculate a broad range of hosts. The system failed to detect any AHLs in the intestine of mammals, but small levels were detected in turtles or in the GI tract of mice infected with *Y. enterocolitica,* a known producer of AHLs (Dyszel et al., 2010; Smith et al., 2008). However, AHLs have been detected in the bovine rumen, although the source of the quorum sensing molecule in this environment is yet unknown (Hughes et al., 2010; Sheng et al., 2013). In this context, it has been shown that sensing of AHLs by enterohemorrhagic *Escherichia coli* (EHEC) O157:H7 through SdiA guides it through the bovine rumen environment and activates genes involved in acid resistance, which are important for the survival of the bacteria in bovine acidic stomachs (Hughes et al., 2010). It has also been shown that *Salmonella* strains that lack *sdiA* have increased virulence, reported by increased fecal shedding and translocation to systemic sites (Volf et al., 2002). Together, these data show that in EHEC and *Salmonella enterica,* SdiA is a regulator of virulence. SdiA has also been shown to sense and respond to other molecules besides AHLs. Recently, it has been shown to bind and sense 1-octanoyl-rac-glycerol that is broadly used as a signaling molecule throughout the tree of life (Nguyen et al., 2015).

Although AHLs can facilitate interspecies communications, they are mostly involved in intraspecies interactions. There are other quorum sensing systems that allow primarily for interspecies communication—the most studied of which is the bacterial signaling molecule known as autoinducer-2 (AI-2). AI-2 molecules are furanones derived from 4,5-dihydroxy-2,3-pentanedione (DPD), which is derived from the SAM metabolism (Schauder et al., 2001). The *luxS* gene encodes an S-ribosylhomocysteine lyase that is required for AI-2 synthesis and is conserved in both Gram-positive and negative bacteria (Pereira et al., 2013).

AI-2 has been shown to be present in the human GI tract (Sperandio et al., 2003). In the gut, most of the AI-2 is produced by the two dominating phyla in the GI, the Bacteroidetes and Firmicutes (Thompson et al., 2015). These phyla, particularly the Clostridia within the Firmicutes, have been implicated in protecting against incoming pathogens in a process known as *colonization resistance* (Itoh and Freter, 1989). Antibiotic treatment leads to a dramatic change in the composition of the microbiota. Streptomycin treatment causes a shift toward

Bacteroidetes as it depletes the spore-forming bacteria that belong to the Clostridia class (Sekirov et al., 2008). It has been recently shown that AI-2 promotes reexpansion of Firmicutes postantibiotic treatment, suggesting a role for AI-2 and quorum sensing in the composition of microbiota (Thompson et al., 2015). The authors engineered strains of *E. coli* that either depleted or increased the extracellular levels of AI-2 in the gut. They showed that the mice that were colonized with the *E. coli* strain that increases AI-2 showed a marked increase in the Firmicutes levels, suggesting that AI-2 can revert antibiotic induced dysbiosis. Also, it has been suggested that epithelial cells respond to AI-2 by inducing the production of inflammatory cytokines like interleukin-8 (Zargar et al., 2015). A recent study also showed that AI-2 production by the microbiota protects against *Vibrio cholerae* infection (Hsiao et al., 2014). *V. cholerae* causes acute and voluminous watery diarrhea. To do so, *V. cholerae* uses quorum sensing mechanisms to regulate its ability to colonize and control the deployment of its virulence factors. Using metagenomics analysis, Hsiao et al. (2014) compared the composition of the gut microbiota of patients infected with *V. cholerae* with that of healthy adults and found that the colonization resistance correlated with the presence of *Ruminococcus obeum*. Virulence gene expression by *V. cholerae* is negatively regulated by quorum sensing; therefore the authors took a metatranscriptomics approach to elucidate the mechanism by which *R. obeum* provided protection against *V. cholerae*. The authors found that the expression levels of a *luxS* homolog in *R. obeum* increase in response to *V. cholerae*. The *R. obeum luxS* gene was cloned into a vector with an inducible promoter in a strain of *E. coli* incapable of producing its own AI-2. Mice that were colonized with this *E. coli* strain were successful at restricting *V. cholerae* colonization, showing that *luxS* from *R. obeum* is sufficient to protect against *V. cholerae* (Hsiao et al., 2014), as depicted in Fig. 1.

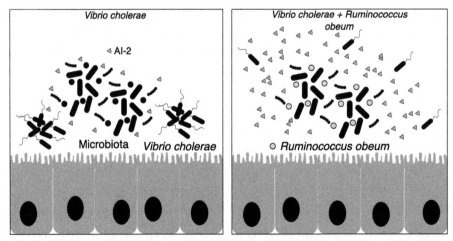

Fig. 1

Ruminococcus obeum restricts *Vibrio cholerae* colonization using quorum sensing. *R. obeum* inhibits *V. cholerae* virulence in vivo. *V. cholerae* uses quorum sensing to regulate the expression of its virulence genes. When the concentration of the quorum sensing molecule AI-2 is low, *V. cholerae* turns on the expression of its virulence genes. *R. obeum* can produce AI-2 and restricts the establishment of *Vibrio cholerae* by downregulating the expression of virulence genes.

Another less well-described signaling molecule is autoinducer-3 (AI-3). AI-3 is a methanol soluble yet uncharacterized signaling molecule that has been shown to be produced by the human intestinal microbiota (Sperandio et al., 2003). The chemical structure or synthetic pathway remain to be elucidated. It is thought to be derived from tyrosine and has been shown to be sensed by the bacterial catecholamine sensor QseC (Clarke et al., 2006; Sperandio et al., 2003). Synthesis of AI-3 was thought to be LuxS dependent. However, it was later shown that the dependency of a functional LuxS is due to metabolic shifts in the *luxS* mutant.

3 Catecholamine Signaling in Bacteria

Host-derived catecholamines control stress responses in mammals. Among the most important catecholamines in the human body are dopamine, epinephrine (E), and norepinephrine (NE) (Molina and Molina, 2006). These host-signaling molecules can be found in the gut, where they contribute to functions such as gut motility, potassium and chloride secretion, epithelial barrier function, and inflammation. The primary sources of E and NE in the gut come from the adrenal medulla and sympathetic adrenergic neurons in the gut, respectively (Asano et al., 2012; Eldrup and Richter, 2000; Hörger et al., 1998).

Salmonella enterica and *E. coli* respond to catecholamines using two-component systems to relay the information across the membrane. Two-component systems are usually composed by a histidine sensor kinase and their cognate response regulator, which is commonly a transcription factor that upon phosphorylation by the histidine kinase induces changes in gene expression (Jung et al., 2012). The catecholamines are sensed in these bacteria using the two-component systems QseC/B and QseE/F (Clarke and Sperandio, 2005a; Clarke et al., 2006; Hughes et al., 2009; Sperandio et al., 2003; Reading et al., 2009). QseC and QseE are membrane-bound histidine kinases that can sense E and NE and relay the information via a phosphorylation event to their cognate response regulators QseB and QseF, respectively (Fig. 2). These response regulators then go on to induce changes in gene expression that control flagella, motility, and virulence in pathogens such as *Salmonella* and EHEC (Clarke and Sperandio, 2005b, a; Clarke et al., 2006; Hughes et al., 2009; Moreira et al., 2010; Moreira and Sperandio, 2012). A variety of animal and plant pathogens have been shown to encode a functional *qseC* gene, and QseC plays a role in virulence gene regulation in these pathogens (Kendall and Sperandio, 2016). Upon binding to E/NE, QseC autophosphorylates itself and transfers the phosphate to QseB. QseC can also phosphorylate the noncognate response regulators QseF and KdpE. QseE can only transfer its phosphate to its cognate response regulator, QseF. QseC has been shown to be essential for in vivo colonization, as EHEC, *Citrobacter rodentium*, and *Salmonella enterica serovar Typhimurium qseC* deletion strains were attenuated in bovine, rabbit, and mice infection models (Clarke et al., 2006; Hughes et al., 2009; Moreira et al., 2010, 2016; Rasko et al., 2008; Sharma and Casey, 2014).

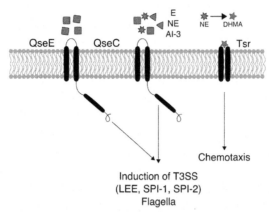

Fig. 2

Epi/NE/AI-3 quorum sensing in EHEC and *S. Typhimurium*. Catecholamines in the gut are sensed by EHEC and *S. Typhimurium* by the QseC and QseE histidine kinases. The sensor kinase QseE senses epinephrine; this leads to an autophosphorylation event and subsequent transfere of the phosphate to its cognate response regulator QseF. QseC autophosphorylates in response to epinephrine, norepinephrine, and AI-3. QseC can transfer its phosphate to its cognate response regulator QseB; this leads to an induction in genes involved in motility, such as an induction in flagellar genes. QseC can also phosphorylate its noncognate response regulators QseF and KdpE, which go on to induce the expression of virulence associated genes. Norepinephrine can be converted to DHMA via a two-step process. DHMA can act as a chemoattractant when sensed by the serine chemoreceptor Tsr.

3,4-Dihydroxymandelic acid (DHMA), a product of NE metabolism by the microbiota, particularly *Enterobacteriaceae*, has also been shown to induce virulence gene expression in EHEC in a QseC-dependent manner. When NE is sensed through QseCE, it induces expression of tyramine oxidase (*tynA*) and aromatic aldehyde dehydrogenase (*feaB*), which transform NE into DHMA in a two-step process. It was also shown that DHMA serves as a chemoattractant molecule in EHEC using the Tsr serine chemoreceptor, perhaps aiding tissue colonization by EHEC (Fig. 2; Pasupuleti et al., 2014).

4 Nutrient Signaling in the Gut

Microbiota and host-derived metabolites have a significant impact on intestinal pathogens and the development of disease. However, how commensal and host metabolites affect the virulence of pathogens is not very well understood. The microbiota is regarded as a barrier against intestinal pathogens in a process known as *colonization resistance* (Bohnhoff et al., 1954). This is thought to be due to intense competition for a limited supply of nutrients. This suggests that enteric pathogens evolve mechanisms to bypass this barrier and outcompete the resident microbiota. Many of these nutrients, whether host, diet, or microbiota-derived, provide chemical cues to incoming pathogens and are used to properly gauge microbiota composition, host physiology, and location within the intestine to deploy their virulence arsenal when appropriate (Bäumler and Sperandio, 2016).

EHEC can sense its nutritional environment and coordinate the production of its virulence repertoire. It has been shown that EHEC preferentially expresses its virulence genes when in nutrient-poor (gluconeogenic) versus nutrient-rich (glycolytic) conditions (Njoroge et al., 2012, 2013). It manages this by the activity of two transcription factors Cra and KdpE. Cra is a regulator of carbon metabolism that gauges fluctuations in the levels of carbon sources in the environment to control target genes. When under glycolytic or nutrient-rich conditions, bacteria accumulate fructose-1-phosphate and fructose-1,6-biphosphate. The accumulation of these metabolic intermediates leads to the inhibition of Cra. These metabolites bind to Cra and decrease its binding affinity to its target genes. KdpE is the response regulator of the KdpDE two-component system involved in responses to osmolarity (Heermann et al., 2003, 2009, 2014; Heermann and Jung, 2010; Jung et al., 2000; Kraxenberger et al., 2012). During poor nutrient conditions, these transcription factors interact with each other to induce virulence gene expression by interaction with the promoter of *ler*, the master regulator of the locus of enterocyte effacement (LEE) pathogenicity island (Carlson-Banning and Sperandio, 2016; Hughes et al., 2009; Njoroge et al., 2012, 2013).

The GI tract is lined with a mucus layer that protects against the microbes by providing a physical separation between the microorganisms and the epithelial cells. A major component of the mucus layer comes in the form of mucin proteins that are highly decorated with O-linked glycans encompassing as far as 80% of their total weight (Johansson et al., 2013). The major sugars that are found in the mucus layer are N-acetylgalactosamine (GalNAc), N-acetylneuraminic acid (NANA), N-acetylglucosamine (GlcNAc), mannose, galactose capped with fucose, and sialic acid (Marcobal et al., 2013). Several members of the microbiota encode hydrolases that can liberate fucose and sialic acid from mucin to access the rest of the mucin derived glycans. *Bacteroides thetaiotaomicron* encodes a variety of loci that are involved in the catabolism of a broad range of polysaccharides, including many of those found in the mucus layer (Koropatkin et al., 2012; Porter and Martens, 2017). *B. thetaiotaomicron* becomes especially efficient at degrading mucin derived sugars when starved of diet-derived complex carbohydrates like in the case of fiber-poor diets. It has been shown that degradation of the mucus layer releases the glycans, which can then be used as carbon sources by other members of the microbiota or by incoming pathogens (Desai et al., 2016).

The importance of mucin-derived sugars as carbon sources for pathogens was addressed in a study using transcriptomics analysis to compare changes in gene expression in *Salmonella* in germ-free mice versus gnotobiotic mice reconstituted with *B. thetaiotaomicron* (Ng et al., 2013). The study showed robust changes in genes involved in the utilization of

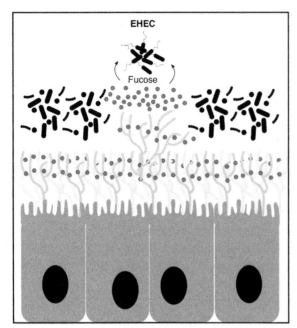
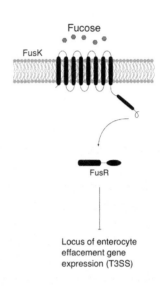

Fig. 3

Microbiota-derived succinate and pathogen colonization. Succinate is one of the most common products excreted by the *Bacteroides* spp. However, it is further metabolized by secondary fermenters in the microbiota and is rarely accumulated in the gut in the absence of an inflammatory stimulus. Inflammation depletes the primary succinate consumers which leads to the accumulation of succinate. Pathogens such as EHEC, *Citrobacter rodentium*, and *Clostridium difficile* take advantage of this microbiota-derived metabolite to expand in the gut environment. *C. difficile* combines the utilization of dietary derived carbohydrates such as sorbitol with the conversion of succinate to butyrate to expand and outcompete the resident microbiota. EHEC can sense the gluconeogenic metabolite succinate to drive the expression of virulence-associated genes using the sensor of carbon metabolism Cra.

sialic acid and fucose. However, *Salmonella* lacks the hydrolases needed to release the sugar from the mucus layer, suggesting that *B. thetaiotaomicron*–derived hydrolases release these sugars. Ng et al. (2013) generated catabolic mutants in *Salmonella* for the *nanA* and *fucI* genes, which are involved in the utilization of sialic acid and fucose, respectively. Elimination of sialic acid and fucose catabolism decreased the ability of this strain to compete against its wild-type parental strain in mice that were reconstituted with *B. thetaiotaomicron*. It was also shown that similar to *Salmonella*, *Clostridium difficile* also utilizes microbiota-derived sialic acid to expand in the gut. Treatment with streptomycin increased the levels of sialic acid in the gut, perhaps explaining one mechanism as to why these pathogens expand postantibiotic treatment (Fig. 3).

Members of the *Bacteroidetes* phylum have been heavily studied for their robust and diverse enzymatic capability. These organisms can degrade a variety of complex carbohydrates.

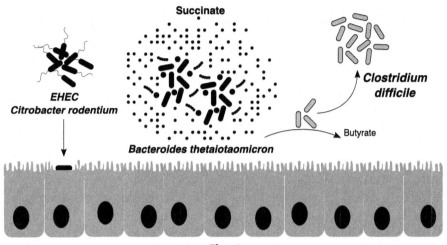

Fig. 4

Regulation of virulence associated genes in EHEC through mucin derived carbohydrates. *Bacteroides thetaiotaomicron* can switch to its metabolisms toward mucin-derived sugars when dietary derived complex carbohydrates are lacking. It is capable of releasing fucose from the mucus layer and makes it available to other bacteria in the gut. Upon the sensing of fucose through the FusKR regulatory system, EHEC represses the expression of genes that encode the T3SS.

Organisms that lack the capacity to degrade these carbohydrates can take advantage of the sugar released from the breakdown by the microbiota (Desai et al., 2016). Fucose is a microbiota-derived sugar, as it was shown that fucosylation of the mucus layer was microbiota dependent and absent in germ-free mice (Bry et al., 1996; Hooper et al., 1999). The release of sugar from the mucus layer is also microbiota dependent from the activity of microbiota-encoded fucosidases. EHEC encodes a two-component system called FusKR that can sense free fucose to repress virulence of EHEC (Fig. 4; Pacheco et al., 2012). FusR is the cognate response regulator of this two-component system, which upon phosphorylation by FusK, goes on to repress the expression of genes within the LEE pathogenicity island in EHEC (Pacheco et al., 2012).

One of the most studied functions of the gut microbiota is its great metabolic capacity. The microbiota can generate a variety of chemical products when utilizing complex carbohydrates. The metabolism of nondigestible carbohydrates such as cellulose or pectin or simple carbohydrates derived from the host from mucin glycoproteins leads to the production of short chain fatty acids (SCFA). These end products are produced by anaerobic fermentation of the complex carbohydrates present in the gut. The metabolic landscape within the intestine can vary greatly and can be affected by shifts in the structure of the microbiota such as diet, antibiotic treatment, or infectious colitis. Therefore, the presence of SCFA in the gut depends on the composition of the microbiota as well as the available carbohydrates. However, in general, the most common SCFA found in the intestine are acetate, propionate, butyrate, and formate (Macfarlane and Macfarlane, 2003). Fermentation of the available carbon sources by the microbiota can lead to the production of intermediates of fermentation, such as lactate and succinate. Intestinal inflammation due to antibiotic treatment or infection by the pathogen *Clostridium difficile* has been shown to increase the levels of succinate in the gut (Lawley et al., 2012). It has been reported that succinate produced by the microbiota can aid in colonization and enhance the expression of virulence genes in intestinal pathogens (Curtis et al., 2014; Ferreyra et al., 2014). The microbiota-derived succinate upregulated the expression of virulence genes in EHEC. Succinate produced by the microbiota can be sensed by EHEC using the regulator of metabolism Cra. Under nutrient-rich conditions, such as when glucose and other simple sugars are abundant, the ability of Cra to interact with the *ler* promoter (ler is the master activator of all LEE genes) is decreased. When the glucose concentration is limiting, *E. coli* can shift to a gluconeogenic metabolism, which allows for efficient utilization of alternative carbon sources, such as succinate, as well as increased expression of virulence associated genes in a Cra-dependent manner. Mice infected with *C. rodentium* that were depleted of their microbiota through antibiotic treatment and reconstituted with *B. thetaiotaomicron* presented increased pathology. Metabolomic profiling showed that mice that were reconstituted with *B. thetaiotaomicron* had higher succinate levels in the gut (Curtis et al., 2014). Similar to EHEC, *C. difficile* is able to utilize succinate derived from the microbiota to aid in its infection process (Fig. 3). To study the effects that the microbiota has in *C. difficile* infection, Ferreyra et al. (2014) monocolonized germ-free mice with *B. thetaiotaomicron* or *C. difficile*, or both. The mice were also fed either a polysaccharide-rich or polysaccharide-deficient diet. It was appreciated that when *C. difficile* was co-colonized with *B. thetaiotaomicron*, the bacterial loads of the pathogen were increased. This expansion was due to the ability of *C. difficile* to utilize succinate as a carbon source using a succinate to butyrate fermentation pathway. Mutants in *C. difficile* that render the strain incapable of transporting succinate are at a competitive disadvantage to their parental strain. This suggests that *C. difficile* depends on microbiota-derived succinate for postantibiotic expansion in the gut (Fig. 3).

The most abundant SCFAs in the intestine are acetate, propionate, butyrate, and formate, the highest concentrations of which are found in the colon, with lower concentrations in the small intestine (Louis et al., 2014; Macfarlane and Macfarlane, 2003). Several pathogens sense SCFAs, and it is thought that they provide a cue for the pathogen to recognize whether it is in the colon or in the small intestine. Acetate and propionate are distributed throughout the small and large intestine in roughly the same concentrations. The SCFA formate and butyrate are thought to provide special cues to bacteria as they are differentially distributed in the GI tract. Formate is found in higher quantities in the small intestine, particularly the ileum, while butyrate is found in higher concentrations in the colon (Louis et al., 2014). Butyrate has been shown to induce the expression of virulence-associated genes in EHEC, such as increased expression of T3SS components and increased bacterial attachment to Caco-2 cells (Nakanishi et al., 2009; Takao et al., 2014). Exposure to butyrate also increases motility in EHEC by increasing expression of flagellar genes in a leucine-responsive protein (Lrp) dependent manner (Tobe et al., 2011). It was also reported that butyrate induced expression of *iha*, which encodes an adhesin (Herold et al., 2009). Sensing of butyrate by EHEC depends on the Lrp transcriptional regulator (Nakanishi et al., 2009; Tobe et al., 2011). In contrast, butyrate and propionate decrease virulence gene expression in *Salmonella* (Gantois et al., 2006; Hung et al., 2013; Lawhon et al., 2002).

Butyrate is one of the major SCFAs found in the GI tract. Its production relies on the metabolism of complex carbohydrates by the microbiota, particularly the *Clostridia*. *Clostridia* are thought to be the members of the intestinal microbiota that are most effective in preventing the colonization of pathogens. Streptomycin treatment is effective at depleting the *Clostridia* populations, which subsequently allows for expansion of commensal and pathogenic strains of *E. coli* as well as *Salmonella*. Depletion of the *Clostridia* community also occurs independent of antibiotic treatment during *Salmonella enterica serovar Typhimurium* infection, due to the activity of neutrophils (Gill et al., 2012). It was recently shown that the depletion of *Clostridia* is dependent on inflammation because of the deployment of virulence factors. Depletion of this population follows a marked decrease in the levels of butyrate (Rivera-Chávez et al., 2016). Butyrate is one of the main energy sources for colonocytes. Colonocytes oxidize butyrate to CO_2 using the β-oxidation pathway, which uses oxygen as the terminal electron acceptor, thereby consuming large quantities of oxygen (Fig. 5; Colgan and Taylor, 2010). When butyrate becomes absent or reduced due to antibiotic treatment or by infection, the colonocytes switch to a fermentative metabolism that is oxygen independent. This causes the leakage of oxygen to the lumen that allows pathogens such as *S. Typhimurium* and *C. rodentium* to use aerobic respiration to outcompete the microbiota (Lopez et al., 2016; Rivera-Chávez et al., 2016). Butyrate can act as an agonist of the butyrate sensor peroxisome proliferator-activated receptor γ (PPAR-γ). The activation of PPAR-γ by butyrate signals colonocytes to engage in β-oxidation. Therefore, PPAR-γ is important for maintaining the hypoxic environment of the colon (Byndloss et al., 2017). It was later shown that the depletion of *Clostridia,* and subsequently butyrate, induces the accumulation of lactate in the gut. This increase in lactate is due to the change in metabolism seen in the colonocytes from β-oxidation to lactate fermentation, as lactate dehydrogenase inhibitors were sufficient to decrease the levels of lactate, suggesting that the lactate is host-derived. Depletion of butyrate producers by antibiotic treatment or infection induces the accumulation of lactate and an increase in the luminal concentration of oxygen. *S. Typhimurium* was shown to take advantage of this environment by aerobically respiring the accumulated lactate to expand in the lumen (Fig. 5; Gillis et al., 2018).

Certain host-derived nutrients can also induce changes in the structure of the microbiota and can aid in the expansion of enteric pathogens. Epithelial cell turnover and shedding provide a supply of host-derived nutrients such as *ethanolamine*. Ethanolamine is one of the main components of cell membranes in both prokaryotes and eukaryotes (Gibellini and Smith, 2010). It is mainly found as phosphatidylethanolamine, and it is the second most abundant

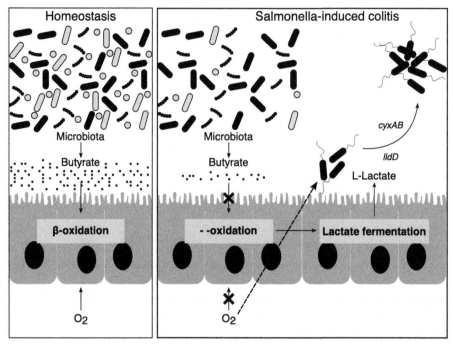

Fig. 5

Salmonella-induced colitis triggers luminal accumulation of oxygen and L-lactate. *S. Typhimurium* uses its two T3SS to induce intestinal inflammation. The initiation of an inflammatory response triggers the release of reactive oxygen and reactive nitrogen species from the epithelium. The induction of inflammation leads to the depletion of commensal *Clostridia,* the major butyrate producers in the gut. Butyrate serves as the major energy source for colonocytes through β-oxidation. β-oxidation utilizes oxygen, which prevents the oxygen from leaking to the lumen of the intestine. When butyrate concentrations drop, the colonocytes switch to lactate fermentation, which leads to the accumulation of both oxygen and L-lactate in the lumen. *S. Typhimurium* can take advantage of this nutrient and aerobically respires it to expand in the lumen.

phospholipid in animal and plant cells. Although ethanolamine is amply found in the GI tract due to the turnover of enterocytes, the resident microbiota does not use ethanolamine (Bertin et al., 2011). Genome mining analysis that attempted to find determinants of enteric infections found that the ethanolamine utilization pathway was important for enteric pathogens such as *Listeria monocytogenes, Clostridium perfringens,* and *Salmonella enterica.* It has been shown that pathogens such as EHEC, *Enterococcus faecalis*, and *Salmonella* take advantage of the available ethanolamine and use it as a noncompetitive source of nitrogen to expand and outcompete the microbiota (Bertin et al., 2011; Maadani et al., 2007; Thiennimitr et al., 2011). The catabolism of ethanolamine in these organisms is possible due to the presence of the ethanolamine utilization (*eut*) locus. This operon encodes genes that are important for the sensing, transport, and catabolism of ethanolamine (Garsin, 2010).

Moreover, ethanolamine can also function as a signal for the induction of virulence genes in *Salmonella* and EHEC. Sensing of ethanolamine is achieved primarily through the EutR transcriptional regulator (Kendall et al., 2012; Luzader et al., 2013). EutR is autoregulated in response to ethanolamine and regulates the expression of the *eut* operon. Apart from regulating the *eut* operon, EutR also directly regulates expression of the LEE in EHEC and the SPI-2 pathogenicity island in *Salmonella* (Anderson et al., 2015). Ethanolamine induces virulence gene expression in a EutR-dependent manner. However, deletion of EutR did not render EHEC or *Salmonella* completely irresponsive to ethanolamine, suggesting that other sensors might exist. Furthermore, deletion of EutR and EutB, a gene important for ethanolamine utilization, had a competitive disadvantage in a *Salmonella* mice model (Anderson et al., 2015; Kendall et al., 2012), suggesting that the sensing and utilization of ethanolamine is important for the establishment and dissemination of the pathogen. Recently, it was shown that in contrast to what is seen in EHEC and *Salmonella*, sensing and utilization of ethanolamine by another pathogen *Clostridium difficile* decreased pathology and morbidity due to downregulation of virulence-associated genes.

5 Concluding Remarks

Here, we have discussed how commensal and pathogenic bacteria sense diverse environmental signals found in the GI tract to promote colonization of its host. We are just beginning to understand the intricate relationship between the host, the microbiota, and incoming enteric pathogens. With advances in technologies in the omics such as genomics, transcriptomics, and metabolomics field, we well on our way to understand the underlying global interactions between these very complex populations. The host-bacterial interactions that take place in the GI tract are reciprocal and flexible. We know that the composition of the microbiota can change in response to changes in diet, host genetics, and immunological processes. However, the physiology of the host can also be altered in response to changes in the microbiota due to metabolic shifts. These changes in the metabolic landscape of the gut can lead to altered susceptibility to incoming pathogens. Here, we have shown examples where changes in the composition of the microbiota lead to fluctuations in metabolites that can be sensed by incoming pathogens to regulate their virulence repertoire. Some pathogens take advantage of the host and microbiota-derived metabolites to better colonize the gut and promote disease, while others are hampered due to decreased virulence or were outcompeted by the resident microbiota due to limiting nutrients.

The metabolite exchange among host, microbiota, and enteric pathogens brings exciting possibilities to prevent or treat enteric infections. An attractive idea is to pursue prebiotic treatments that trigger changes in the interaction between hosts and microbiota to restrict the establishment of incoming enteric pathogens. Understanding these interactions could also set the stage for the development of therapies that focus on ameliorating the infection by concentrating on antivirulence approaches (Rasko et al., 2008).

Glossary

Catecholamine Monoamine compounds derived from tryptophan that are produced primarily by the adrenal gland and serve as hormones involved in the fight or flight response. Epinephrine, norepinephrine, and dopamine are examples of catecholamine.

Colonization resistance Mechanism by which the resident microbes in a habitat inhibit the colonization of new or harmful microorganisms.

Dysbiosis Denotes a change in the structure of the microbiota from a "normal" state.

Germ-free Denotes animals that are devoid of a resident microbiota, sterile or clean.

Interkingdom signaling Refers to the chemical interactions between organisms of different kingdoms such as that between bacteria and mammals.

Interspecies signaling Chemical signaling between organisms of different species but of the same kingdom.

Intraspecies signaling Chemical signaling between members of the same species.

Microbiota Refers to the community of microorganisms including bacteria, archaea, protists, fungi, and viruses in a particular site.

Abbreviations

AHL	acylated homoserine lactone
AI	autoinducer
DHMA	3,4-dihydroxymandelic acid
DPD	4,5-dihydroxy-2,3-pentanedione
E/Epi	epinephrine
EHEC	enterohemorrhagic *Escherichia coli*
GalNAc	*N*-acetylgalactosamine
GI	gastrointestinal
GlcNAc	*N*-acetylglucosamine
LEE	locus of enterocyte effacement
NANA	*N*-acetylneuraminic acid
NE	norepinephrine
PPAR-γ	peroxisome proliferator-activated receptor γ
SAM	S-adenosylmethionine
SCFA	short chain fatty acids
T3SS	type three secretion system

References

Anderson, C.J., Clark, D.E., Adli, M., Kendall, M.M., 2015. Ethanolamine signaling promotes *Salmonella niche* recognition and adaptation during infection. PLoS Pathog. 11, e1005278.

Asano, Y., Hiramoto, T., Nishino, R., Aiba, Y., Kimura, T., Yoshihara, K., Koga, Y., Sudo, N., 2012. Critical role of gut microbiota in the production of biologically active, free catecholamines in the gut lumen of mice. Am. J. Physiol. Gastrointest. Liver Physiol. 303, G1288–G1295.

Bäumler, A.J., Sperandio, V., 2016. Interactions between the microbiota and pathogenic bacteria in the gut. Nature 535, 85–93.

Bertin, Y., Girardeau, J.P., Chaucheyras-Durand, F., Lyan, B., Pujos-Guillot, E., Harel, J., Martin, C., 2011. Enterohaemorrhagic *Escherichia coli* gains a competitive advantage by using ethanolamine as a nitrogen source in the bovine intestinal content. Environ. Microbiol. 13, 365–377.

Bohnhoff, M., Drake, B.L., Miller, C.P., 1954. Effect of streptomycin on susceptibility of intestinal tract to experimental Salmonella infection. Proc. Soc. Exp. Biol. Med. 86, 132–137.

Bry, L., Falk, P.G., Midtvedt, T., Gordon, J.I., 1996. A model of host-microbial interactions in an open mammalian ecosystem. Science 273, 1380–1383.

Byndloss, M.X., Olsan, E.E., Rivera-Chávez, F., Tiffany, C.R., Cevallos, S.A., Lokken, K.L., Torres, T.P., Byndloss, A.J., Faber, F., Gao, Y., Litvak, Y., Lopez, C.A., Xu, G., Napoli, E., Giulivi, C., Tsolis, R.M., Revzin, A., Lebrilla, C.B., Baumler, A.J., 2017. Microbiota-activated PPAR-γ signaling inhibits dysbiotic Enterobacteriaceae expansion. Science 357, 570–575.

Carlson-Banning, K.M., Sperandio, V., 2016. Catabolite and oxygen regulation of Enterohemorrhagic *Escherichia coli* virulence. MBio 7, e01852-16.

Case, R.J., Labbate, M., Kjelleberg, S., 2008. AHL-driven quorum-sensing circuits: Their frequency and function among the Proteobacteria. ISME J. 2, 345–349.

Clarke, M.B., Sperandio, V., 2005a. Transcriptional autoregulation by quorum sensing *Escherichia coli* regulators B and C (QseBC) in enterohaemorrhagic *E. coli* (EHEC). Mol. Microbiol. 58, 441–455.

Clarke, M.B., Sperandio, V., 2005b. Transcriptional regulation of flhDC by QseBC and sigma (FliA) in enterohaemorrhagic *Escherichia coli*. Mol. Microbiol. 57, 1734–1749.

Clarke, M.B., Hughes, D.T., Zhu, C., Boedeker, E.C., Sperandio, V., 2006. The QseC sensor kinase: a bacterial adrenergic receptor. Proc. Natl. Acad. Sci. U. S. A. 103, 10420–10425.

Colgan, S.P., Taylor, C.T., 2010. Hypoxia: an alarm signal during intestinal inflammation. Nat. Rev. Gastroenterol. Hepatol. 7, 281–287.

Curtis, M.M., Hu, Z., Klimko, C., Narayanan, S., Deberardinis, R., Sperandio, V., 2014. The gut commensal Bacteroides thetaiotaomicron exacerbates enteric infection through modification of the metabolic landscape. Cell Host Microbe 16, 759–769.

Desai, M.S., Seekatz, A.M., Koropatkin, N.M., Kamada, N., Hickey, C.A., Wolter, M., Pudlo, N.A., Kitamoto, S., Terrapon, N., Muller, A., Young, V.B., Henrissat, B., Wilmes, P., Stappenbeck, T.S., Núñez, G., Martens, E.C., 2016. A dietary fiber-deprived gut microbiota degrades the colonic mucus barrier and enhances pathogen susceptibility. Cell 167, 1339–1353.e21.

Dyszel, J.L., Smith, J.N., Lucas, D.E., Soares, J.A., Swearingen, M.C., Vross, M.A., Young, G.M., Ahmer, B.M., 2010. *Salmonella enterica serovar Typhimurium* can detect acyl homoserine lactone production by *Yersinia enterocolitica* in mice. J. Bacteriol. 192, 29–37.

Eldrup, E., Richter, E.A., 2000. DOPA, dopamine, and DOPAC concentrations in the rat gastrointestinal tract decrease during fasting. Am. J. Physiol. Endocrinol. Metab. 279, E815–E822.

Engebrecht, J., Silverman, M., 1984. Identification of genes and gene products necessary for bacterial bioluminescence. Proc. Natl. Acad. Sci. U. S. A. 81, 4154–4158.

Engebrecht, J., Nealson, K., Silverman, M., 1983. Bacterial bioluminescence: Isolation and genetic analysis of functions from *Vibrio fischeri*. Cell 32, 773–781.

Ferreyra, J.A., Wu, K.J., Hryckowian, A.J., Bouley, D.M., Weimer, B.C., Sonnenburg, J.L., 2014. Gut microbiota-produced succinate promotes *C. difficile* infection after antibiotic treatment or motility disturbance. Cell Host Microbe 16, 770–777.

Galloway, W.R., Hodgkinson, J.T., Bowden, S.D., Welch, M., Spring, D.R., 2011. Quorum sensing in Gram-negative bacteria: small-molecule modulation of AHL and AI-2 quorum sensing pathways. Chem. Rev. 111, 28–67.

Gantois, I., Ducatelle, R., Pasmans, F., Haesebrouck, F., Hautefort, I., Thompson, A., Hinton, J.C., Van Immerseel, F., 2006. Butyrate specifically down-regulates salmonella pathogenicity island 1 gene expression. Appl. Environ. Microbiol. 72, 946–949.

Garsin, D.A., 2010. Ethanolamine utilization in bacterial pathogens: roles and regulation. Nat. Rev. Microbiol. 8, 290–295.

Gibellini, F., Smith, T.K., 2010. The Kennedy pathway–De novo synthesis of phosphatidylethanolamine and phosphatidylcholine. IUBMB Life 62, 414–428.

Gill, N., Ferreira, R.B., Antunes, L.C., Willing, B.P., Sekirov, I., Al-Zahrani, F., Hartmann, M., Finlay, B.B., 2012. Neutrophil elastase alters the murine gut microbiota resulting in enhanced Salmonella colonization. PLoS One 7, e49646.

Gillis, C.C., Hughes, E.R., Spiga, L., Winter, M.G., Zhu, W., Furtado de Carvalho, T., Chanin, R.B., Behrendt, C.L., Hooper, L.V., Santos, R.L., Winter, S.E., 2018. Dysbiosis-associated change in host metabolism generates lactate to support Salmonella growth. Cell Host Microbe 23, 54–64.e6.

Gordon, J.I., Klaenhammer, T.R., 2011. A rendezvous with our microbes. Proc. Natl. Acad. Sci. U. S. A. 108, 4513–4515. Suppl. 1.

Grenham, S., Clarke, G., Cryan, J.F., Dinan, T.G., 2011. Brain–gut–microbe communication in health and disease. Front. Physiol. 2, 94.

Heermann, R., Jung, K., 2010. The complexity of the 'simple' two-component system KdpD/KdpE in *Escherichia coli*. FEMS Microbiol. Lett. 304, 97–106.

Heermann, R., Altendorf, K., Jung, K., 2003. The N-terminal input domain of the sensor kinase KdpD of *Escherichia coli* stabilizes the interaction between the cognate response regulator KdpE and the corresponding DNA-binding site. J. Biol. Chem. 278, 51277–51284.

Heermann, R., Weber, A., Mayer, B., Ott, M., Hauser, E., Gabriel, G., Pirch, T., Jung, K., 2009. The universal stress protein UspC scaffolds the KdpD/KdpE signaling cascade of *Escherichia coli* under salt stress. J. Mol. Biol. 386, 134–148.

Heermann, R., Zigann, K., Gayer, S., Rodriguez-Fernandez, M., Banga, J.R., Kremling, A., Jung, K., 2014. Dynamics of an interactive network composed of a bacterial two-component system, a transporter and K+ as mediator. PLoS One 9, e89671.

Herold, S., Paton, J.C., Srimanote, P., Paton, A.W., 2009. Differential effects of short-chain fatty acids and iron on expression of iha in Shiga-toxigenic *Escherichia coli*. Microbiology 155, 3554–3563.

Hooper, L.V., Xu, J., Falk, P.G., Midtvedt, T., Gordon, J.I., 1999. A molecular sensor that allows a gut commensal to control its nutrient foundation in a competitive ecosystem. Proc. Natl. Acad. Sci. U. S. A. 96, 9833–9838.

Hörger, S., Schultheiss, G., Diener, M., 1998. Segment-specific effects of epinephrine on ion transport in the colon of the rat. Am. J. Physiol. 275, G1367–G1376.

Hsiao, A., Ahmed, A.M., Subramanian, S., Griffin, N.W., Drewry, L.L., Petri Jr., W.A., Haque, R., Ahmed, T., Gordon, J.I., 2014. Members of the human gut microbiota involved in recovery from *Vibrio cholerae* infection. Nature 515, 423–426.

Hudaiberdiev, S., Choudhary, K.S., Vera Alvarez, R., Gelencsér, Z., Ligeti, B., Lamba, D., Pongor, S., 2015. Census of solo LuxR genes in prokaryotic genomes. Front. Cell. Infect. Microbiol. 5, 20.

Hughes, D.T., Clarke, M.B., Yamamoto, K., Rasko, D.A., Sperandio, V., 2009. The QseC adrenergic signaling cascade in Enterohemorrhagic *E. coli* (EHEC). PLoS Pathog. 5, e1000553.

Hughes, D.T., Terekhova, D.A., Liou, L., Hovde, C.J., Sahl, J.W., Patankar, A.V., Gonzalez, J.E., Edrington, T.S., Rasko, D.A., Sperandio, V., 2010. Chemical sensing in mammalian host-bacterial commensal associations. Proc. Natl. Acad. Sci. U. S. A. 107, 9831–9836.

Hung, C.C., Garner, C.D., Slauch, J.M., Dwyer, Z.W., Lawhon, S.D., Frye, J.G., McClelland, M., Ahmer, B.M., Altier, C., 2013. The intestinal fatty acid propionate inhibits Salmonella invasion through the post-translational control of HilD. Mol. Microbiol. 87, 1045–1060.

Itoh, K., Freter, R., 1989. Control of *Escherichia coli* populations by a combination of indigenous clostridia and lactobacilli in gnotobiotic mice and continuous-flow cultures. Infect. Immun. 57, 559–565.

Johansson, M.E., Sjövall, H., Hansson, G.C., 2013. The gastrointestinal mucus system in health and disease. Nat. Rev. Gastroenterol. Hepatol. 10, 352–361.

Jung, K., Veen, M., Altendorf, K., 2000. K+ and ionic strength directly influence the autophosphorylation activity of the putative turgor sensor KdpD of *Escherichia coli*. J. Biol. Chem. 275, 40142–40147.

Jung, K., Fried, L., Behr, S., Heermann, R., 2012. Histidine kinases and response regulators in networks. Curr. Opin. Microbiol. 15, 118–124.

Kendall, M.M., Sperandio, V., 2016. What a dinner party! Mechanisms and functions of interkingdom signaling in host-pathogen associations. MBio 7, e01748.

Kendall, M.M., Gruber, C.C., Parker, C.T., Sperandio, V., 2012. Ethanolamine controls expression of genes encoding components involved in interkingdom signaling and virulence in enterohemorrhagic *Escherichia coli* O157:H7. MBio 3, e00050-12.

Koropatkin, N.M., Cameron, E.A., Martens, E.C., 2012. How glycan metabolism shapes the human gut microbiota. Nat. Rev. Microbiol. 10, 323–335.

Kraxenberger, T., Fried, L., Behr, S., Jung, K., 2012. First insights into the unexplored two-component system YehU/YehT in *Escherichia coli*. J. Bacteriol. 194, 4272–4284.

Lawhon, S.D., Maurer, R., Suyemoto, M., Altier, C., 2002. Intestinal short-chain fatty acids alter *Salmonella typhimurium* invasion gene expression and virulence through BarA/SirA. Mol. Microbiol. 46, 1451–1464.

Lawley, T.D., Clare, S., Walker, A.W., Stares, M.D., Connor, T.R., Raisen, C., Goulding, D., Rad, R., Schreiber, F., Brandt, C., Deakin, L.J., Pickard, D.J., Duncan, S.H., Flint, H.J., Clark, T.G., Parkhill, J., Dougan, G., 2012. Targeted restoration of the intestinal microbiota with a simple, defined bacteriotherapy resolves relapsing *Clostridium difficile* disease in mice. PLoS Pathog.. 8e1002995.

Lopez, C.A., Miller, B.M., Rivera-Chávez, F., Velazquez, E.M., Byndloss, M.X., Chávez-Arroyo, A., Lokken, K.L., Tsolis, R.M., Winter, S.E., Bäumler, A.J., 2016. Virulence factors enhance *Citrobacter rodentium* expansion through aerobic respiration. Science 353, 1249–1253.

Louis, P., Hold, G.L., Flint, H.J., 2014. The gut microbiota, bacterial metabolites and colorectal cancer. Nat. Rev. Microbiol. 12, 661–672.

Luzader, D.H., Clark, D.E., Gonyar, L.A., Kendall, M.M., 2013. EutR is a direct regulator of genes that contribute to metabolism and virulence in enterohemorrhagic Escherichia coli O157:H7. J. Bacteriol. 195, 4947–4953.

Maadani, A., Fox, K.A., Mylonakis, E., Garsin, D.A., 2007. *Enterococcus faecalis* mutations affecting virulence in the Caenorhabditis elegans model host. Infect. Immun. 75, 2634–2637.

Macfarlane, S., Macfarlane, G.T., 2003. Regulation of short-chain fatty acid production. Proc. Nutr. Soc. 62, 67–72.

Marcobal, A., Southwick, A.M., Earle, K.A., Sonnenburg, J.L., 2013. A refined palate: Bacterial consumption of host glycans in the gut. Glycobiology 23, 1038–1046.

Michael, B., Smith, J.N., Swift, S., Heffron, F., Ahmer, B.M., 2001. SdiA of *Salmonella enterica* is a LuxR homolog that detects mixed microbial communities. J. Bacteriol. 183, 5733–5742.

Molina, P.E., Molina, P.E., 2006. Endocrine Physiology. Lange Medical Books/McGraw-Hill, New York.

Moreira, C.G., Sperandio, V., 2012. Interplay between the QseC and QseE bacterial adrenergic sensor kinases in *Salmonella enterica serovar Typhimurium* pathogenesis. Infect. Immun. 80, 4344–4353.

Moreira, C.G., Weinshenker, D., Sperandio, V., 2010. QseC mediates *Salmonella enterica serovar Typhimurium* virulence in vitro and in vivo. Infect. Immun. 78, 914–926.

Moreira, C.G., Russell, R., Mishra, A.A., Narayanan, S., Ritchie, J.M., Waldor, M.K., Curtis, M.M., Winter, S.E., Weinshenker, D., Sperandio, V., 2016. Bacterial adrenergic sensors regulate virulence of enteric pathogens in the gut. MBio 7, e00826-16.

Nakanishi, N., Tashiro, K., Kuhara, S., Hayashi, T., Sugimoto, N., Tobe, T., 2009. Regulation of virulence by butyrate sensing in enterohaemorrhagic *Escherichia coli*. Microbiology 155, 521–530.

Ng, K.M., Ferreyra, J.A., Higginbottom, S.K., Lynch, J.B., Kashyap, P.C., Gopinath, S., Naidu, N., Choudhury, B., Weimer, B.C., Monack, D.M., Sonnenburg, J.L., 2013. Microbiota-liberated host sugars facilitate post-antibiotic expansion of enteric pathogens. Nature 502, 96–99.

Nguyen, Y., Nguyen, N.X., Rogers, J.L., Liao, J., MacMillan, J.B., Jiang, Y., Sperandio, V., 2015. Structural and mechanistic roles of novel chemical ligands on the SdiA quorum-sensing transcription regulator. MBio 6, e02429-14.

Njoroge, J.W., Nguyen, Y., Curtis, M.M., Moreira, C.G., Sperandio, V., 2012. Virulence meets metabolism: Cra and KdpE gene regulation in enterohemorrhagic *Escherichia coli*. MBio 3, e00280-12.

Njoroge, J.W., Gruber, C., Sperandio, V., 2013. The interacting Cra and KdpE regulators are involved in the expression of multiple virulence factors in enterohemorrhagic *Escherichia coli*. J. Bacteriol. 195, 2499–2508.

Pacheco, A.R., Curtis, M.M., Ritchie, J.M., Munera, D., Waldor, M.K., Moreira, C.G., Sperandio, V., 2012. Fucose sensing regulates bacterial intestinal colonization. Nature 492, 113–117.

Pasupuleti, S., Sule, N., Cohn, W.B., MacKenzie, D.S., Jayaraman, A., Manson, M.D., 2014. Chemotaxis of *Escherichia coli* to norepinephrine (NE) requires conversion of NE to 3,4-dihydroxymandelic acid. J. Bacteriol. 196, 3992–4000.

Pereira, C.S., Thompson, J.A., Xavier, K.B., 2013. AI-2-mediated signalling in bacteria. FEMS Microbiol. Rev. 37, 156–181.

Porter, N.T., Martens, E.C., 2017. The critical roles of polysaccharides in gut microbial ecology and physiology. Annu. Rev. Microbiol. 71, 349–369.

Rasko, D.A., Moreira, C.G., Li de, R., Reading, N.C., Ritchie, J.M., Waldor, M.K., Williams, N., Taussig, R., Wei, S., Roth, M., Hughes, D.T., Huntley, J.F., Fina, M.W., Falck, J.R., Sperandio, V., 2008. Targeting QseC signaling and virulence for antibiotic development. Science 321, 1078–1080.

Reading, N.C., Rasko, D.A., Torres, A.G., Sperandio, V., 2009. The two-component system QseEF and the membrane protein QseG link adrenergic and stress sensing to bacterial pathogenesis. Proc. Natl. Acad. Sci. U. S. A. 106, 5889–5894.

Rivera-Chávez, F., Zhang, L.F., Faber, F., Lopez, C.A., Byndloss, M.X., Olsan, E.E., Xu, G., Velazquez, E.M., Lebrilla, C.B., Winter, S.E., Bäumler, A.J., 2016. Depletion of butyrate-producing clostridia from the gut microbiota drives an aerobic luminal expansion of Salmonella. Cell Host Microbe 19, 443–454.

Schauder, S., Shokat, K., Surette, M.G., Bassler, B.L., 2001. The LuxS family of bacterial autoinducers: biosynthesis of a novel quorum-sensing signal molecule. Mol. Microbiol. 41, 463–476.

Sekirov, I., Tam, N.M., Jogova, M., Robertson, M.L., Li, Y., Lupp, C., Finlay, B.B., 2008. Antibiotic-induced perturbations of the intestinal microbiota alter host susceptibility to enteric infection. Infect. Immun. 76, 4726–4736.

Sharma, V.K., Casey, T.A., 2014. *Escherichia coli* O157:H7 lacking the qseBC-encoded quorum-sensing system outcompetes the parental strain in colonization of cattle intestines. Appl. Environ. Microbiol. 80, 1882–1892.

Sheng, H., Nguyen, Y.N., Hovde, C.J., Sperandio, V., 2013. SdiA aids enterohemorrhagic *Escherichia coli* carriage by cattle fed a forage or grain diet. Infect. Immun. 81, 3472–3478.

Smith, J.N., Ahmer, B.M., 2003. Detection of other microbial species by Salmonella: Expression of the SdiA regulon. J. Bacteriol. 185, 1357–1366.

Smith, J.N., Dyszel, J.L., Soares, J.A., Ellermeier, C.D., Altier, C., Lawhon, S.D., Adams, L.G., Konjufca, V., Curtiss 3rd, R., Slauch, J.M., Ahmer, B.M., 2008. SdiA, an N-acylhomoserine lactone receptor, becomes active during the transit of *Salmonella enterica* through the gastrointestinal tract of turtles. PLoS One. 3e2826.

Sperandio, V., 2010. SdiA sensing of acyl-homoserine lactones by enterohemorrhagic *E. coli* (EHEC) serotype O157:H7 in the bovine rumen. Gut Microbes 1, 432–435.

Sperandio, V., Torres, A.G., Jarvis, B., Nataro, J.P., Kaper, J.B., 2003. Bacteria-host communication: The language of hormones. Proc. Natl. Acad. Sci. U. S. A. 100, 8951–8956.

Stevens, A.M., Dolan, K.M., Greenberg, E.P., 1994. Synergistic binding of the *Vibrio fischeri* LuxR transcriptional activator domain and RNA polymerase to the lux promoter region. Proc. Natl. Acad. Sci. U. S. A. 91, 12619–12623.

Swearingen, M.C., Sabag-Daigle, A., Ahmer, B.M., 2013. Are there acyl-homoserine lactones within mammalian intestines? J. Bacteriol. 195, 173–179.

Takao, M., Yen, H., Tobe, T., 2014. LeuO enhances butyrate-induced virulence expression through a positive regulatory loop in enterohaemorrhagic *Escherichia coli*. Mol. Microbiol. 93, 1302–1313.

Thiennimitr, P., Winter, S.E., Winter, M.G., Xavier, M.N., Tolstikov, V., Huseby, D.L., Sterzenbach, T., Tsolis, R.M., Roth, J.R., Bäumler, A.J., 2011. Intestinal inflammation allows Salmonella to use ethanolamine to compete with the microbiota. Proc. Natl. Acad. Sci. U. S. A. 108, 17480–17485.

Thompson, J.A., Oliveira, R.A., Djukovic, A., Ubeda, C., Xavier, K.B., 2015. Manipulation of the quorum sensing signal AI-2 affects the antibiotic-treated gut microbiota. Cell Rep. 10, 1861–1871.

Tobe, T., Nakanishi, N., Sugimoto, N., 2011. Activation of motility by sensing short-chain fatty acids via two steps in a flagellar gene regulatory cascade in enterohemorrhagic *Escherichia coli*. Infect. Immun. 79, 1016–1024.

Volf, J., Sevcik, M., Havlickova, H., Sisak, F., Damborsky, J., Rychlik, I., 2002. Role of SdiA in *Salmonella enterica serovar Typhimurium* physiology and virulence. Arch. Microbiol. 178, 94–101.

von Bodman, S.B., Willey, J.M., Diggle, S.P., 2008. Cell-cell communication in bacteria: united we stand. J. Bacteriol. 190, 4377–4391.

Zargar, A., Quan, D.N., Carter, K.K., Guo, M., Sintim, H.O., Payne, G.F., Bentley, W.E., 2015. Bacterial secretions of nonpathogenic *Escherichia coli* elicit inflammatory pathways: a closer investigation of interkingdom signaling. MBio 6, e00025.

Zhang, R.G., Pappas, K.M., Brace, J.L., Miller, P.C., Oulmassov, T., Molyneaux, J.M., Anderson, J.C., Bashkin, J.K., Winans, S.C., Joachimiak, A., 2002. Structure of a bacterial quorum-sensing transcription factor complexed with pheromone and DNA. Nature 417, 971–974.

Zhu, J., Winans, S.C., 1999. Autoinducer binding by the quorum-sensing regulator TraR increases affinity for target promoters in vitro and decreases TraR turnover rates in whole cells. Proc. Natl. Acad. Sci. U. S. A. 96, 4832–4837.

PART 3

Quorum Sensing Inhibition

CHAPTER 7

Enzymatic Quorum Quenching in Biofilms

Jan Vogel, Wim J. Quax

Groningen Research Institute of Pharmacy, Department of Chemical and Pharmaceutical Biology, University of Groningen, Groningen, The Netherlands

1 Introduction

In the beginning of the 20th century, the German scientist Paul Ehrlich was trying to find a drug he referred to as a *magic bullet*, aimed at a specific target to eliminate the cause of a disease without harming the human body. This research resulted in a better understanding of the human immune system and also led to a better understanding of the pathogens involved. In 1928, Alexander Fleming found exactly a magic bullet on one of his agar plates: a *Penicillium* strain producing penicillin, thus effectively killing *Staphylococcus aureus* bacteria. This event marked the start of a medical revolution. Nevertheless, upon accepting his Nobel Prize in Medicine in 1945 Alexander Fleming stated that "a misuse of the drug could result in selection for resistant bacteria" (Rosenblatt-Farrell, 2009). Indeed, in the past decades the number of multidrug resistant bacteria has raised alarmingly. Furthermore, hardly any novel antibiotics have been introduced onto the market for the past 40 years. This lack of treatment options is one of the most dangerous threats to humanity (Gill et al., 2015; Infectious Diseases Society of America (IDSA) et al., 2011). Thus the demand for novel antimicrobials is high.

The discovery of the luminescent bacterium *Vibrio fischerii* in the early 1960s revealed that light emission by these bacteria correlates with their population density. It was shown that genes responsible for light emission are only switched on when the cell density of the bacterial culture reaches a particular threshold. Sensing of the correct quorum (quorum sensing, QS) was shown to be dependent on the secretion of certain signaling molecules. Interestingly, at that time this phenomenon was put aside as an oddity of deep-water bacteria. Today, however, it is clear that bacterial populations in nearly every ecological niche communicate with a variety of diffusible molecules. This opens up novel possibilities to exploit interference with the QS system for novel, much needed antimicrobial therapies. To this purpose, understanding the QS system is crucial.

In this review, we reflect on interference with the bacterial QS system as a promising approach to battle bacterial infections. A broad variety of enzymes has been shown to be capable of

Quorum Sensing. https://doi.org/10.1016/B978-0-12-814905-8.00007-1

degrading QS signaling molecules with a concomitant reduction of virulence (Grandclement et al., 2015).

Translating knowledge of QQ enzymes into a practical application offers great potential in diverse research areas, ranging from the prevention of biofouling on structures exposed to seawater to the functionalization of medical devices to hinder biofilm formation (Bzdrenga et al., 2017). The emphasis of this chapter, however, is directed toward the medical perspective and opportunities offered by quorum quenching (QQ) enzymes.

2 Life on Surfaces: The Bacterial Biofilm

Bacterial biofilms are a perfect example of a coordinated behavior of bacteria in response to various external factors, leading to a change of phenotype. In turn, this phenotype is responsible for the increased attachment to and persistence of the bacteria on surfaces. The biofilm is described as a multicellular community of bacteria, offering bacteria many benefits and therefore an evolutionary advantage in special environments.

Sessile bacteria differ from planktonic individuals in a variety of ways. A common characteristic of bacteria living in biofilms is their ability to produce an extracellular matrix. This matrix is not only providing structural support, but is also a key element in tolerating antimicrobial compounds. This matrix is composed of different macromolecules, proteins, lipids, extracellular DNA, and polysaccharides (Billings et al., 2013). Formation of the biofilm is divided into five fundamental stages (Fig. 1).

First, the cells attach to a surface, followed by aggregation and recruitment of planktonic cells from the environment. For example, the aggregation of *Pseudomonas aeruginosa* on surfaces takes place with a type IV pilus-mediated twitching motility. Binary cell division is crucial for

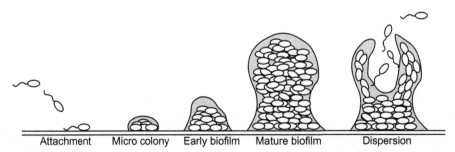

Attachment   Micro colony   Early biofilm   Mature biofilm   Dispersion

Fig. 1

Scheme of biofilm formation. In the first stage, the bacteria attach to the surface. Subsequently, the bacteria form micro colonies and produce eDNA and EPS compounds depicted in *dark gray*. The colonies develop from an early biofilm to a mature biofilm showing complex three-dimensional structures. The latest stage is the biofilm dispersion, in which individual cells segregate from the biofilm.

initial biofilm aggregation and also a subsequent differentiation of the structure (Stoodley et al., 2002). In a later stage, the biofilm is changing from a surface-attached micro colony into a three-dimensional structure through a phenotypical differentiation of the individual bacteria. For example, upon initial attachment, genes expressing extracellular polymeric substances (EPS) in *P. aeruginosa* are upregulated within 15 min. These components are most notably the matrix component named PsI and alginate (Billings et al., 2013; Chang, 2018). This marks the transition from a reversible to an irreversible attachment, and is also described as the transition from weak interactions to a permanent bonding with the substrate (Stoodley et al., 2002). The dispersal of microorganisms from the biofilm, which is marking the final stage, can be actively regulated or can be a consequence of mechanical influences like the erosion through fluid flow (Nadell et al., 2016). Newly dispersed *P. aeruginosa* cells were shown to be more virulent than their planktonic counterparts (Chua et al., 2014).

The opportunistic pathogen *P. aeruginosa* is a good model organism for studying bacterial biofilms. The *P. aeruginosa* PAO1 strain was shown to be dependent on the 3-oxo-C12 homoserine lactone (HSL)-mediated *lasI* QS circuit to be able to form a differentiated biofilm. The LasI/LasR QS system consists of the AHL synthase LasI and its cognate response regulator LasR. Upon binding of 3-oxo-C12 HSL, LasR acts as transcription factor, activating various genes including virulence factors and also the expression of LasI in a positive feedback loop. A Δ*lasI*-mutant formed a biofilm, which is much thinner than the *P. aeruginosa* PAO1 wild type. The wild type-phenotype could be restored by the addition of exogenous 3-oxo-C12 HSL (De Kievit and Iglewski, 2000). Nevertheless, the *lasI* gene expression level was found to decrease with the growing thickness of the biofilm over time. The expression level of *rhlI* remained constant (Stoodley et al., 2002).

3 Biofilms and Antibiotic Resistance

The formation of a biofilm is a highly complex and coordinated interaction between microorganisms. This communal effort requires a high degree of communication, which is mediated by QS (Solano et al., 2014). It makes the bacteria less accessible to antibiotic treatment and increases the ability of bacteria to persist on surfaces (Flemming et al., 2016). Indeed, bacteria organized in a biofilm are considered to be their predominant lifeform found in all ecological niches (Solano et al., 2014). Accordingly, fighting bacterial infectious diseases means understanding and ultimately controlling biofilms.

It was previously suggested that 60% of all infections treated by physicians are related to bacteria organized in biofilms (Fux et al., 2005). Yet the treatment of bacterial biofilms represents a major problem due to various reasons. Most of the research on the activity of antimicrobials is performed on planktonic and dividing bacteria. Accordingly, determined minimal inhibitory concentration values (MIC) may not be representative of the real disease state when the bacteria persist in biofilms (Fux et al., 2005). As the accessibility of these bacteria differs in biofilms, an inadequate dose and prolonged use of antibiotics could be the result. Further taking into account that within the bacterial biofilm, horizontal gene transfer including resistance genes is promoted, the problem of the emerging antibiotic resistance becomes obvious. Bacteria organized in a biofilm are thus less susceptible to antibiotics and can tolerate higher concentrations of antibiotics. In the majority of cases, the infection is not cleared completely by the end of the treatment.

In this context, the biofilm matrix plays an important role in its tolerance toward antibiotics and is protecting the bacteria from the host's immune system (Fux et al., 2005). In some bacteria, exposure to subinhibitory concentrations of antibiotics even induces mucoid phenotypes. This results in thicker biofilms due to additional matrix components (Fux et al., 2005). In biofilms produced by *P. aeruginosa*, secreted alginate has a protective effect against leukocyte phagocytosis and additionally serves as an absorbing shield toward aminoglycoside antibiotics (Bayer et al., 1991).

In contrast to the common belief that the extracellular matrix offers a strict barrier toward diffusion of antibiotics, there is no evidence to confirm this hypothesis. Diffusion studies show that antibiotics are indeed able to penetrate bacterial biofilms and that the protective effect of the extracellular matrix is due to interactions with special properties of the antibiotic, such as its charge (Tseng et al., 2013). Furthermore, sometimes enzymes are incorporated into the outer matrix layer—for example, in the *P. aeruginosa* biofilm, β-lactamases are incorporated, actively degrading β-lactam antibiotics and thus contributing to antibiotic resistance (Heydari and Eftekhar, 2015).

Bacteria within the inner layers of a biofilm show many similarities with bacteria in their stationary phase. Compared with planktonic bacteria, stationary phase cells were shown to be more tolerant to antibiotics due to their different metabolic state (Spoering and Lewis, 2001). The antibiotic tolerance of bacteria in biofilms is thus the result of their phenotype rather than an intrinsic resistance mediated through mutations, which can be the case in planktonic cells (Fux et al., 2005). Interestingly, the expression of efflux pumps, which are considered a key element of intrinsic resistance in planktonic *P. aeruginosa*, was believed to have only little influence on the antibiotic tolerance of the mature biofilm as a whole (Ciofu and Tolker-Nielsen, 2010). However, more recent findings in *P. aeruginosa* have shown that efflux pumps MexAB-OprM and MexCD-oprJ are indeed involved in biofilm tolerance toward azithromycin (Ciofu and Tolker-Nielsen, 2010).

3.1 Multispecies Biofilms

In Nature, mono-species biofilms are rather an exception. This is exemplified by the lungs of cystic fibrosis patients: their disturbed electrolyte balance and resulting viscous, dehydrated mucus offers a perfect habitat for microorganisms. The lungs of young CF patients are commonly colonized by *S. aureus* and *Haemophilus influenza*, and in the further course of the infection, 60%–70% of the CF patients show a colonialization with *P. aeruginosa* that in 50% represents the dominating species (O'Brien and Fothergill, 2017). This is in most cases followed by secondary infection of the *Burkholderia cepacia* complex. Chronic bacterial infections involving many different species will subject the CF patients to a lifelong medical treatment. In addition, prominent pathogens are fungi and yeasts such as *Aspergillus fumigatus* and *Candida albicans*. Studies confirm that in 50% of CF patients *Aspergillus* spp. and *Candida* spp. are present. Needless to say, chronic infections in the lung tissue are a great danger to the host, often resulting in the death of the patient (O'Brien and Fothergill, 2017).

Porphyromonas gingivalis and *Streptococcus gordonii* inhabit the human oral cavity, where they can form a mixed biofilm upon autoinducer-2 (AI-2) signaling (McNab et al., 2003). AI-2 is mediating communication between different species of Gram negative and Gram positive bacteria and is synthesized by LuxS type synthases (McNab et al., 2003; Xavier and Bassler, 2003). Research showed that LuxS synthase-deficient mutants of both strains did not form biofilms on polystyrene surfaces (McNab et al., 2003). In addition, a synergistic way of living in a mixed community was observed: the bio-volume of mixed communities was reported to be higher than when both species were cultivated separately (Elias and Banin, 2012). This hypothesis was confirmed by an experiment showing that a mixed biofilm consisting of *Actinomyces naeslundii* and a LuxS-deficient *Streptococcus oralis* could be more easily dispersed in contrast to the mutualistic biofilm of the wild type strains (Rickard et al., 2006).

3.2 Communication in P. aeruginosa *Biofilms*

The bacterial biofilm plays a crucial part in the persistence of *P. aeruginosa* infections in the lungs of cystic fibrosis (CF) patients. Moreover, as QS plays an important role in the formation of biofilms, using QQ enzymes to disturb biofilm formation could help make these pathogens more susceptible to treatment and eventually also to the host immune system. To highlight how the disruption of QS system might affect the pathogens, it is important to elucidate several QS mechanisms important for virulence and persistence. As previously described, alginate is a part of the structural elements of the biofilm that can act as a protective layer against antibiotics and also save the bacteria against leukocyte phagocytosis of the immune system (Leid et al., 2005). The EPS production was shown to be regulated by QS; various experiments show that QQ impairs the biofilm formation. However, it is also known that the *las*I dependent QS signal decreases in a maturating biofilm (De Kievit et al., 2001). Moreover, studies suggest that QS provides a benefit for the bacterial population as a whole by increasing the production of common goods that can favor growth in certain environmental conditions. The researchers simulated an environment in which the expression of proteases is required for growth (quorum sensing growth medium, QSM). The same group found that an impairment of the QS system leads to a way slower growth, suggesting that in certain conditions QQ can reduce the growth of pathogenic bacteria (Diggle et al., 2007).

3.3 Quorum Sensing Signals

To inhibit biofilm formation through bacterial communication, it is crucial to understand how QS networks work and which signals are used. Bacteria use a broad spectrum of different molecules for communication. At the same time, to individual bacteria it is crucial to recognize signals from the same species and distinguish foreign molecules.

QS molecules or autoinducers (AIs) include peptides, *N*-acyl homoserine lactones (AHLs), and quinolones. These molecules do not only have different properties, but also have different metabolic costs. For example, the production of AgrD, a signaling peptide of *S. aureus*, requires 184 ATP molecules, whereas a C4-HSL requires a mere 8 ATP molecules for its synthesis (Keller and Surette, 2006). In Gram positive bacteria, signaling molecules are mostly small peptides, actively secreted into the environment. They are recognized by membrane-bound receptors or internalized and recognized in the cytoplasm (Monnet and Gardan, 2015).

The most studied signaling molecules are AHLs used by Gram negative bacteria. Here, the chain length of the acyl chain determines the specificity of the whole molecule (Papenfort and Bassler, 2016). Different bacteria use different signaling molecules for their QS systems.

Autoinducer 2 (AI-2) is part of a set of signaling molecules all derived from 4,5-dihydroxy-2, 3-pentanedione (DPD), produced by the LuxS synthase. This protein is reported to be present in more than 500 bacterial species, which makes AI-2 to this day the most common signaling molecule (Pereira et al., 2013). It is thought that AI-2 is used for interspecies communication among microorganisms (Papenfort and Bassler, 2016). Interestingly, the genome of *P. aeruginosa* does not encode a *luxS* gene, though it is reported that this bacterium responds to AI-2 (Duan et al., 2003).

Approximately 76% of the annotated LuxR-type receptor proteins are catalogued as LuxR-solo class receptors, indicating that these are AHL-dependent response regulators without a corresponding LuxI type synthase. Perhaps the best described solo-LuxR receptor is QscR from *P. aeruginosa*. This receptor controls its target genes in the presence of nanomolar concentrations of C8-HSL, C10-HSL, 3-oxo-C10-HSL, C12-HSL, 3-oxo-C12-HSL, and C14-HSL (Lee et al., 2006). In this context, it is tempting to draw the conclusion that a large variety of autoinducers could be synthesized by non-LuxI type synthases. More likely, these solo class receptors allow bacteria to intercept the communication between microorganisms they are competing or collaborating with.

3.4 Quorum Quenching Enzymes

The enzymatic degradation of QS molecules has been a focus of research for many years. The best researched QQ enzymes in this context are: (i) lactonases, which catalyze the opening of the homoserine lactone ring; (ii) acylases hydrolyzing the amide bond in between the

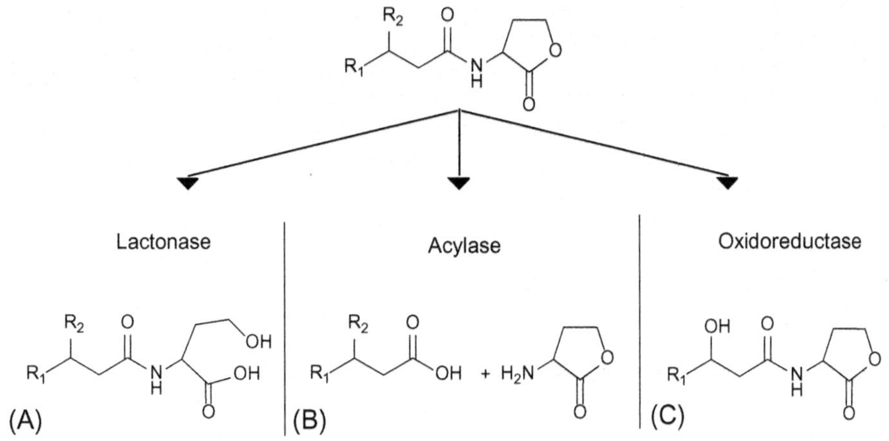

Fig. 2

Three classes of enzymes are categorized as AHL QQ enzymes: (A) lactonase that opens the homoserine lactone ring; (B) acylases that hydrolyze the amide bond between the acyl chain and the homoserine lactone ring; and (C) oxidoreductases that modify the AHLs by oxidizing or reducing the acyl chain without degrading the AHLs.

homoserine lactone ring and the fatty acid chain; and (iii) oxidoreductases that can convert 3-oxo-AHLs to 3-hydroxy-AHLs, which do not show the same binding affinity to their corresponding response regulator (Fig. 2).

3.4.1 Lactonases

AHL lactonases are active toward the homoserine lactone ring of AHLs. A prominent example of AHL lactonases is the metallo β-lactamase AiiA from *Bacillus* sp. 240B1. These enzymes show a characteristic Zn^{2+}-binding HXHXDH motif and interestingly belong to the same superfamily of enzymes, which is mediating a resistance to β-lactam antibiotics (Wang et al., 1999). When expressing the *aiia* gene *of B. cereus* A24 in *P. aeruginosa*, researchers reported a decrease in the production of virulence factors as well as reduced motility (Michel et al., 2002).

Phosphotriesterase-like lactonases (PLLs) are a group of lactonases with a broad substrate spectrum. They were first described as paraoxonases, but later on it was discovered that they are indeed natural AHL-lactonases with a promiscuous activity toward organophosphate pesticides (Afriat-Jurnou et al., 2012)

A very promising candidate for a QQ lactonase is SsoPox from the extremophile archaeon *Sulfolobus solfataricus*. Its extreme stability and an activity over a broad range of temperatures and pH levels are of interest. Furthermore, SsoPox shows a remarkable tolerance against organic solvents, surfactants, as well as proteases (Hiblot et al., 2012). These traits make it especially interesting for biotechnological usage (Bzdrenga et al., 2017; Rémy et al., 2016). The protein is crystalized, and protein engineering efforts have been performed to increase the catalytic efficiency (Bzdrenga et al., 2017). Studies confirm the efficiency and the feasibility of a protein-engineered variant of SsoPox. The protein shows a change in one amino acid, in particular residue W263, which is located in the active center of the protein. The Trp was altered to an Ile resulting in a lactonase more efficient to 3-oxo-C12 HSL. In the presence of this QQ lactonase in a culture of *P. aeruginosa* PAO1, the researchers reported that the protease production was reduced 90% and the pyoverdine production was nearly completely abolished. In addition, the formation of biofilm was reduced also 90%. The same group also showed that the quenching ability of SsoPox showed comparable results when the enzyme was immobilized in a polyurethane coating using glutaraldehyde as a cross-linking compound (Guendouze et al., 2017).

In the context of QQ, lactonases have a very beneficial trait, since the enzymes recognize and cleave the lactone ring of the AHLs, whereas the acyl chain length plays a minor role in the recognition of the molecule. This means that lactonases can degrade a broad spectrum of AHL molecules with various chain lengths (Fetzner, 2015). These experiments underline the feasibility of QQ enzymes to fight infectious diseases. In this context, it is important to mention that the opened lactone ring can recyclize under acidic conditions (Fetzner, 2015; Grandclement et al., 2015). In contrast, QQ mediated by acylases provides an irreversible degradation of the signaling molecule.

As an example, when fighting bacterial infections, QQ enzymes are shown to reduce the biofilm of the human pathogen *Acinetobacter baumannii* (Chow et al., 2014). In the past years *A. baumannii* was put in focus of medical research, because of its emergence as a nosocomial or hospital-acquired infection with a high morbidity (Chow et al., 2014). The biofilm formation of a clinical isolate of *Acinetobacter* was shown to be dependent on QS (Niu et al., 2008). Unfortunately, the QS regulation of *A. baumannii* genes is not well characterized. It is reported that 63% of all isolated *A. baumannii* strains produce more than one type of AHL (Peleg et al., 2008). However, the completed genome sequence of *A. baumannii* ATCC 17978 suggests that only the LuxI type synthase AbaI is involved in the synthesis of AHLs. Scientists discovered at least five AHLs next to the putative main signaling molecule 3-hydroxy-C12 HSL in *A. baumannii* M2, whereas it could be shown that in the clinical isolate *A. baumannii* S1 the major QS molecule is 3-hydroxy-C10 HSL (Chow et al., 2014). A Δ*abaI* deletion mutant of *A. baumannii* M2 shows a significant reduction in biofilm formation (30%–40%) when compared with the same strain which is able to produce AHLs. When the signaling molecule was added exogenously to the *A. baumannii* M2 culture, the biofilm maturation could be restored (Niu et al., 2008). Furthermore, it could be shown that the biofilm formation could be significantly reduced by the enzymatic degradation of the QS signal by an engineered QQ lactonase (Chow et al., 2014). According to the fact that biofilms are harder to treat than planktonic bacteria, these results offer great chances for a successful treatment strategy of *A. baumannii*.

Researchers performed a directed evolution of a thermostable QQ lactonase from *Geobacillus kaustophilus* (GKL) and explored its capability to quench the QS system dependent biofilm formation of *A. baumannii* S1 (Chow et al., 2014). The enzyme belongs to the phosphotriesterase-like lactonase (PLL) family of the amidohydrolase superfamily. As different clinical isolates of *A. baumannii* use AHLs with different chain lengths, the authors highlighted that a promiscuous activity toward a broader spectrum of AHLs has obvious advantages to fight biofilms (Chow et al., 2014).

3.4.2 QQ acylases

Bacteria are extremely flexible and adaptable, and it does not surprise that some are able to use AHLs as an energy and carbon source. *Variovorax paradoxus* cleaves AHLs into the homoserine lactone ring and its corresponding fatty acid, and uses AHLs as a nitrogen source (Leadbetter and Greenberg, 2000). The opportunistic pathogen *P. aeruginosa* possesses several acylases capable of degrading AHLs, the most prominent enzyme being PvdQ (Sio et al., 2006). Most of the identified AHL acylases belong to the super family of Ntn-Hydrolases (Utari et al., 2017). This superfamily is very versatile and stands out by an interesting and very characteristic αββα-fold in which the active center is located.

Research has focused on Ntn-hydrolases, and some members are used in industrial applications. Penicillin acylases, for instance, cleave the acyl chain of penicillin molecules separating 6-amino penicillanic acid (6-APA) and the corresponding organic acid. 6-APA is an important intermediate toward many semisynthetic β-lactam antibiotics (Hewitt et al., 2000). The reaction of Ntn-hydrolases is dependent on an N-terminal nucleophile, which gives this superfamily its name. Within the active center, the nucleophile can be an N-terminal serine, threonine or cysteine, which is launching a nucleophilic attack on the amide bond of the AHL (Utari et al., 2017). A notable members of this family, penicillin G acylases (PGA) from *Kluyvera citrophila,* was recently shown also to be able to degrade AHLs with a chain length of 6–8 carbons (Fetzner, 2015).

The acylase PvdQ was identified in the genome of *P. aeruginosa* next to two others, namely HacB and QuiP (Wahjudi et al., 2011). The physiological functions of these acylases is not completely resolved; however, PvdQ plays a key role in the production of pyoverdine in *P. aeruginosa* (Drake and Gulick, 2011). PvdQ cleaves long-chain AHLs with side chains longer than 10 carbons (Koch et al., 2014). In a *C. elegans* infection model with *P. aeruginosa*, it could be shown that PvdQ is potent enough to degrade the 3-oxo-C12 HSL signal to such an extent that QS dependent pathogenicity factors were significantly reduced. This was visible by an increased survival rate of *C. elegans* within this assay (Papaioannou et al., 2009). Furthermore, the substrate specificity of PvdQ could be altered from cleaving long-chain AHLs to also accepting C8-HSLs as a substrate. Two single point mutations (Lα146W;Fβ24Y) conferred the activity toward the AHL used by *B. cenocepacia*. In a subsequent experiment, it could be shown that this engineered enzyme can protect *Galleria mellonella* larva in a *B. cenocepacia* infection scenario (Koch et al., 2014).

3.4.3 Oxidoreductases

QQ oxidoreductases target the sidechain of the AHL molecules by oxidizing or reducing the carboxyl group of the third carbon or are able to oxidize ω-1, ω-2, and ω-3 carbons. This class of enzymes does not destroy the signal but rather modifies it and is thereby modulating the efficiency by which the signaling molecule is recognized by its cognate receptor (Chen et al., 2013). *Rhodococcus erythropolis* strain W2, for example, possesses (in addition to a lactonase and an acylase) an oxidoreductase, which is able to reduce 3-oxo-HSLs with a chain length of 8–12 carbons to its 3-hydroxy-HSL form (Phane Uroz et al., 2005). BpiB09 is an oxidoreductase, which catalyzes the reduction of 3-oxo-HSL to 3-hydroxy-HSL, and when expressed in *P. aeruginosa* had an effect on the QS system and lead to a reduced production of pyocyanin as well as a change in motility and the biofilm formation. In addition, a reduced virulence in a *C. elegans* infection model with *P. aeruginosa* PAO1 was observed (Bijtenhoorn et al., 2011).

3.4.4 Quorum quenching enzymes against AI-2

LuxS is not only the synthesizing enzyme for the AI-2 signaling molecule, but also is a part of the activated methyl cycle (AMC). This has led to the thought that AI-2 molecules could just be side products of AMC. Notably the autoinducers of the AHL class as well as CAI1 are also synthesized with an *S*-Adenosyl methionine (SAM) molecule as a substrate. An interesting finding is the fact that different bacteria can sense AI-2 without synthesizing it themselves. For example, *P. aeruginosa* strains upregulate the genes *rhl*A (rhamnolipid biosynthesis), *las*B (elastase), *exo*T (exotoxin), *phzA1* and *phzA2* (phenazine synthesis), and *fliC* (flagellar component) upon AI-2 sensing (Pereira et al., 2013). In the case of *Escherichia coli* cells, the AI-2 signal is taken up by the bacterial cell actively though a membrane transporter and is phosphorylated by the receptor kinase LsrK to phosphor-AI-2. The phosphorylated autoinducer can now bind to the transcriptional repressor LsrR, leading to activation of target genes (Pereira et al., 2013). Interestingly, the kinase LsrK, once added exogenously to cultures of *V. harveyi*, *Salmonella typhimurium*, and *E. coli*, reduced their QS response. The mechanism behind the QQ activity of LsrK is assumed to be due to the phosphorylation of AI-2 outside of the cell, prohibiting transport into the cell. In addition, the modification of AI-2 makes the molecule less stable.

3.5 Quenching the Quinolone Dependent QS Signal

In *P. aeruginosa*, the PQS signal was shown to play in important role in the production of virulence factors as well as biofilm formation (Pustelny et al., 2009). Besides *P. aeruginosa*, only *Burkholderia* spp. is known to utilize a 2-alkyl-4(1H)-quinolone (AQ) signal for QS (Fetzner, 2015). The possibility of enzymatically degrading this signal underlines that QQ enzymes can be used in a very selective way, targeting especially pathogenic bacteria. So far the only QQ enzyme identified that is able to degrade AQ is the dioxygenase Hod (1H-3-hydroxy-4-oxoquinaldine 2,4-dioxogenase). The enzyme catalyzes the cleavage of the heterocyclic PQS ring to *N*-octanylanthranilic acid and carbon monoxide. The exogenous addition of Hod to a *P. aeruginosa* culture led to a reduction in the expression of various virulence factors (Pustelny et al., 2009). The enzyme was isolated from the soil bacterium *Arthrobacter* sp. Rue61a. In soil environments, AQs are produced by higher organisms such as plants as secondary metabolites. Fetzner and colleagues (2015) hypothesize that bacteria sharing this ecological niche could have evolved AQ degrading enzymes. This was supported by the finding of PQS degrading activity by a yet uncharacterized soil isolate.

3.6 Applications of Quorum Quenching Enzymes

One of the best characterized organisms when it comes to the characterization of QS as well as QQ is the Gram negative bacterium *P. aeruginosa*. In 2017 this bacterium was listed as number two on the World Health Organization list of priority pathogens for research and development of new antimicrobials. Data collected in medical environments in the United States highlighted that 7.5% of all general health care-associated infections are connected to *P. aeruginosa* (Sievert et al., 2013). It has two hierarchically ordered QS circuits, which are controlled via *N*-acyl homoserine lactones (AHLs) signaling molecules. The Las system is dependent on *N*-3-oxo dodecanoyl-homoserine lactone (3-oxo-C12 HSL), and the second system, namely the Rlh regulon, utilizes *N*-butyryl-homoserine lactone (C4 HSL). Both signals can be degraded by QQ enzymes, which has a significant impact on the production of virulence factors, including biofilm formation (Papaioannou et al., 2009). More than that, QQ acylases discriminate between different chain lengths of different AHLs, which opens the possibility to target QS circuits more specifically.

3.7 Functionalization of Surfaces

Biofilms on surfaces are often problematic in industrial applications such as in membrane bio reactors membranes. Strategies including QQ enzymes are already tested for an application to counteract the formation of biofilms in that respect (Kim et al., 2011). Furthermore, biofilm formation on medical devices is shown to be the cause of a multitude of severe diseases. In the clinic, a huge effort is undertaken to fight bacteria on surfaces. The introduction of any foreign material in the human body causes a big health threat and predisposes the body to infections. An important example of this situation comes in the form of medical implants. Orthopedic implants serve as a surface for many bacteria, which causes an increasing problem in the background of a raising antibiotic resistance. These infections are hard to treat (Gbejuade et al., 2015), emphasizing the strong need for a nonantibiotic-dependent approach to clear or disrupt bacterial biofilms on surfaces.

The common materials usually used for orthopedic implants are stainless steel, cobalt-chromium, polymethylmethacrylate (PMMA), and a spectrum of polymeric biomaterials (Gbejuade et al., 2015). All the surfaces of these materials are susceptible to colonization by biofilms. The possible infections occur immediately after the operation in early infections (<3 months postoperatively) or as delayed infections (3–24 months postoperatively), and even as very late infections (>24 months). The management of these infections poses a major problem for physicians. The standard procedures in this case include an intravenous antimicrobial therapy over a longer period of time. If these measures show no effect, this can result in the worst case scenario of limb amputation or the death of the patient (Osmon et al., 2013). Needless to say, there is a demand for strategies to prevent biofilm formation, thereby prophylactically counteracting infections. Recently, several approaches have been explored to enzymatically coat various surfaces to interfere with the biofilm formation of bacteria. Swartjes et al. (2013) immobilized DNaseI on a PDMMA surface and thus created a coating with antibiofilm properties. Extracellular DNA is an integral part of the extracellular polymeric substances. The molecules are reported to be the longest EPS molecule and support the formation of stable filamentous network structures in biofilms (Flemming et al., 2016). In *S. aureus*, the release of eDNA is crucial for the initial attachment during the biofilm formation (Swartjes et al., 2013). The PMMA was coated with a coupling layer of dopamine, which in turn was immersed in a DNaseI solution. The functionality of the immobilized enzyme with this method was tested with plasmid DNA. The scientists reported a full digestion of the plasmid DNA after 30 min. In addition, it could be shown that there was no loss of activity within 8 h after the coating procedure. When the researchers observed the initial attachment of bacteria to the coated surface, they found that after 60 min of incubation, there was a reduction of 99% in the case of *P. aeruginosa* and 95% for *S. aureus*, compared with uncoated PMMA. This result clearly indicates the feasibility of enzymes for biotechnological applications, as well as the possibility of functionalized surfaces with biofilm inhibiting properties.

The convergence of enzymatically biofilm inhibition and the functionalization of surfaces can be seen in the efforts to functionalize surfaces with QQ enzymes in order to give them antibiofilm properties. Recently such a surface coating with a QQ acylase was achieved, showing a significant reduction of a *P. aeruginosa* biofilm. The acylase was targeting the AHL signal molecule and was applied to a silicone surface of a urinary catheter (Ivanova et al., 2015b). Urinary catheters are usually made from silicone for its hydrophobic properties, which was shown to counteract bacterial attachment on its own (Ivanova et al., 2015b). However, the clinical use of catheters poses a risk for the patient, leading to urinary tract infections and in turn to a longer duration of hospitalization (Jacobsen et al., 2008). In this study, the commercially available AHL acylase of *Aspergillus melleus* was used. The enzyme was shown to degrade the AHL signaling molecules. The silicone in this study was washed with sodium dodecyl sulfate (SDS) and pretreated with amino-propyl-tri-ethoxysilane solution (APTES) with the purpose of the introduction of amino groups. Subsequently the surface was coated with the acylase using linear polyethyleneimine (PEI) as a polycation. The pH conditions were chosen to give the enzyme polyanionic properties. The coating procedure was repeated in a layer by layer (LbL) fashion, creating a multilayer surface coating. A cytotoxicity assay with human fibroblast cells confirmed that the coated material does not exhibit toxic effects. Notably, PEI is considered as a toxic, but the study showed that the compound within the coating does not cause any adverse effects. Researchers suggest that this could be due to interaction with the acylase, neutralizing the positive charge. In addition, the stability was tested by exposing the coated silicone to artificial urine. After an incubation period of 7 days, an antibiofilm activity of the surface still could be observed. All the results together confirm the robustness of the coating and also highlight the potential of QQ enzymes. Moreover, the loss of enzyme activity during the deposition of the coating process was only 20%. The immobilized enzyme showed activity toward several AHL model compounds, including C4 HSL, C6 HSL, 3-oxo-C10 HSL, and 3-oxo-C12 HSL. Comparing the biomass of *P. aeruginosa* biofilms after 24 h on untreated and coated silicone surfaces showed a reduction of 75%. The same group went a step further by showing the first integration of two enzymatic inhibition approaches (Ivanova et al., 2015a). As in the preceding study, an acylase from *Aspergillus melleus* was utilized as a QQ enzyme, and in addition, the coating approach was combined with the α-amylase of *Bacillus amyloliquefaciens*. Both enzymes are commercially available. The choice of an amylase as a biofilm preventing enzyme was based on the fact that, as described, EPS components consist, among other things, of polysaccharides, and the enzyme is catalyzing endohydrolysis of 1,4-α-glycosidic bonds in said polysaccharides. Thus the α-amylase is targeting structural components of the biofilms by depolymerization of the EPS components (Craigen et al., 2011). In total, 10 layers of acylase and α-amylase were applied in turns. This leads to two different architectures—one with an acylase as the terminal layer and one with amylase at the outmost layer. The immobilized α-amylase showed activity toward purified *P. aeruginosa* EPS components measured by the amount of released reducing sugars. Under static conditions, individually immobilized acylase and α-amylase reduced the biofilm formation of *P. aeruginosa* by 30% in each setup compared with untreated silicone. The hybrid coating with the terminal acylase reduced biofilm formation to 60%, whereas for the terminal amylase, a reduction of 38% was measured. This underlines, on the one hand, the synergistic effect of the hybrid layer coating, but on the other hand shows that the architecture of its assembly plays an important role on the biofilm reducing effect. Using an acylase for QQ has been proven to be a good approach; however, AHLs represent only one kind of messenger molecule, which is exclusively reported in Gram negative QS systems. The EPS components, however, do vary in their composition, but the polysaccharides with an 1,4-α-glycosidic bond can be cleaved by α-amylase (Ivanova et al., 2015a). In a dynamic biofilm inhibition test of *P. aeruginosa* with a coated catheter in a physical model of a human bladder, the results were comparable to the results in the static environment. The biofilm inhibition was reported to be 70% by the acylase-terminated hybrid LbL coating. Notably the coating with the amylase as an outmost layer showed only a reduction of 10%. The hybrid coating with the terminal acylase was subsequently studied in an animal model, and in a catheterized rabbit, the spontaneous bacterial infection was monitored. The researchers reported 7 days of catheterization, and the biofilm formation was varying, depending on the catheter sections. In the balloon part, a biofilm reduction of 70% and in the shaft of 30% could be reported as compared with the control silicone catheter (Ivanova et al., 2015b).

3.8 Membrane Bioreactor

Nowadays an important field of use for QQ enzymes is wastewater treatment, precisely in the maintenance of membrane bioreactors (MBR). These machines work with a combination of activated sludge and micro- or nanofiltration. MBRs are a promising technology, but parameters still need to be improved in order to maximize their potential. Over the duration of their use, the filtration membrane is exposed to biofouling (Lade et al., 2014). To ensure long-term performance and reduce the costs due to the reduction of biofouling, QQ enzymes offer a solution to make this technology effective. The formation of a bacterial biofilm plays a crucial role in the biofouling of MBR membranes. A broad variety of antimicrobial compounds and antibiotics was tested to solve this problem. An MBR nanofiltration (NF) membrane was coated with porcine kidney acylase I; the membrane was subsequently put into a flow cell and incubated for 5 days. The results showed that the relative volume of the *P. aeruginosa* biofilm was only 24% compared with an untreated NF membrane (Kim et al., 2011). This is only a small example of the technical application of QQ enzymes outside of the medical field.

3.9 Inhaled Lactonase

Hraiech et al. (2014) researched the impact of the lactonase SsoPox-I on rat pneumonia caused by *P. aeruginosa*. As mentioned here, many in vitro studies confirm the potency of enzymatic QQ; however, in vivo experiments are very limited. Lactonases have been shown to efficiently decrease the production of virulence factors as well as the production of biofilm in *P. aeruginosa* (Hiblot et al., 2012). Preceding experiments by Migiyama et al. (2013) showed that the *P. aeruginosa* PAO1 expressing the lactonase AiiM caused a far lower amount of lung injury and greatly improved the survival rate in an acute pneumonia mouse model. In addition, it could be shown that the co-expression of AiiM in *P. aeruginosa* reduced the production of AHL-mediated virulence factors and attenuated cytotoxicity against epithelial cells of the human lung.

The goal of Hraiech et al. (2014) was to explore the possibilities of an exogenous administration of the QQ lactonase SsoPox-I in a *P. aeruginosa* PAO1 lung infection model in rats and thus if the enzyme was able to reduce the mortality caused by the infection. The bioengineered lactonase SsoPox from the extremophile archaeon *Sulfolobus solfataricus*, as described in this chapter, shows an increased affinity toward 3-oxo-C12 HSL, which is the signal molecule in the LasI/LasR QS circuit, playing an important role in the expression of virulence factors

(Guendouze et al., 2017; Hiblot et al., 2012). The rats were infected with 2.5×10^8 colony forming units (CFU) of *P. aeruginosa* PAO1 in a total volume of 250 μL. However, in this model the impact of the biofilm per se was not closer elucidated. After the infection, the animals were observed for 2 days. The administration of SsoPox-I as well as the infection of the bacteria was performed using a catheter inserted through the exposed trachea of the animals. The administration of the enzyme followed the same protocol as the infection, 250 μL of 1 mg/mL SsoPox-I in PBS, through the catheter, directly after the infection, and in another group with a delay of 3 h. Notably in this study, it could be shown that the administration of an enzyme into the lung environment of the animals was tolerated by the tissue. None of the test animals exhibited a bad tolerance toward the enzyme, and macroscopic examinations of the lungs showed no injuries. The histological assessment also confirmed that the tissue was not damaged. This finding marks an important step in research toward the application of QQ enzymes as a medical treatment. It was shown in a preceding in vitro assay that the addition of SsoPox-I in a *P. aeruginosa* PAO1 culture led to a decreased protease activity as well as a reduced biofilm formation. The infection model showed that the rats treated immediately after infection showed a reduced mortality rate within the monitored 50 h. A very important observation was also that among the tested groups of animals treated, untreated, and with a delayed treatment, the bacterial count was on a comparable level but virulence differed.

These examples illustrate that interference with bacterial communication can have a significant influence on the pathogenicity of bacteria.

4 Discussion and Perspectives

The focus of this review is set on the role of enzymes in QQ and also on their potential applications. Various examples of QQ enzymes point out the amazing potential this field offers. As mentioned previously, QQ is not only achieved by the enzymatic degradation of the signaling molecules, but also small molecules inhibiting the receptor or the synthase are being considered.

Bacteria live in mixed communities and are exposed to a variety of signaling molecules by a variety of different bacterial species. There is increasing evidence for bacterial crosstalk that is not limited to the same genus, but even overcomes the barrier between kingdoms (Tang and Zhang, 2014). In recent years, considerable research has been performed to elucidate the symbiotic role of the microbiome in the rhizosphere of plants dependent on QS interkingdom communication (Sanchez-Contreras et al., 2007). In the past 15 years, considerable research has been performed on QQ enzymes and the possibility of applying these enzymes (Fetzner, 2015). The use of the enzymatic degradation of AI in an extracellular environment offers many advantages that make this field of research extremely versatile for different applications and promising in the medical field (Gill et al., 2015). Some of these enzymes can discriminate different chain lengths of AHLs and thus be used selectively on QS circuits dependent on different signaling molecules (Bokhove et al., 2010). In addition, it is possible to immobilize these enzymes on surfaces, which makes it possible to bring a QQ effect to a locally defined area (Ivanova et al., 2015a). It is thought that the external degradation of AIs is not likely to lead to a formation of resistance because it does not generate a direct selective pressure for bacteria. QQ, in general, does not kill cells, as it interferes with gene expression in a nonlethal way and thus does not expose them to a strong selective pressure (Bzdrenga et al., 2017). Assessing the risk of a formation of resistance against QQ enzymes is very difficult since the interference of bacterial communication poses no immediate evolutionary pressure on the organisms per se (Maeda et al., 2012). Diggle et al. (2007) showed that, when choosing conditions in which the bacterial culture was dependent on QS regulated proteases, this leads to a severe impairment of growth compared with the same strain without a QQ enzyme. In this model, QQ showed a direct effect on the populations fitness (Diggle et al., 2007). On the other hand, this can be also seen as a disadvantage, since possible treatment strategies do not lead to a clearance of the bacterial infection, but rely on the host immune system or a combination of treatments in acute cases. There is exemplary research on β-lactamase inhibitors in combination with antibiotics, which are already in use (Gill et al., 2015). Proposing enzymes as a treatment option also means that the compounds have to be administered to a patient's body, thus exposing the patient to various factors which are unfavorable for the stability and functionality of enzymes. *P. aeruginosa*, for instance, expresses the protease LasA. Thus the question arises: Is it feasible to consider enzymes as a medical treatment? As a matter of fact, DNase I is already clinically used as an aerosol to support patients in their CF management (Bakker and Tiddens, 2007). Furthermore, the previously presented application of QQ enzymes as a surface coating underscores the stability of immobilized enzymes and the option to use them in the clinic (Ivanova et al., 2015b).

Glossary

Autoinducers (AI) In the literature, bacterial signaling molecules in a QS background are generally described as *autoinducers* due to the early finding that the molecules induce their own production.

Cystic fibrosis A genetic disorder that causes the lungs to produce large amount of mucus, which favors the colonialization of pathogenic bacteria causing chronic lung infections. Other symptoms can include may include sinus infections, poor growth, fatty stool, and clubbing of the fingers and toes.

Layer by layer coating A technique of depositing alternating layers of oppositely charged materials on surfaces.

Membrane bioreactor (MBR) A system used in wastewater treatment utilizing a filtration step in order filter particles. A major advantage is the high volumetric load, which is only limited by membrane pollution.

Quorum quenching (QQ) The disruption of the bacterial quorum sensing system.

Abbreviations

AHL	*N*-acyl homoserine lactone
AI	autoinducer
AI-2	autoinducer-2
AQ	2-alkyl-4(1H)-quinolone
CAI-1	cholerae autoinducer-1
CF	cystic fibrosis
CFU	colony forming unit
C4-HSL	*N*-butanoyl homoserine lactone
C6-HSL	*N*-hexanoyl homoserine lactone
C8-HSL	*N*-octanoyl-l-homoserine lactone
3-oxo-C6-HSL	*N*-(3-oxo-hexanoyl)-L-homoserine lactone
3-oxo-C8-HSL	*N*-(3-oxo-octanoyl)-L-homoserine lactone
3-oxo-C12-HSL	*N*-(3-oxo-dodecanoyl)-L-homoserine lactone
DSF	diffusible signaling factor
LbL	layer by layer
MIC	minimal inhibitory concentration
PQS	pseudomonas quinolone signal
QS	quorum sensing
QQ	quorum quenching

Acknowledgment

The authors would like to thank Dr. Ykelien L. Boersma for critically reading the manuscript and Dr. Robbert H. Cool for helpful discussions.

References

Afriat-Jurnou, L., Jackson, C.J., Tawfik, D.S., 2012. Reconstructing a missing link in the evolution of a recently diverged phosphotriesterase by active-site loop remodeling. Biochemistry 51, 6047–6055.

Bakker, E., Tiddens, H., 2007. Pharmacology, clinical efficacy and safety of recombinant human DNase in cystic fibrosis. Expert Rev. Respir. Med. 1 (3), 317–329.

Bayer, A.S., Speert, D.P., Park, S., Tu, J., Witt, M., Nast, C.C., Norman, D.C., 1991. Functional role of mucoid exopolysaccharide (alginate) in antibiotic-induced and polymorphonuclear leukocyte-mediated killing of *Pseudomonas aeruginosa*. Infect. Immun. 59, 302–308.

Bijtenhoorn, P., Mayerhofer, H., Müller-Dieckmann, J., Utpatel, C., Schipper, C., Hornung, C., Szesny, M., Grond, S., Thürmer, A., Brzuszkiewicz, E., Daniel, R., Dierking, K., Schulenburg, H., Streit, W.R., 2011. A novel metagenomic short-chain dehydrogenase/reductase attenuates *Pseudomonas aeruginosa* biofilm formation and virulence on *Caenorhabditis elegans*. PLoS One 6, e26278.

Billings, N., Ramirez Millan, M., Caldara, M., Rusconi, R., Tarasova, Y., Stocker, R., Ribbeck, K., 2013. The extracellular matrix component Psl provides fast-acting antibiotic defense in *Pseudomonas aeruginosa* biofilms. PLoS Pathog. 9e1003526.

Bokhove, M., Nadal Jimenez, P., Quax, W.J., Dijkstra, B.W., 2010. The quorum-quenching N-acyl homoserine lactone acylase PvdQ is an Ntn-hydrolase with an unusual substrate-binding pocket. Proc. Natl. Acad. Sci. U. S. A. 107, 686–691.

Bzdrenga, J., Daudé, D., Rémy, B., Jacquet, P., Plener, L., Elias, M., Chabrière, E., 2017. Biotechnological applications of quorum quenching enzymes. Chem. Biol. Interact. 267, 104–115.

Chang, C.-Y., 2018. Surface sensing for biofilm formation in *Pseudomonas aeruginosa*. Front. Microbiol. 8, 2671.

Chen, F., Gao, Y., Chen, X., Yu, Z., Li, X., 2013. Quorum quenching enzymes and their application in degrading signal molecules to block quorum sensing-dependent infection. Int. J. Mol. Sci. 14, 17477–17500.

Chow, J.Y., Yang, Y., Tay, S.B., Chua, K.L., Yew, W.S., 2014. Disruption of biofilm formation by the human pathogen *Acinetobacter baumannii* using engineered quorum-quenching lactonases. Antimicrob. Agents Chemother. 58, 1802–1805.

Chua, S.L., Liu, Y., Yam, J.K.H., Chen, Y., Vejborg, R.M., Tan, B.G.C., Kjelleberg, S., Tolker-Nielsen, T., Givskov, M., Yang, L., 2014. Dispersed cells represent a distinct stage in the transition from bacterial biofilm to planktonic lifestyles. Nat. Commun. 5, 4462.

Ciofu, O., Tolker-Nielsen, T., 2010. Antibiotic tolerance and resistance in biofilms. In: Biofilm Infections. Springer New York, New York, NY, pp. 215–229.

Craigen, B., Dashiff, A., Kadouri, D.E., 2011. The use of commercially available alpha-amylase compounds to inhibit and remove *Staphylococcus aureus* biofilms. Open Microbiol. J. 5, 21–31.

De Kievit, T.R., Iglewski, B.H., 2000. Bacterial quorum sensing in pathogenic relationships. Infect. Immun. 68, 4839–4849.

De Kievit, T.R., Gillis, R., Marx, S., Brown, C., Iglewski, B.H., 2001. Quorum-sensing genes in *Pseudomonas aeruginosa* biofilms: their role and expression patterns. Appl. Environ. Microbiol. 67, 1865–1873.

Diggle, S.P., Griffin, A.S., Campbell, G.S., West, S.A., 2007. Cooperation and conflict in quorum-sensing bacterial populations. Nature 450, 411–414.

Drake, E.J., Gulick, A.M., 2011. Structural characterization and high-throughput screening of inhibitors of PvdQ, an NTN hydrolase involved in pyoverdine synthesis. ACS Chem. Biol. 6, 1277–1286.

Duan, K., Dammel, C., Stein, J., Rabin, H., Surette, M.G., 2003. Modulation of *Pseudomonas aeruginosa* gene expression by host microflora through interspecies communication. Mol. Microbiol. 50, 1477–1491.

Elias, S., Banin, E., 2012. Multi-species biofilms: living with friendly neighbors. FEMS Microbiol. Rev. 36, 990–1004.

Fetzner, S., 2015. Quorum quenching enzymes. J. Biotechnol. 201, 2–14.

Flemming, H.-C., Wingender, J., Szewzyk, U., Steinberg, P., Rice, S.A., Kjelleberg, S., 2016. Biofilms: an emergent form of bacterial life. Nat. Rev. Microbiol. 14, 563–575.

Fux, C.A., Costerton, J.W., Stewart, P.S., Stoodley, P., 2005. Survival strategies of infectious biofilms. Trends Microbiol. 13, 34–40.

Gbejuade, H.O., Lovering, A.M., Webb, J.C., 2015. The role of microbial biofilms in prosthetic joint infections. Acta Orthop. 86, 147–158.

Gill, E.E., Franco, O.L., Hancock, R.E.W., 2015. Antibiotic adjuvants: diverse strategies for controlling drug-resistant pathogens. Chem. Biol. Drug Des. 85, 56–78.

Grandclement, C., Tannieres, M., Morera, S., Dessaux, Y., Faure, D., 2015. Quorum quenching: role in nature and applied developments. FEMS Microbiol. Rev. 40, 86–116.

Guendouze, A., Plener, L., Bzdrenga, J., Jacquet, P., Rémy, B., Elias, M., Lavigne, J.-P., Daudé, D., Chabrière, E., 2017. Effect of quorum quenching lactonase in clinical isolates of *Pseudomonas aeruginosa* and comparison with quorum sensing inhibitors. Front. Microbiol. 8, 227.

Hewitt, L., Kasche, V., Lummer, K., Lewis, R.J., Murshudov, G.N., Verma, C.S., Dodson, G.G., Wilson, K.S., 2000. Structure of a slow processing precursor penicillin acylase from *Escherichia coli* reveals the linker peptide blocking the active-site cleft. J. Mol. Biol. 302, 887–898.

Heydari, S., Eftekhar, F., 2015. Biofilm formation and β-lactamase production in burn isolates of *Pseudomonas aeruginosa*. Jundishapur J. Microbiol. 8, e15514.

Hiblot, J., Gotthard, G., Chabriere, E., Elias, M., 2012. Characterisation of the organophosphate hydrolase catalytic activity of SsoPox. Sci. Rep. 2, 779.

Hraiech, S., Hiblot, J., Lafleur, J., Lepidi, H., Papazian, L., Rolain, J.-M., Raoult, D., Elias, M., Silby, M.W., Bzdrenga, J., Bregeon, F., Chabriere, E., 2014. Inhaled lactonase reduces *Pseudomonas aeruginosa* quorum sensing and mortality in rat pneumonia. PLoS ONE 9 (10), e107125.

Infectious Diseases Society of America (IDSA), Spellberg, B., Blaser, M., Guidos, R.J., Boucher, H.W., Bradley, J.S., Eisenstein, B.I., Gerding, D., Lynfield, R., Reller, L.B., Rex, J., Schwartz, D., Septimus, E., Tenover, F.C., Gilbert, D.N., 2011. Combating antimicrobial resistance: policy recommendations to save lives. Clin. Infect. Dis. 52 (Suppl. 5), S397–S428.

Ivanova, K., Fernandes, M.M., Francesko, A., Mendoza, E., Guezguez, J., Burnet, M., Tzanov, T., 2015a. Quorum-quenching and matrix-degrading enzymes in multilayer coatings synergistically prevent bacterial biofilm formation on urinary catheters. ACS Appl. Mater. Interfaces 7, 27066–27077.

Ivanova, K., Fernandes, M.M., Mendoza, E., Tzanov, T., 2015b. Enzyme multilayer coatings inhibit *Pseudomonas aeruginosa* biofilm formation on urinary catheters. Appl. Microbiol. Biotechnol. 99, 4373–4385.

Jacobsen, S.M., Stickler, D.J., Mobley, H.L.T., Shirtliff, M.E., 2008. Complicated catheter-associated urinary tract infections due to *Escherichia coli* and *Proteus mirabilis*. Clin. Microbiol. Rev. 21, 26–59.

Keller, L., Surette, M.G., 2006. Communication in bacteria: an ecological and evolutionary perspective. Nat. Rev. Microbiol. 4, 249–258.

Kim, J.-H., Choi, D.-C., Yeon, K.-M., Kim, S.-R., Lee, C.-H., 2011. Enzyme-immobilized nanofiltration membrane to mitigate biofouling based on quorum quenching. Environ. Sci. Technol. 45, 1601–1607.

Koch, G., Nadal-Jimenez, P., Reis, C.R., Muntendam, R., Bokhove, M., Melillo, E., Dijkstra, B.W., Cool, R.H., Quax, W.J., 2014. Reducing virulence of the human pathogen *Burkholderia* by altering the substrate specificity of the quorum-quenching acylase PvdQ. Proc. Natl. Acad. Sci. U. S. A. 111, 1568–1573.

Lade, H., Paul, D., Kweon, J.H., 2014. Quorum quenching mediated approaches for control of membrane biofouling. Int. J. Biol. Sci. 10, 550–565.

Leadbetter, J.R., Greenberg, E.P., 2000. Metabolism of acyl-homoserine lactone quorum-sensing signals by *Variovorax paradoxus*. J. Bacteriol. 182, 6921–6926.

Lee, J.H., Lequette, Y., Greenberg, E.P., 2006. Activity of purified QscR, a *Pseudomonas aeruginosa* orphan quorum-sensing transcription factor. Mol. Microbiol. 59, 602–609.

Leid, J.G., Willson, C.J., Shirtliff, M.E., Hassett, D.J., Parsek, M.R., Jeffers, A.K., 2005. The exopolysaccharide alginate protects pseudomonas aeruginosa biofilm bacteria from IFN-γ-mediated macrophage killing. J. Immunol. 175 (11), 7512–7518.

Maeda, T., García-Contreras, R., Pu, M., Sheng, L., Garcia, L.R., Tomás, M., Wood, T.K., 2012. Quorum quenching quandary: resistance to antivirulence compounds. ISME J. 6, 493–501.

McNab, R., Ford, S.K., El-Sabaeny, A., Barbieri, B., Cook, G.S., Lamont, R.J., 2003. LuxS-based signaling in Streptococcus gordonii: autoinducer 2 controls carbohydrate metabolism and biofilm formation with *Porphyromonas gingivalis*. J. Bacteriol. 185, 274–284.

Michel, L., Harms, H., Heurlier, K., Michaux, P., Zala, M., Ginet, N., Haas, D., Défago, G., Keel, C., Reimmann, C., Triandafillu, K., Krishnapillai, V., 2002. Genetically programmed autoinducer destruction reduces virulence gene expression and swarming motility in *Pseudomonas aeruginosa* PAO1. Microbiology 148, 923–932.

Migiyama, Y., Kaneko, Y., Yanagihara, K., Morohoshi, T., Morinaga, Y., Nakamura, S., Miyazaki, T., Hasegawa, H., Izumikawa, K., Kakeya, H., Kohrogi, H., Kohno, S., 2013. Efficacy of AiiM, an N-acylhomoserine lactonase, against *Pseudomonas aeruginosa* in a mouse model of acute pneumonia. Antimicrob. Agents Chemother. 57, 3653–3658.

Monnet, V., Gardan, R., 2015. Quorum-sensing regulators in Gram-positive bacteria: "cherchez le peptide" Mol. Microbiol. 97, 181–184.

Nadell, C.D., Drescher, K., Foster, K.R., 2016. Spatial structure, cooperation and competition in biofilms. Nat. Rev. Microbiol. 14, 589–600.

Niu, C., Clemmer, K.M., Bonomo, R.A., Rather, P.N., 2008. Isolation and characterization of an autoinducer synthase from *Acinetobacter baumannii*. J. Bacteriol. 190, 3386–3392.

O'Brien, S., Fothergill, J.L., 2017. The role of multispecies social interactions in shaping Pseudomonas aeruginosa pathogenicity in the cystic fibrosis lung. FEMS Microbiol. Lett. 364 (15). fnx128.

Osmon, D.R., Berbari, E.F., Berendt, A.R., Lew, D., Zimmerli, W., Steckelberg, J.M., Rao, N., Hanssen, A., Wilson, W.R., 2013. Diagnosis and Management of Prosthetic Joint Infection: Clinical Practice Guidelines by the Infectious Diseases Society of America. Clin. Infect. Dis. 56, e1–e25.

Papaioannou, E., Wahjudi, M., Nadal-Jimenez, P., Koch, G., Setroikromo, R., Quax, W.J., 2009. Quorum-quenching acylase reduces the virulence of *Pseudomonas aeruginosa* in a *Caenorhabditis elegans* infection model. Antimicrob. Agents Chemother. 53, 4891–4897.

Papenfort, K., Bassler, B.L., 2016. Quorum sensing signal-response systems in Gram-negative bacteria. Nat. Rev. Microbiol. 14 (9), 576–588.

Peleg, A.Y., Seifert, H., Paterson, D.L., 2008. *Acinetobacter baumannii*: emergence of a successful pathogen. Clin. Microbiol. Rev. 21, 538–582.

Pereira, C.S., Thompson, J.A., Xavier, K.B., 2013. AI-2-mediated signalling in bacteria. FEMS Microbiol. Rev. 37, 156–181.

Phane Uroz, S., Chhabra, S.R., Cá Mara, M., Williams, P., Oger, P., Dessaux, Y., Uroz, S., Chhabra, S.R., Cámara, M., Williams, P., Oger, P., Dessaux, Y., 2005. N-Acylhomoserine lactone quorum-sensing molecules are modified and degraded by *Rhodococcus erythropolis* W2 by both amidolytic and novel oxidoreductase activities. Microbiology 151, 3313–3322.

Pustelny, C., Albers, A., Büldt-Karentzopoulos, K., Parschat, K., Chhabra, S.R., Cámara, M., Williams, P., Fetzner, S., 2009. Dioxygenase-mediated quenching of quinolone-dependent quorum sensing in *Pseudomonas aeruginosa*. Chem. Biol. 16, 1259–1267.

Rémy, B., Plener, L., Poirier, L., Elias, M., Daudé, D., Chabrière, E., 2016. Harnessing hyperthermostable lactonase from *Sulfolobus solfataricus* for biotechnological applications. Sci. Rep. 6, 37780.

Rickard, A.H., Palmer, R.J., Blehert, D.S., Campagna, S.R., Semmelhack, M.F., Egland, P.G., Bassler, B.L., Kolenbrander, P.E., 2006. Autoinducer 2: a concentration-dependent signal for mutualistic bacterial biofilm growth. Mol. Microbiol. 60, 1446–1456.

Rosenblatt-Farrell, N., 2009. The landscape of antibiotic resistance. Environ. Health Perspect. 117, A244–A250.

Sanchez-Contreras, M., Bauer, W.D., Gao, M., Robinson, J.B., Allan Downie, J., 2007. Quorum-sensing regulation in rhizobia and its role in symbiotic interactions with legumes. Philos. Trans. R. Soc. Lond. Ser. B Biol. Sci. 362, 1149–1163.

Sievert, D.M., Ricks, P., Edwards, J.R., Schneider, A., Patel, J., Srinivasan, A., Kallen, A., Limbago, B., Fridkin, S., National Healthcare Safety Network (NHSN) Team and Participating NHSN Facilities, 2013. Antimicrobial-resistant pathogens associated with healthcare-associated infections summary of data reported to the National Healthcare Safety Network at the Centers for Disease Control and Prevention, 2009-2010. Infect. Control Hosp. Epidemiol. 34, 1–14.

Sio, C.F., Otten, L.G., Cool, R.H., Diggle, S.P., Braun, P.G., Bos, R., Daykin, M., Camara, M., Williams, P., Quax, W.J., 2006. Quorum quenching by an N-acyl-homoserine lactone acylase from *Pseudomonas aeruginosa* PAO1. Infect. Immun. 74, 1673–1682.

Solano, C., Echeverz, M., Lasa, I., 2014. Biofilm dispersion and quorum sensing. Curr. Opin. Microbiol. 18, 96–104.

Spoering, A.L., Lewis, K., 2001. Biofilms and planktonic cells of *Pseudomonas aeruginosa* have similar resistance to killing by antimicrobials. J. Bacteriol. 183, 6746–6751.

Stoodley, P., Sauer, K., Davies, D.G., Costerton, J.W., 2002. Biofilms as complex differentiated communities. Annu. Rev. Microbiol. 56, 187–209.

Swartjes, J.J.T.M., Das, T., Sharifi, S., Subbiahdoss, G., Sharma, P.K., Krom, B.P., Busscher, H.J., van der Mei, H.C., 2013. A functional DNase I coating to prevent adhesion of bacteria and the formation of biofilm. Adv. Funct. Mater. 23, 2843–2849.

Tang, K., Zhang, X.H., 2014. Quorum quenching agents: resources for antivirulence therapy. Mar. Drugs 12, 3245–3282.

Tseng, B.S., Zhang, W., Harrison, J.J., Quach, T.P., Song, J.L., Penterman, J., Singh, P.K., Chopp, D.L., Packman, A.I., Parsek, M.R., 2013. The extracellular matrix protects *Pseudomonas aeruginosa* biofilms by limiting the penetration of tobramycin. Environ. Microbiol. 15, 2865–2878.

Utari, P.D., Vogel, J., Quax, W.J., 2017. Deciphering physiological functions of AHL quorum quenching acylases. Front. Microbiol. 8, 1123.

Wahjudi, M., Papaioannou, E., Hendrawati, O., van Assen, A.H.G., van Merkerk, R., Cool, R.H., Poelarends, G.J., Quax, W.J., 2011. PA0305 of *Pseudomonas aeruginosa* is a quorum quenching acylhomoserine lactone acylase belonging to the Ntn hydrolase superfamily. Microbiology 157, 2042–2055.

Wang, Z., Fast, W., Valentine, A.M., Benkovic, S.J., 1999. Metallo-β-lactamase: structure and mechanism. Curr. Opin. Chem. Biol. 3, 614–622.

Xavier, K.B., Bassler, B.L., 2003. LuxS quorum sensing: more than just a numbers game. Curr. Opin. Microbiol. 6, 191–197.

CHAPTER 8

Effect of Polyphenols on Microbial Cell-Cell Communications

Filomena Nazzaro*, Florinda Fratianni*, Antonio d'Acierno*, Vincenzo De Feo[†], Fernando Jesus Ayala-Zavala[‡], Adriano Gomes-Cruz[§], Daniel Granato[¶], Raffaele Coppola[‖]

*Institute of Food Science, ISA-CNR, Avellino, Italy, [†]Department of Pharmacy, University of Salerno, Salerno, Italy, [‡]Center for Research in Nutrition and Development, A. C (CIAD AC), Hermosillo, Mexico, [§]Federal Institute of Education, Science and Technology of Rio de Janeiro (IFRJ), Department of Food, Rio de Janeiro, Brazil, [¶]State University of Ponta Grossa (UEPG), Department of Food Engineering, Ponta Grossa, Brazil, [‖]DiAAA, Department of Agricultural, Environmental and Food Sciences, University of Molise, Campobasso, Italy

1 What is Quorum Sensing?

Microorganisms coordinate both microbe-microbe interactions and associations with higher organisms through intercellular communication systems, known as quorum sensing (QS) systems, in which behavior takes place only when microorganisms reach a specific cell density. Such behavior, unproductive when carried out by a singular organism, becomes operative when a group of microorganisms (bacteria-bacteria, bacteria-fungi, fungi-fungi) simultaneously performs the action indeed. A QS system is capable of controlling a certain number of activities, including sporulation, bioluminescence, virulence factor expression, biofilm formation, and mating. In addition, some QS systems can represent virulence factors by themselves, since they are toxic to host cells and/or can modulate host immunity. The system is realized throughout the microbial production of signaling molecules, of chemical nature, which are also known as autoinducers, produced in a particular stage of growth by microorganisms, above a threshold concentration, and that give rise to the activation or repression of specific genes. The accumulation of a stimulatory amount of such molecules can take place only when a specific and sufficient number of cells, referred to as a *quorum*, are present (Bassler, 2002). In general, the QS system is based on the following key elements: (a) autoinducers, (b) signal synthase, (c) signal receptor, (d) signal response regulator, and (e) regulated genes (which form the so-called *QS regulon*; Nazzaro et al., 2013). The QS systems used by bacteria can be classified into three main types: the LuxR/I-type systems, primarily used by Gram negative bacteria; the peptide signaling systems used primarily by Gram positive bacteria; and the *lux*S/AI-2

Quorum Sensing. https://doi.org/10.1016/B978-0-12-814905-8.00008-3

signaling used for interspecies communication. In addition, AI-3/epinephrine/norepinephrine is used as an interkingdom signaling system (Reading and Sperandio, 2006). Gram negative bacteria generally produce acylated homoserine lactones (AHLs) as autoinducers, which are synthesized by a LuxI-type enzyme (signal synthase), encoded by the first gene of the lux operon. A low level of transcription of the lux operon—caused by a concurrent low level of bacterial cell density—is not enough for the activation of LuxR, and it can be activated only when the cell density increases and signal levels reach a specified threshold level. The LuxR/3-oxo-C6-HSL complex thereby activates transcription via the *lux* operon promoter, giving rise to the expression of other genes, including (in *Vibrio fischeri*) *lux* AB genes encoding luciferase and *lux* CDE, which encodes the enzymes that produce the substrate for luciferase and hence bioluminescence. These molecules passively diffuse through the bacterial membrane and accumulate both intra- and extracellularly in proportion to the cell density. QS circuits have been identified in more than 25 species of Gram negative bacteria that can efficiently couple gene expression to fluctuations in cell population density through the QS mechanism (Fig. 1A).

V. fischeri, *Pseudomonas aeruginosa*, *P. fluorescens*, *P. chlororaphis*, *Erwinia carotovora*, and *Agrobacterium tumefaciens* systems can be considered the best understood. Furthermore, the quorum system mechanism of *Chromobacterium violaceum* is usually used to evaluate the QS activity and QS inhibitory activity of several natural molecules. In this microorganism, the phenotypic response to AHLs involves the production of a variety of factors, including antibiotics, hydrogen cyanide, proteases, chitinase, and particularly violacein, a water-insoluble purple pigment with antibacterial activity (Chernin et al., 1998). Gram positive bacteria also regulate a variety of processes in response to an increasing cell population density, through the employment of secreted peptides that are processed from precursors and used as autoinducers for QS (Fig. 1B). Such signals are dynamically carried outside the cell, where they interact with the external domains of membrane-bound sensor proteins. The transduction of a signal generated by a phosphorylation cascade culminates in the activation of a DNA-binding protein: this event affects the transcription of specific genes, so that each sensor protein is extremely selective for a particular peptide signal. In Gram positive bacteria, a peptide-signal precursor locus is present within the sequence of a precursor protein; this is then cleaved, and the processed peptide autoinducer signal produced is usually transported out of the cell via an ATP-binding cassette (ABC) transporter. Similar to Gram negative bacteria, Gram positive bacteria can employ multiple autoinducers and sensors. However, some peptides can exclusively act externally to the bacterial cell, causing a specific set of gene expression changes, which are translated in the activation of a different set of behavioral changes. Making and responding to combinations of these and other types of chemical signals could lead bacteria to take a sort of census of their own population, as well as of the population density of other species in the vicinity. A distinct response to each signal, or a response based on a combinatorial sampling of a variety of signals, could enable bacteria to continuously modulate

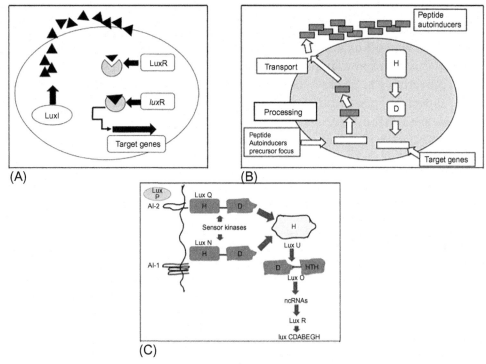

Fig. 1

Mechanism of quorum sensing in Gram negative (A), Gram positive (B) bacteria, and a mix between components of Gram positive and Gram negative systems (as an e.g., herein is shown the LuxS/AI-2 signaling system in the marine bacterium *Vibrio harveyi*, C). (A) Quorum sensing system in Gram negative bacteria. (B) Quorum sensing system in Gram positive bacteria H and D phosphorylated residues of histidine and aspartate, respectively, on the signalling proteins. (C) *V. harveyi* QS system represents a typical example of a mix between components of Gram positive and Gram negative systems. *(A and B) Modified from Nazzaro, F., Fratianni, F., Coppola, R., 2013. Quorum sensing and phytochemicals. Int. J. Mol. Sci. 14, 12607–12619; (C) Modified from Reading, N.C., Sperandio, V., 2006. Quorum sensing: the many languages of bacteria. FEMS Microbiol. Lett. 254, 1–11.*

their behavior according to the species present in a consortium. In lactic acid bacteria (LAB), QS implicates peptides directly sensed by membrane-located histidine kinases, after which the signal is transmitted to an intracellular response regulator. This regulator in sequence activates the transcription of target genes that commonly include the structural gene for the inducer molecule. The two-component signal-transduction machinery is essential for transcription activation and production of several autoinducers found in LAB, which are predominantly bacteriocins or bacteriocin-like peptides, which can act as the response regulator and activate transcription of target genes, which, for example, can lead them to the adhesion process (Sturme et al., 2005). The LuxS/AI-2 signaling system is exhibited by the marine bacterium *Vibrio harveyi*, which controls bioluminescence through QS. The *V. harveyi* QS system represents a typical example of a mix between components of Gram positive and Gram negative systems. The bacterium shows two QS systems: system 1, in which the autoinducer (AI-1) is an AHL, principally involved in intraspecies signaling; and system 2, in which the autoinducer (AI-2) is a furanosyl borate diester, associated with interspecies signaling (Fig. 1C). *V. harveyi* uses two sensor kinases, LuxN and LuxQ, to recognize AI-1 and AI-2, respectively. LuxQ recognizes AI-2 complexed with its periplasmic receptor LuxP. Upon sensing these signals, the kinases convert to phosphatases, and this complex phosphorelay system—through LuxU and LuxO—is dephosphorylated. The process no longer activates transcription of not-codifying (nc)RNAs, so that LuxR mRNA are no longer degraded, and LuxR activates transcription of the luciferase operon (Reading and Sperandio, 2006).

In general, QS is reported for several microorganisms involved in the production of different fermented foods such as sourdough, dairy products, fermented vegetables, and wine, suggesting that QS plays a role in the fermentation of these foods (Lebeer et al., 2007). A promising application of this system is the development of autolytic starter LAB, which will lyse in an early stage during cheese ripening, thereby easing the liberation of intracellular enzymes that can contribute to flavor formation (Kuipers et al., 1998). Our knowledge of the mammalian gut microbiome reveals how QS could affect microbiome species composition. The same knowledge of QS might lead also to the study of new ways to control infectious diseases, such as cholera, and how, in the host organism, different mechanisms have evolved to influence bacterial QS and thus shape their microbiomes. For example, the AI-2 molecule, produced by many bacterial species, is capable of promoting gut colonization by *Firmicutes* over *Bacteroidetes* (Thompson et al., 2015). Furthermore, the production of such a molecule by a gut commensal bacterium can limit *Vibrio cholerae* infections (Bäumler and Sperandio, 2016). Although the interactions of bacteria in the mammalian gut are much more complex, it is now established that the involvement of QS in these microbiomes can offer an opportunity to intervene in the so-called gut dysbiosis (Whiteley et al., 2017). In eukaryotic organisms, including yeasts and molds, QS is a relatively recent field of study and application. In fact, it was practically unknown until the discovery of *E*-farnesol as a QS molecule (QSM) in the pathogenic fungus *Candida albicans* (Hornby et al., 2001). This is an exogenous molecule, which inhibits biofilm formation when provided early during adherence. Farnesol also accumulates in supernatants of mature biofilms, where the stimulation of yeast cell production might also stimulate their dispersal, and indeed acts as an inhibitor of hyphal formation (Peleg et al., 2010). Besides, aromatic alcohols, dodecanol, tyrosol, and γ-butyrolactone are other molecules identified as mediators of QS processes in eukaryotic organisms, such as filamentous fungi *Aspergillus* and *Penicillium* species.

2 Polyphenols

2.1 Structure and Distribution

Polyphenols are secondary metabolites of plants that, due to their structural chemistry, contribute significantly to defending plants from ultraviolet radiation, pathogens, and predators, and overall are crucial for their growth and survival. Polyphenols can be categorized as phenolic acids (with single rings), stilbenes, flavonoids, and lignans (which show multiple rings), depending on the rings and the structural groups binding one ring to another (Fig. 2).

The structural differences between individual compounds influence the differences among polyphenols also in terms of bioavailability and biological properties (Manach et al., 2009). In nature, polyphenols are typically conjugated to sugars and organic acids. They can be classified as two major types: flavonoids and nonflavonoid phenolics, which are distributed in fruit and vegetables, as well as in herbs and spices.

Flavonoid phenolics have a basic structure, based on two benzene rings (A and B) linked through a heterocyclic pyrone C ring. Nonflavonoid phenolics result from a more heterogeneous group of compounds, ranging from the simplest of the class, such as hydroxycinnamates and benzoic acids (Selma et al., 2009), to more complex compounds that include stilbenes, lignans hydrolysable tannins, as well as gallotannins and ellagitannins, with main components gallic and hexahydroxydiphenic acids and, upon their hydrolysis, ellagic acid. The cleavage of glycosyl- or glucuronosyl-moiety from the phenolics basis, known as deconjugation, gives rise to the formation of aglycones (Rechner et al., 2004). Flavonols, a class of flavonoids at planar structure with different positions of the phenolic-OH groups, are present predominantly in Brassicaceae, tea, red wine, and apples (Hollman and Katan, 1999). They are actively degraded by gut microbiota and substituted with simpler, smaller phenolic compounds derived from A and B rings after break-up of the flavonoid C ring (Aura, 2008; Quirós Sauceda et al., 2017). In the intestine, flavonols could be transformed firstly by a C-ring fission, then by dehydroxylation reactions. The type of glycosylation can also affect the stability of the flavonoids in the intestine. Flavanones, essentially present in the genus *Citrus*, show a 2,3-dihydro-2-phenylchromen-4-one structure, with the pyran ring of nonplanar flavanones. They have a bigger bioavailability than flavonols or flavan-3-ols (Selma et al., 2009). This can be related to a minor degradation by colonic microbiota respect to other flavonoids; thus they can be more suitable for absorption, even in the distal part of the intestine. Flavanones are present as glycosides—generally rutinosides and neohesperidosides. Under such form, they can also be found in some medicinal herbs, such as menthe. Flavanone glycosides, such as naringin, are firstly subjected to a deglycosylation, giving the aglycone naringenin; naringenin is therefore split and gives rise to phloroglucinol and with 3-(4-hydroxyphenyl) propionic acid. This can be additionally dehydroxylated to give 3-phenyl propionic acid (Rechner et al., 2004). Flavan-3-ols and procyanidins are a very complex group of polyphenols produced from simple flavan-3-ols (catechin, epicatechin, gallocatechin, epigallocatechin, and the corresponding gallate esters), which, when metabolized, give rise to polymeric procyanidins or condensed tannins. They are among the main constituents of the dietary phenolic intake, and are chiefly provided by fruits, tea, and wine. During digestion and transfer across the small intestine, and from this to the liver, flavan-3-ols are rapidly metabolized to different *o*-sulphated, *o*-glucuronidated, and *o*-methylated forms (Kuhnle et al.,

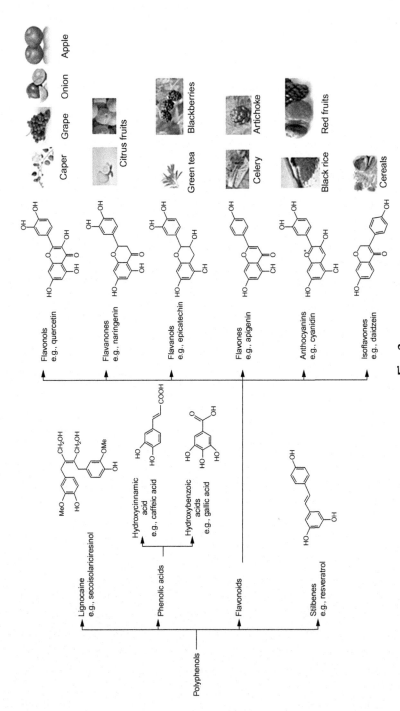

Fig. 2

Polyphenols. Classification and presence in fruit, vegetables, and herbs.

2000). A combined action of bacterial and human metabolisms degrades catechin and epicatechin (Meselhy et al., 1997; Rechner et al., 2004). Anthocyanins, one of the most important classes of flavonoids, are plant pigments widely distributed in colored fruits and flowers, who are responsible for their purple, violet, blue, and red colors. The major sources of anthocyanins in edible plants are Vitaceae (grape) and Rosaceae (raspberry, strawberry, cherry, blackberry, plum, apple, peach, etc.) and, to a less extent, Solanaceae (eggplant, tamarillo), Saxifragaceae (red and blackcurrant), Cruciferae (red cabbage), and Ericaceae (blueberry and cranberry). The aglycone forms of anthocyanins are commonly found in fruits and vegetables (Granato et al., 2015; Pascual-Teresa and Sanchez-Ballesta, 2007). They are much more stable in the gut with respect to the respective glucosides, probably because they are attacked by glucosidase in the small intestine. The protective action by *p*-coumaric acid (Nazzaro et al., 2015) or the entrapment of the molecules into food and safe grade polymers, such alginate, can stabilize the molecules, protecting them during gut transit (Fratianni et al., 2010). Syringic acid (3,5-dimethoxy-4-hydroxybenzoic acid), vanillic acid (3-methoxy-4-hydroxybenzoic acid), phloroglucinol aldehyde (2,4,6-trihydroxybenzaldehyde), phloroglucinol acid (2,4,6-trihydroxybenzoic acid), gallic acid (3,4,5-trihydroxybenzoic acid), and 3-*O*-methylgallic acid are the most common anthocyanins metabolites produced by gut microbiomes (Keppler and Humpf, 2005). Isoflavones, nonsteroidal estrogens with a chemical structure similar to estrogens, are particularly abundant in soy nuts, tempeh, and red clover. The majority of them are not absorbed as such across the enterocytes, owing to their molecular weight and high hydrophilicity. The activity of isoflavones is subsequent to the conversion of the glucosides into the main bioactive aglycones (glycitein, daidzein, genistein, and equol) through the action of the intestinal β-glucosidase of microbiome (Setchell et al., 2002). Nonflavonoid phenolics include hydrolysable tannins, such as gallotannins and ellagitannins, which in hydrolysis gives rise to gallic acid and ellagic acid, respectively, which generate urolithin A and urolithin B (Cerdá et al., 2004). Gallotannins, typically found in berries, are responsible for their so-called astringent taste. Ellagitannins are usually present in some berries, nuts, a few fruits, and oak-aged wines. All these molecules are transformed by microbiomes. Lignans, recognized as phytoestrogens, include secoisolariciresinol, matairesinol, pinoresinol, lariciresinol, isolariciresinol, and syringaresinol. They are primarily contained in fruits and vegetables, and, to a minor extent, in tea, cereals, coffee, and alcoholic drinks (Scalbert and Williamson, 2000). Their metabolism involves both mammalian and gut microbial enzymatic activities (Possemiers et al., 2007). The bioactivation of lignans to phytoestrogens is mostly performed by the microbiome (Borriello et al., 1985). Hydroxycinnamates are generally located in many plant-derived food products. *p*-Coumaric acid (4-hydroxycinnamic acid), caffeic acid (3,4-dihydroxycinnamic acid), ferulic acid (4-hydroxy-3-methoxycinnamic acid), and sinapic acid (4-hydroxy-3,5-dimethoxycinnamic acid) and their esters with quinic and tartaric acids (chlorogenic, caftaric, etc.) are the most abundant hydroxycinnamates. Ferulic and *p*-coumaric acids occur in spinach and in the arabinoxylans of cereal brans. Chlorogenic acid is very abundant in coffee, but it represents one of the key rings of several metabolic pathways toward the formation of other secondary metabolites (http://www.genome.jp/kegg/pathway.html). Hydroxycinnamic acids are very important vegetal components with biological properties, including chemopreventive action (Karlsson et al., 2005), exhibited after their absorption in the human gut and microbial transformation (Gonthier et al., 2006; Kroon and Williams, 1999). Bacterial esterases release free acids, which are subsequently metabolized to other compounds such as phenylpropionic acid and then to phenylacetic acids. Stilbenes, compounds present in wood or in soft vegetal tissues, have health benefits. They are not transformed by microbiome, with the exception of the hydrogenation of *trans*-resveratrol to di-hydroresveratrol (Walle et al., 2004). Benzoic acids, benzoates, and benzoic acid esters are common in fruits, in particular berries (cranberries). Such molecules also represent the most common metabolites deriving from the microbial degradation of different metabolites, both flavonoids and nonflavonoid phenolics. The major molecules belonging to such classes are gallic acid, *p*-hydroxybenzoic acid, vanillic acid, syringic acid, and protocatechuic acid; they can even be transformed by the microbiome to simpler molecules, including pyrogallol, catechol, and *o*-methylcatechol, which are more easily absorbed in the gut.

2.2 Antimicrobial Activity of Polyphenols

The structural differences in the phytochemicals give rise to marked differences in their bioavailability (Manach et al., 2009). Microbial action on polyphenols, essentially on their core, gives rise to simpler metabolites, which often affect their bioactivities (Van Duynhoven et al., 2011), such as phenolic acids and short-chain fatty acids. Some of the phenolic acids can be absorbed across the intestinal mucosa. At same time, polyphenols and their metabolites can affect the intestinal ecology-modulating microbiota (Selma et al., 2009). Many phenolic compounds have demonstrated potential in vitro antimicrobial activity, acting as inhibitors of infection-causing bacteria, suggesting that some phenolic compounds might be applied as antimicrobial agents against human infections (Fratianni et al., 2013; Selma et al., 2009). Thus we can assume that there is a strict link between the composition of microbiota and phytochemicals, with a great impact on human health and, subsequently, on the potential role that such molecules can exhibit on the microbial cell-cell mechanisms of communication. Several polyphenols act as inhibiting agents against different pathogens. Naringenin and quercetin demonstrated a total and dose-dependent inhibitory effect on the growth of *Escherichia coli*. A higher inhibitory effect of the relative aglycones has been demonstrated. This effect could be explained by the presence/absence of the sugar moiety in such molecules. Pure polyphenols, except rutin, tested at different concentrations on *E. coli, Staphylococcus aureus*, and *Salmonella typhimurium*, caused a decrease in bacterial growth—mainly quercetin (flavonol) and naringenin (flavanone; Parkar et al., 2008). Some phenolic aglycones, epicatechin, catechin, 3-*O*-methylgallic acid, gallic acid, and caffeic acid, were capable of inhibiting the growth of different pathogens (Lee et al., 2006), such as *Clostridium difficile, C. perfringens*, and some *Bacteroides* spp. Caffeic acid acted as a powerful inhibitor against

Salmonella, Pseudomonas, E. coli, Clostridia, and *Bacteroides*. Proanthocyanidins (condensed tannins) affected the growth of different bacteria belonging to *Enterobacteriaceae*, *Bacteroides, Prevotella*, and *Porphyromonas* (Smith and Mackie, 2004). Nonflavonoids, such as resveratrol, showed in vitro antimicrobial activity against several pathogens. Ellagitannins—that are hydrolyzed in vivo, releasing ellagic acid and transformed by gut microbiota in urolithin (Espin et al., 2013)—are known to exert a strong inhibitory effect against some pathogens such as *Clostridia* and *S. aureus* (Bialonska et al., 2009).

2.3 Polyphenols as Quorum Quenching Players

The food industry's consumers commonly demand natural products because of the therapeutic values in traditional applications. In recent years, many researches focused their attention on the role of QS among the microbial strains in the food industry; according to their reports, some signaling molecules have been detected in different concentrations in food products such as meat, vegetables, and milk products. The time of food processing as well as the storage conditions might influence the secretion of the QSMs. Therefore, food constitutes an available medium for microbial growth and biofilm formation due to its richness in nutrients. Concurrently, the growth of microorganisms in the food matrix might be also influenced by different parameters, including communication among the different microbial strains found in the food, and the subsequent ability of signal molecule production. Thus more knowledge about the microflora present in a biological matrix and the relative QS mechanism has the potential to help formulate a new approach to prevent not only the growth of pathogen and spoilage microorganisms but also their biofilm matrix, which could be detected in food products. In this regard, it could be useful to deepen the research on the therapeutic properties of plants, and the quorum quenching (QQ) mechanism should be better studied for biotechnological application. In the last years, plant food extracts and phytochemicals were highlighted as QS inhibitors (QSIs; Koh et al., 2013). One of the elements of their success can be ascribed to their likeness to the ideal QSI, which involves, for instance, a chemical stability and highly effectiveness, without being harmful for human health. Although the molecular mechanisms are still poorly understood, it is thought that some phytochemicals can interfere with the key components in QS mechanism (Dong et al., 2007). A diet rich in plant-derived foods and vegetables might affect, as previously described, the intestinal microbiota, preventing the colonization of the intestinal system and biofilm formation by pathogens (El-Hamid, 2016; Yang et al., 2018). The mechanism of QQ is to prevent small signal molecules from reaching the relevant signal receptor, and/or to control the gene expression in the microbial metabolism, so that the cumulative behaviors could not be regulated by the signaling mechanism. Essentially, the capacity of plant extracts to interrupt QS systems may provide a defense instrument to fight against bacterial invasion. The biologically active components naturally located in plants opened a new door for the exploration and exploitation of newly generated drugs and antimicrobial agents, helpful in the food industry as natural preservatives. Many studies highlighted the capacity of plant extracts and phytochemicals to interfere in intra- and interspecies QS communication systems (Al-Refi, 2016; Yang et al., 2018). Phytochemicals present in several fruits and vegetables show the capability to inactivate the QS mechanism (Korukluoglu and Gulgor, 2017). Therefore, the basis of the QQ mechanism is a condition for the plant to protect itself against pathogen attacks (Bacha et al., 2016). The first studies on anti-QS activities of plant extracts in bacteria were concentrated mainly to clarify how they were capable of inhibiting the expression of well-established specific QS-induced gene(s) (Schuster et al., 2013), in which reduction is related to the attenuation of bacterial virulence and can be a support in prevention of bacterial adverse effects (Suga and Smith, 2003). The microbial activities could be regulated by the QQ mechanism instead of the QS mechanism because the signaling molecules are inhibited (Chen et al., 2013). Overall, the target of biological components is the microbial QS system, and the effect of the compounds goes through the inactivation, degradation, or modification of the signal molecules. DNA microarray studies and quantitative real-time PCR (qRT-PCR) data demonstrated the effect of plant extracts in the downregulation of the QS-related genes of various bacterial species, which is thought to be one of the main motives responsible for the decrease of their virulence (Tolmacheva et al., 2014). The plant components can block the signaling molecules from being synthesized by the *lux*I encoded AHL synthase, degrading the signaling molecules and/or targeting the *lux*R signal receptor. The most commonly identified mechanisms of action of plant food extracts and phytochemicals are correlated with their similarity in chemical structure to QS signals (AHL), as well as to their ability to destroy signal receptors (LuxR/LasR; Rasmussen and Givskov, 2006). Lastly, the reduction of QS gene expression and signaling molecule levels, which affects the production of virulence factors, provides more insight into why plant extracts were used in the past and in what way they might be used in the future as support to combat the bacterial infections (Vandeputte et al., 2010). QS-antagonist components, isolated from plants, are well-known flavonoids found in almost all parts of the plants. Flavonoids, which give the relevant colors to the flowers and fruits, also play a fundamental role in the protection of the plant from insect pests. In addition, they can act as an important QQ mechanism against communication among microorganisms (Bacha et al., 2016; Koh et al., 2013). Andrographolide, a labdane diterpenoid isolated from *Andrographis paniculata*, and curcumin, a yellow chemical component produced by some plants belonging to the ginger family, can act as repressors able to affect the QS-regulated virulence of *P. aeruginosa*. Other plant-derived active ingredients, such as ellagic acid and fisetin, have some attenuation effects on the biofilm formation of *Streptococcus dysgalactiae*. Salicylic acid can repress AHL production in *P. aeruginosa*; on the other hand, tannic acid and *trans*-cinnamaldehyde might act, as the salicylic acid, to inhibit AHL synthase (Chang et al., 2014). Extracts of *Vaccinium macrocarpon*, *Rubus idaeus*, *Origanum vulgare*, *Rosmarinus officinalis*, *Ocimum basilicum*, *Brassica oleracea*, *Curcuma longa*, and *Zingiber officinale* show QQ mechanisms against different microorganisms, such as *P. aeruginosa* and *E. coli* O157:H7, two of the most important foodborne pathogens known (Vattem et al., 2007). *Rosa canina, Ballota* *nigra, Juglans regia, Castanea sativa*, and *R. officinalis* extracts have QQ ability against *S. aureus* indeed, and the orange extract could prevent the communication among *Yersinia enterocolitica* strains (Korukluoglu and Gulgor, 2017). Tannin-rich fractions of some plants have the capability to inhibit swarming and act against virulence factors, especially in pathogen bacterial strains. In the future, the plant extracts capable of exhibiting a QQ activity might replace drugs in medicine as well as all (or lot of) synthetic molecules used as preservatives and/or additives in the food industry. The use of such agents might improve the advancement for new biocontrol strategies (Koh et al., 2013; Mahmoudi et al. 2014; Shukla and Bhathena, 2016). Most antagonists exhibit narrow spectrum activities, which only target specific pathogens. This aspect can represent an interesting way when, for instance, we should combat against a specific type of pathogen in a polymicrobial environment, but such a narrow antagonistic action has limited clinical values. Therefore, a cocktail therapy, which includes both antibiotics and QS inhibitors, may supply synergistic effects (Rutherford and Bassler, 2012). Many fruit and vegetables exhibit inhibitory activities against both adhesion and biofilm formation. An extract of *Helichrysum italicum* inhibits the adherence of *Streptococcus mutans* cells on the glass surface; cranberry juice inhibited the biofilm formation of *S. sobrinus* and *S. sanguinis* indeed (Abachi et al., 2016). Ethanol and ethyl acetate extracts from *Hypericum connatum* exhibit QQ activity against *C. violaceum*, limiting its production of violacein (Fratianni et al., 2013). Ethanolic extracts of several plants are capable of reducing the biofilm formation of different pathogens. The extracts of *Cinnamomum zeylanicum, Marrubium vulgare, Tamarix aphylla, Cuminum cyminum, Pelargonium hortorum, Lawsonia inermis, Salvia officinalis, Triticum aestivum, Artemisia absinthium, Hibiscus sabdariffa, Thymus vulgaris, Punica granatum, Agave sisalana, Mentha longifolia*, and *Portulaca oleracea* reduce the *P. aeruginosa* biofilm formation in the range 20%–80% (Korukluoglu and Gulgor, 2017). It has been reported that any biofilm inhibition or reduction was observed after the application of ethanolic extracts of *Pelargonium graveolens, Foeniculum vulgare, Solenostemon scutellarioides, R. officinalis, Urtica dioica, Matricaria recutita, Erythrina crista-galli*, and *Momordica charantia* (Al-Refi, 2016). The oak bark is rich in the hydrolysable tannins, pyrogallol tannins, and condensed tannins-proanthocyanidins, which act as QQ molecules (Deryabin and Tolmacheva, 2015). *Conocarpus erectus* contains vescalagin and castalagin, which are hydrolysable tannins, with QQ activity (Adonizio et al., 2008; Vandeputte et al., 2010). An extract of *Allium sativum* contains at least three different QS inhibitors, whose activity depends on their concentration. Several phytochemicals act also as inhibitors of the QS mechanism in fungi: the phytoalexin, resveratrol (3,5,4′-trihydroxystilbene), naturally present in grapes and other vegetables, is an antifungal agent; *Coriandrum sativum* extracts are rich in α-pinene, β-bisabolene, *p*-cymene, hexanal, and linalool with QQ activities against fungi (Savoia, 2012). Fruit, leaves, and stems of *Salvadora persica* possess QQ activity, due to the presence and the synergistic effect of some polyphenols, such as gallic, caffeic, and chlorogenic acids, which are particularly strong against the biofilm formed by *S. aureus* and the production of violacein by *C. violaceum* (Noumi et al., 2017). Several plants secrete substances that mimic bacterial AHLs and subsequently affect QS-regulated behaviors in the bacteria associated with these plants (Brackman et al., 2009), *Medicago truncatula* is capable of modulating AhyR, CviR, and LuxR reporter activities in different organisms (Gao et al., 2003). *C. longa*, through the production of curcumin, can block the expression of virulence genes in *P. aeruginosa* PA01 (Rudrappa and Bais, 2008). Extracts from some varieties of apples and their constituents exhibit QSI activity, essentially due to the presence of different polyphenols, such as rutin, epicatechin, and hydroxycinnamic acids, capable of showing a QQ behavior in a synergistic way against *C. violaceum* (Fratianni et al., 2011). QQ activities have also been observed for extracts of *Laurus nobilis*, *Sonchus oleraceus*, *R. officinalis*, *Tecoma capensis*, *Jasminum sambac*, *P. alba*, and *P. nigra*, all capable of breaking down the amount of violacein (Al-Hussaini and Mahasneh, 2009). Flavanones are capable of interfering with the QS system, subsequently affecting the related physiological processes (Truchado et al., 2015). Such molecules can be also capable of significantly reducing the production of pyocyanin and elastase in *P. aeruginosa* without affecting bacterial growth. Flavonoids, such as quercetin, naringenin, apigenin, and kaempferol, inhibit HAI-1- or AI-2-mediated bioluminescence in *V. harveyi* BB886 and MM32. Naringenin and taxifolin also reduce the expression of several QS-controlled genes (i.e., *las*I, *las*R, *rhl*I, *rhl*R, *las*A, *las*B, *phz*A1, and *rhl*A) in *P. aeruginosa* PAO1. Naringenin also markedly reduces the production of the acylhomoserine lactones *N*-(3-oxododecanoyl)-L-homoserine lactone (3-oxo-C12-HSL) and *N*-butanoyl-L-homoserine lactone (C4-HSL), driven by the *las*I and *rhl*I gene products, respectively (Vandeputte et al., 2011). Quercetin, sinensetin, apigenin, and naringenin present antibiofilm formation activity against *V. harveyi* BB120 and *E. coli* O157:H7 (Truchado et al., 2015; Vikram et al., 2010). Flavan-3-ol catechin is capable of reducing the production of QS-mediated virulence factors, such as pyocyanin and elastase, and biofilm formation by *P. aeruginosa* PAO1 (Rasamiravaka et al., 2013; Vandeputte et al., 2010). Biofilm formation by *E. coli* can even be disrupted by grapefruit juice and by rosmarinic acid produced by the roots of *O. basilicum* (sweet basil) (Vattem et al., 2007). Extracts of edible plants and fruits, such as *Musa paradisiaca*, *Manilkara zapota*, *Ananas comosus*, and *O. sanctum*, show QQ activity against violacein production by *C. violaceum* and against pyocyanin pigment, staphylolytic protease, and elastase production in *P. aeruginosa* PAO1, as well as its biofilm formation ability (Musthafa et al., 2010). *Brassica* extracts and constituents can inhibit the expression of QS-associated genes, in that way downregulating the virulence attributes of *E. coli* O157:H7 both in vitro and in vivo; this suggests that they have potentiality as antiinfective agents (Lee et al., 2011). Polyphenol compounds containing a gallic acid moiety (i.e., epigallocatechin gallate, ellagic acid, and tannic acid) can interfere with AHL-mediated signaling through the block of AHL-mediated communication between bacteria (Sarabhai et al., 2013). Ellagitanins, such as punicalagin and ellagic acid, contained mainly in pomegranates and berries, are hydrolyzed in the gut to ellagic acid by the microbiota, and subsequently metabolized to form urolithin-A and urolithin-B. They are capable of decreasing up to 40% the QS-associated processes and decreasing the levels of AHLs produced by the enteropathogen *Y. enterocolitica* (Giménez-Bastida et al., 2012). 4′,5′-*O*-Dicaffeoylquinic acid (4′,5′-ODCQA) can act as a pump inhibitor with a potential of targeting efflux systems in a wide panel of Gram positive human pathogenic bacteria (Fiamegos et al., 2011). Chlorogenic and vanillic acids, as well as rutin, kynurenic acid, and dimethyl-esculetin, can all be used as positive controls for QS inhibition (Leach et al., 2007). Flavanols and proanthocyanidins form a complex with the spores and hyphae of pathogenic fungi of fruit crops (Feucht et al., 2000), and phenolic polymer deposition is related to a decrease in bacterial multiplication rates. Plant extracts are promising instruments for the control of bacterial pathogenesis and microbial modulation. The knowledge of these activities provides more insight into why these plant extracts can be used in the future to combat bacterial infections (Vandeputte et al., 2010) and to best preserve food (Hwa In Lee et al., 2017; Vazquez-Armenta et al., 2017), safeguarding human health.

3 Polyphenols From Marine Organisms With Antimicrobial and QQ Activity

Marine and terrestrial organisms produce secondary metabolites capable of protecting them from bacterial infection. It is unexpected that such molecules also exhibit activity against many pathogens of humans and animals. Marine algal extracts have been used since ancient times in folk medicine and pharmacognosy. Because marine and terrestrial organisms are facing extremely different environmental challenges, their structural characteristics, as well as the resulting pharmacological activity of their metabolites, vary significantly. Some antibacterial compounds available from marine sources, such as algae, are capable of exhibiting better antibacterial efficacy with respect to those from terrestrial sources; this is due to the greater number of bacterial cells in seawater compared with that present in the air and the need for sessile organisms to prevent surface biofouling in the ocean (Bixler and Bhushan, 2012). The search for bioactive compounds from marine organisms in recent decades has produced an abundance of extracts with pharmaceutical and industrial applications. In the last decade, more than one thousand active compounds have been studied; many of them have been demonstrated to support treatments to combat hypertension, high cholesterol, and other diseases, while also exhibiting potential efficiency as antimicrobial, antifungal, anticancer agents. This last aspect is of particular importance: in fact, the dramatic increase of the bacterial resistance to existing antibiotics has necessarily stimulated and induced the study to discover new naturally occurring candidate bioactive molecules from marine sources, which might be capable of acting as antifouling agents and food preservation agents. The sea and ocean represent an incredible array of biodiversity, and organisms such as macro- and microalgae, cnidarians, phytoplankton, mollusks, sponges, corals, tunicates, and bryozoans have been studied to characterize their primary and secondary metabolites, aiming to identify those compounds with antimicrobial activity and capable of blocking or limiting the microbial mechanism of communication. In the course of evolution, some marine organisms, such as micro and macroalgae (such as diatoms and seaweeds), have matured indigenous systems to oppose resistance against pathogenic bacteria and other environmental microorganisms. Thus many of these marine organisms should be considered not only for food purposes or as producing hydrocolloids, but also as having the potential to provide bioactive molecules with healthy activity. Modern screening methods have led to the identification of secondary metabolites with antibacterial property in algal classes such as Rhodophyceae (red), Chlorophyceae (green), Phaeophyceae (brown), Chrysophyceae (golden), and Bacillariophyceae (diatoms; Fig. 3; Shannon and Abu-Ghannam, 2016).

Their functional groups exhibiting antibacterial activity include phlorotannins, peptides, terpenes, fatty acids, polyacetylenes, polysaccharides, indole alkaloids, sterols, aromatic organic acids, shikimic acid, polyketides, hydroquinones, alcohols, aldehydes, ketones, halogenated furanones, alkanes, and alkenes (Shannon and Abu-Ghannam, 2016).

What could be their ecological function in algae is still far from being completely understood. All phenomena accomplishing algae growth and reproduction, as well as their occurrence, could be seen as a factor—probably the main factor—related to the environmental, often adverse, conditions: impediments to algal survival including grazing by herbivores, competition for space from other organisms, thallus injury, biofilm formation, osmotic stress,

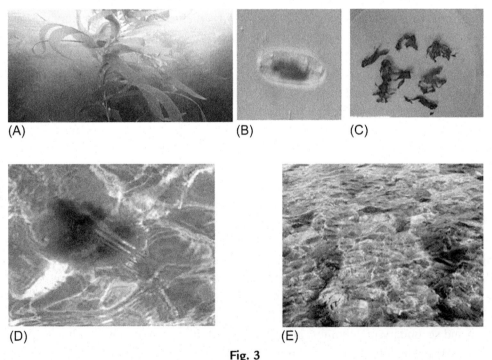

Fig. 3
Different types of seaweed: (A) Green algae; (B) Diatoms; (C) Red algae; (D) Brown algae; (E) Golden algae.

high levels of UV light, oxygen, and salinity. In addition, about 1 million bacterial cells are present in each milliliter of seawater. Some of the chemicals produced by algae to fight such stresses show antibacterial potential; a lot of them have high biological activity due to the diluting effect of seawater and the harsh environment. For instance, ethanolic extracts of the green algae *Ulva fasciata* and red algae *Gracilaria salicornia*, rich in polyphenols, are capable of inhibiting the growth of different pathogens, such as *Enterococcus faecalis*, *Vibrio alginolyticus*, *V. cholerae*, *S. aureus*, *S. typhimurium*, and *E. coli* (Vijayavel and Martinez, 2010).

3.1 Structure, Characterization, and Biological Activity of Phlorotannins

Naturally occurring phenolic compounds found in seaweeds include one or more hydroxyl group, directly bonded to an aromatic hydrocarbon group. For the phenolic compounds to chelate metal ions, a structure with two neighboring OH groups (*o*-diphenol) is required (Hermund, 2018). Their aptitude to react with radicals is related to the number of phenolic rings and catecholic structures (*o*-diphenol; Capitani et al., 2009). Brown algae are the most studied in terms of content and biological activity of phlorotannins, and only few recent studies have explored chemical defenses in red and green macroalgae.

Brown algae, especially *Fucus vesiculosus*, produce high amounts of polyphenolic secondary metabolites, phlorotannins. These polyphenols exhibit the chemical characteristics of tannins; like tannins, they are likely to be bound to proteins and carbohydrates (Hermund, 2018). Owing to their complex structure, phlorotannins have potential as multifunctional natural antioxidants with both primary and secondary antioxidant properties. Phlorotannins are water soluble, oligomeric, or polymeric molecules of phloroglucinol (1,3,5-trihydroxybenzene) and are formed in the acetate-malonate pathway in marine algae, biosynthesized through the acetate-malonate pathway in the Golgi apparatus, in the perinuclear area of the cell. They are highly hydrophilic components, and are stored in in vesicles called *physodes*, where they perform different biological tasks within algal cells and comprise 1%–15%, and in some cases until 25%, of the thallus dry mass (Hermund, 2018).

As they are not normally secreted outside the cell, their release occurs only after cell damaging. The amount of phlorotannins is affected by different factors, such as seaweed size, age, tissue type, nutrient levels, as well as by salinity and light intensity. Water temperature, season, and intensity of herbivory influence their presence too.

In brown seaweeds, this can vary among species, and species of the order *Fucales* are generally richer in such type of compounds (Shibata et al., 2004). Their concentration can reach the maximum temperate in the tropical Atlantic Ocean (where they represent up to 20% of brown seaweed dry mass), and the minimum in tropical Pacific and Indo-Pacific regions. Naturally occurring phlorotannins exhibit a molecular weight ranging from 0.126 to 650 kDa, even if the most common observed range varies between 10 and 100 kDa (Hermund, 2018).

Following the studies of Martínez and Castañeda (2013), phlorotannins can be classified in three main groups, depending on their linkage of phloroglucinol units (PGUs): (1) fucols, (2) phloroethols, and (3) fucophloroethols. The structural diversity and complexity of these molecules increases by increasing the number and linkages (linear or branched or both) of PGUs. The method of extraction of phlorotannins affects both yields and the resulting biological properties, including the antimicrobial one. The so-called *environmentally friendly techniques*, such as supercritical water extraction (SWE; Plaza et al., 2010), can be used to extract phenolic compounds from brown algae, obtaining a high yield of phlorotannins.

Several phlorotannins purified from brown seaweeds, such as *Ecklonia cava*, *E. kurome*, *E. stolonifera*, *Eisenia arborea*, *E. bicyclis*, *Ishige okamurae*, and *Pelvetia siliquosa*, show medicinal and pharmaceutical activities (Eom et al., 2012; Gupta and Abu-Ghannam, 2011).

Sargassum muticum contains mainly fuhalols, hydroxyfuhalols, and phlorethols. Other species, like *Fucus serratus*, *F. vesiculosus*, *Himanthalia elongate*, and *Cystoseira nodicaulis*, contain low molecular weight phlorotannins (corresponding to 4–12 monomers of phloroglucinol with different extents of positional isomerism), recognized as the main components (Panzella and Napolitano, 2017).

E. cava also contains phlorotannins such as 6,60-bieckol, 8,80-bieckol, 8400 o-dieckol, dioxinodehydroeckol, fucodiphlorethol G, phlorofucofuroeckol-A, and triphlorethol-A. Other compounds, such as eckol, phlorofucofuroeckol, dieckol, and 8,80-bieckol, have been isolated from *E. kurome* and *E. bicyclis*; 6,60-bieckol diphlorethohydroxycarmalol and phloroglucinol have been isolated from brown algae *I. okamurae* (Eom et al., 2012). Phlorotannin mixtures, as well as the polymers dieckol, phlorofucofuroeckol A, and 8,8-bieckol, can be capable of strongly inhibiting digestive enzymes from the viscera of the herbivorous turban snail *Turbo cornutus* (Shibata et al., 2004).

Phlorotannins and their derivatives offer a potentially useful source of natural antibacterial agents for food and medical applications.

In general, phlorotannins purified extracts exhibit bacteriostatic activity against both Gram positive and Gram negative strains, with Gram positive bacteria more sensitive respect to Gram negative ones, and this effect tends to increase with the polymerization of phloroglucinol (Nagayama et al., 2002). The physical differences between Gram negative and Gram positive bacteria can be the basis of the behavior of phlorotannins extracts. Gram negative bacteria are surrounded by an external membrane with high lipopolysaccharide content, which can confer to bacteria a greater resistance to several synthetic and natural antibiotics. Lopes et al. (2012) demonstrated that the extracts were most active against among Gram positive bacteria using as tester strains *Staphylococcus,* and that *S. epidermidis* and in a lesser way *S. aureus* were the most susceptible strains, in particular to the phlorotannins extracted from *Fucus spiralis* and *C. nodicaulis*, which show high content in phlorotannins. Such result is of particular meaning:

S. aureus and *S. epidermidis* are more often associated with human infections, *S. aureus* being the major cause of morbidity and mortality, involved also in immunocompromised hosts and in the formation of biofilm, a signal of pathogenicity. The extracts were also active, although in lesser extent, against Gram negative strains. In addition, *C. nodicaulis* was the only seaweed with some ability, in particular, to inhibit the growth of *S. typhimurium*, *Proteus mirabilis*, and *E. coli,* three of the main infectious agents of the gastrointestinal tract, the development of which causes cystitis, kidney, and bladder stones, as well as diarrhea, catheter obstruction due to stone encrustation, acute pyelonephritis, and fever. Some of these strains produce one or more types of toxins that can seriously damage the mucous membranes of the digestive system and kidneys, with dysentery and renal failure, especially in young children or patients with compromised immune systems.

Species belonging to the genus *Cystoseira* and *Fucus* are probably the ones with phlorotannins of higher molecular weight; concurrently, they tend to have higher amounts of phlorotannins with more hydroxyl groups in the free form, and they show lower MICs for all the studied bacteria. Among *Cystoseira* genus, although *C. tamariscifolia* shows the highest phlorotannins amount, its extract is less effective against bacteria than that of *C. nodicaulis* and *C. usneoides*, probably due to the presence of polyphenols with lower molecular weight or fewer hydroxyl groups free to react. The interactions between bacterial proteins and phlorotannins indeed play an important role in the bactericidal action of phlorotannins (Eom et al., 2012).

3.2 Antifungal and QQ Activities Exhibited by Phlorotannins

The cell membrane and cell wall of fungi are the most important targets for antifungal drugs. They are responsible for the communication with the environment and therefore play a key role in metabolic processes. Ergosterol represents the predominant sterol in fungal cell membranes; it is responsible for preserving cell integrity and viability, as well as its function and regular growth. For this reason, the fungal cell wall is a target for antifungals action, which affects the fungal cell wall, as well as the synthesis of chitin and β-glucans, essential cell wall components responsible for fungal structure and normal cell growth. Antifungals can also affect the germ tube formation and adhesion capability of yeasts, and interact with the respiratory chain processes in mitochondria. The inhibition of the germ tube formation of yeast is believed to be the mechanism by which several antifungal compounds reduce the microorganism's virulence, also inhibiting or at least reducing the QS mechanisms of communication (Nazzaro et al., 2017). In fact, by affecting the yeast dimorphic transition, fungistatic compounds reduce microorganisms' adhesion to target epithelial cells, decreasing the progression of infection and making it easier to overcome. Some antifungal compounds also have the capacity to affect the mitochondrial respiratory chain, acting as potential cell growth inhibitors and capable of triggering cell death. Seaweeds are particularly attractive, not only for the abundance of substances with industrial interest, but also for the diversity on secondary metabolites with interesting pharmaceutical properties (Thomas and Kim, 2011). Phlorotannins are particularly interesting because they present important biological activities, without giving toxicity to eukaryotic cells. Phlorotannin extracts seem to be less active against fungi than bacteria. The purified extracts displayed antifungal properties against the fungal strains, *T. rubrum* and *C. albicans*. Under the tested conditions, *Aspergillus fumigatus*, commonly responsible for allergic symptoms associated with inhalation of spores, is, on the contrary, resistant to the action of all extracts. *T. rubrum* is one of the most sensitive to the phlorotannins purified extracts, in particular to the phlorotannins extracted by *F. spiralis* and *C. nodicaulis*, and to those obtained by *C. usneoides*, *S. vulgare*, and *C. tamariscifolia*. The activity of *C. nodicaulis* phlorotannins is remarkable against *C. albicans*, a commensal yeast that generally colonizes the mucosa of the majority of healthy humans without causing tissue damage. However, as opportunistic pathogens, *Candida* species can establish disease in a variety of permissive circumstances: its cells can disseminate from mucosa and gut, being the origin of invasive infections and formation of biofilm at level of catheters. Biofilm-forming capacity greatly increases the potency of *Candida* to convert from the commensal stage into a virulent pathogen (Nazzaro et al., 2017). The mechanism of action of phlorotannins is probably ascribable to the alteration of ergosterol composition of the yeast cell membrane. Phlorotannins seem capable of increasing the activity of mitochondria dehydrogenases, and inhibiting the dimorphic transition of *C. albicans*, leading to the formation of pseudo hyphae with diminished ability to adhere to epithelial cells associated with a decline of *C. albicans* virulence and the ability to attack host cells (Lopes et al., 2013). Few studies have been performed about the QQ activity exhibited by marine seaweed polyphenols. Liu et al. (2014) evaluated the QS inhibitors activity of marine algae polyphenols, using the bacterial model of *C. violaceum* CV026, and identifying and detecting the presence of phenolic compounds. The study, performed on 25 algae, demonstrated that 12 among the 25-polyphenolic extracts possessed QS inhibitors activity. In particular, polyphenols extracted from *Ulva lactuca*, *Sargassum vachellianum*, and *Undaria pinnatifida* significantly reduced violacein production of *C. violaceum* CV026 at the concentration of 1.0 mg/mL, causing a diameter of inhibitory violacein of 25.18, 21.61, and 20.43 mm, respectively. Other extracts, from *S. thunbergii*, *Ectocarpus lyngbye*, and *Spirulina platensis*, also reduced the violacein production, although to a lesser extent. The activity was due to the presence, within the polyphenol pattern, also of catechin and epicatechin. These results suggested that the inhibiting QS system of bacteria using algae polyphenols could lead to the development of new QS inhibitors.

4 Lichens Secondary Metabolites: Generalities, Antimicrobial and Antifungal Activities

Lichens are symbiotic organisms resultant from the association between a fungus, named the mycobiont (typically an ascomycete), photoautotrophic partners, which can be green algae,

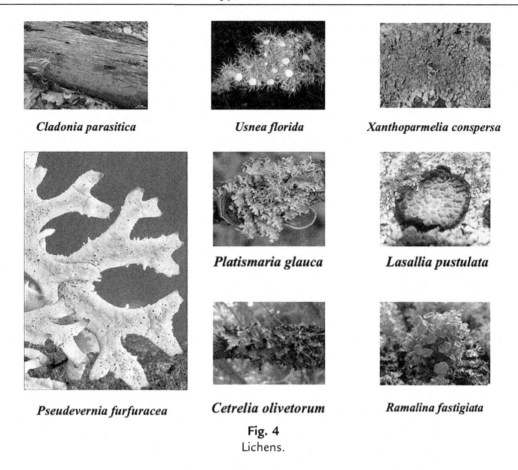

Cladonia parasitica   *Usnea florida*   *Xanthoparmelia conspersa*

Platismaria glauca   *Lasallia pustulata*

Pseudevernia furfuracea   *Cetrelia olivetorum*   *Ramalina fastigiata*

Fig. 4
Lichens.

and/or cyanobacteria, and that are called the photobiont, and a third partner embedded in the lichen cortex, which is identified as a basidiomycete yeast (Fig. 4; Millot et al., 2017)

Lichens are precious plant resources with great versatility of use. They have been known and used since ancient times not only in folk medicine, but also in food, cosmetics, spices, and dyes in several countries in the world.

The lichen thallus represents a support for other microorganisms that live inside and outside the thallus, and that include *endo-* and epi-lichenic fungi and bacteria. Interactions within this complex ecosystem give rise to the synthesis of many molecules, which are involved in the interactions between the community members and influence overall community homeostasis and survival. Lichens have approximately more than 1000 secondary metabolites (Shukla et al., 2010).

Most of them originate from different biosynthetic pathways, which give rise to a variety of compounds. The acetyl-polymalonyl pathway gives rise to secondary aliphatic acids, esters, and related derivatives, and polyketide derived aromatic compounds, synthesized by fungal partners, only if associated with algae (lichen symbiosis), and that play an important role for the success of the lichen association. Through this pathway, different compounds are synthesized, formed by two to three orcinol and β-orcinol type phenolics units joined by different ether, ester, and C—C bonds. These molecules are known essentially both to establish a better interaction between the symbionts and the outer environment and to defend the lichens against the oxidative stress (Shukla et al., 2010). Others compounds, in particular triterpenes such as zeorin, rarely diterpenes, take place from the mevalonic acid pathway. The shikimic acid pathway led to the formation of terphenylquinones and pulvinic acid derivates, occurring mainly in the lichen of the family Stictaceae.

A diverse group of polyketide secondary metabolites includes monocyclic phenols (orsellinic and β-orsellinic acids), bicyclic or tricyclic phenols joined by ester bonds (depsides, tridepsides, and benzyl esters), both ester and ether bonds (depsidones, depsones, and diphenyl ethers), or a furan heterocycle (dibenzofurans, usnic acids and derivates), antraquinones, chromones, naphtaquinones, and xanthones (Nomura et al., 2013). Other phenolics include lichexanthone, diffractaic, divaricatic, perlatolic, psoromic, protocetraric, and norstictic acids (Boustie et al., 2011).

Metabolites from lichens can represent until 20% of the dry weight of thallus, more commonly 5%–10%. Most of them are accumulated as crystals on the outer surfaces of hyphae (extrolites) in the upper cortex or in the inner thallus medulla, and generally, one to two of the cortex "lichen metabolites" are amassed in higher amount, ranging, as previously indicated, from 5% to 20% of the extractable material in the organic solvents.

The cost in terms of energy and carbon source used by lichens to produce their phytochemicals suggests that such compounds might take part in the defense mechanisms lichens use against environmental stresses, especially biotic ones, and to sustain and protect the lichenic association. Thus such a composite community of molecules can represent a potential source for new drugs. The enormous number of compounds isolated from lichens have shown different biologic activities. Several lichenic phytochemicals exhibit antiviral, antioxidant, antitumor, and enzyme inhibiting properties (Gökalsın and Sesal, 2016).

Lichen extracts and their metabolites are known for their antimicrobial and antifungal properties too.

Kosanic and Rankovic (2011) demonstrated the antimicrobial activity of methanolic extracts from the lichens *C. furcata*, *H. physodes*, and *U. polyphylla* against several Gram positive bacteria, such as *Bacillus mycoides*, *B. subtilis*, and *S. aureus*, as well as the Gram negative bacteria *Enterobacter cloacae*, *E. coli*, and *Klebsiella pneumoniae*. The extracts also demonstrated effective activity against *Aspergillus flavus*, *A. fumigatus*, *Botrytis cinerea*, *C. albicans*, *Fusarium oxysporum*, *Mucor mucedo*, *Paecilomyces variotii*, *P. purpurescens*, *P. verrucosum*, and *Trichoderma harzianum*. The antimicrobial activity of the extracts, which showed a concurrent high content in polyphenols, was of particular relevance, taking into account the importance of blocking the growth of some species, such as *S. aureus*, *E. coli*, and *C. albicans*, capable of activating the QS system of cell-to-cell communication that triggers the majority of events leading to the their virulence. Phenolic compounds isolated from lichens, such as orcinol derivatives and ethyl-β-orsellinate hydrolytic derivative from some depsides, isolated from *U. vellea*, *Peltigera leucophlebia*, *Stereocaulon* sp., and *Umbilicaria* sp., exhibited antifungal and antibacterial activity (Boustie et al., 2011).

4.1 QS and Antibiofilm Activity by Lichenic Polyphenols

Lichen extracts and their metabolites have been widely studied for their antimicrobial properties, but their QQ and antibiofilm potential is still poorly explored.

Some common lichens, such as *Platismatia glauca* and *Pseudevernia furfuracea* (Fig. 4), and more particularly their acetone, ethyl acetate, or methanol extracts, demonstrated the capability to inhibit the biofilm capability by some pathogens, such as *S. aureus*, with a biofilm inhibitory concentration ranging from 0.63 to 1.25 mg/mL (Mitrovic et al., 2014). Chang et al. (2012) demonstrated the antibiofilm activity of retigeric acid B, a secondary metabolite isolated from the lichen *Lobaria kurokawae*, inhibiting *C. albicans* yeast-to-hypha transition in infected mice, and by acting synergistically in conjunction with azoles to block not only the fungal growth but also the biofilm formation by this microorganism. Effective activities have been described for curcumin and pyrogallol against *C. albicans* biofilm; on the other hand, purpurin, a natural anthraquinone pigment, has demonstrated an effect against *C. albicans* hyphal formation and biofilm development. In addition, several acetone extracts of lichens (belonging to nine families, principally to Parmeliaceae and Cladoniaceae) are effective against planktonic and sessile *C. albicans* cells (Millot et al., 2017). The control of development of *C. albicans* biofilms is clinically important. *C. albicans* is a well-known commensal colonizer of the human gastrointestinal tract and acts as an opportunistic pathogen, especially through its capacity to form biofilms. This lifestyle is frequently involved in infections and increases the yeast resistance to antimicrobials and immune defenses, because, as it is proven, yeast cells within biofilm are more resistant to traditional antifungal agents and host immune factors, so that the majority of the biofilm-associated *C. albicans* infections are much more difficult to treat.

The biological action of the lichens polyphenols seems different depending also both on their structure and on the amount in the lichenic extracts tested: in fact, some metabolites do not appear to be involved in the antimaturation effect observed. Lichens with a chemical profile dominated by terpenoids (for instance in the genus *Peltigera*) can be mostly inactive. The dibenzofurans placodiolic, pannaric, or didymic acids (predominant in *Leprocaulon microscopicum*, *Lepraria membranacea*, and *Cladonia incrassata*, respectively); the depsides atranorin and thamnolic acid (abundant in *P. furfuracea* and *Cetrelia olivetorum*, and in

Cladonia parasitica, C. squamosa, and *Usnea florida* (Fig. 4), respectively); the depsides divaricatic and perlatolic acids (present in *Neofuscellia pulla*); the tridepsides gyrophoric acid and tenuiorin (abundant in *Lasallia pustulata* and *Peltigera* species); the depsidone norstictic acid (predominant in *Pleurosticta acetabulum*); caperatic acid (predominant in *Flavoparmelia caperata* and *P. glauca*); rocellic acid (*L. membranacea*); and finally the triterpene zeorin (present in *L. microscopicum*) were reported in inactive extracts. Some compounds (usnic acid, fumarprotocetraric, protocetraric, salazinic, and squamatic acids) were found both in active and inactive extracts, and in this case, the biological activity or inactivity of the extracts can be due to the amount of singular metabolites. For instance, a weak activity has been shown for lichenic extracts with high proportions of usnic acid this metabolite. Usnic acid represents 40% of the dried weight of the extract in *U. florida* and 20% of the dried weight of the extract of *F. caperata*; the relative extracts showed poor activity against the development and maturation of *C. albicans* biofilms. In other cases, the activity can be related to other factors, such as to the rate of isomerization. For example, the activity of usnic acid can be associated with its isomerization: in fact, depending on the structure, usnic acid can act against bacterial biofilms (e.g., *Staphylococcus*) as well as against fungal biofilms (e.g., by inhibiting the yeast to hyphal switch and reducing the thickness of mature biofilms, or reducing different sugars present in the exopolysaccharide layer). Furthermore, it can provide a significant biofilm inhibition against azole-resistant and azole-sensitive *C. albicans* strains (71.08% and 87.84%, respectively, Millot et al., 2017; Nithyanand et al., 2015; Pompilio et al., 2016). Some compounds, such as stictic acid, evernic acid, and parietin, are predominant only in active extracts of *Xanthoparmelia conspersa, Ramalina fastigiata,* and *X. parietina*; this suggests their implication in the antimaturation process of *C. albicans* biofilms. In particular, the extracts of *E. prunastri* and *R. fastigiata* possessed the highest activity against both growing biofilms and 24-h-old biofilms. They are dominated by the presence of evernic acid (Culberson, 1963; Ristic et al., 2016). The effect of evernic acid on bacterial biofilm has been evaluated, and this depside has demonstrated a capacity to inhibit *P. aeruginosa* QS systems (Gökalsın and Sesal, 2016). Mitrovic et al. (2014) also observed the activity of stictic acid isolated from *X. conspersa* against *C. albicans* biofilm. In other cases, only a possible synergistic mechanism between usnic acid and salazinic acid, in *Xanthoparmelia tinctina*, could explain the antibiofilm activity of the extract. Another depside, squamatic acid, associated with usnic acid in *C. uncialis* extract, showed a structure similar to that of evernic acid could be involved in antibiofilm activities (Millot et al., 2017). Phenols and dipesides isolated from two species of lichens, belonging to the *Parmeliaceae* family (*P. glauca* and *P. furfuracea*), show antibiofilm activity against *Legionella*, as they can block the mechanisms of cell-cell communication of such bacteria. Lichens also contain other well-known polyphenols with ascertained QQ and antibiofilm activity. Quercetin, isolated from an edible lichen (*Usnea longissima*), has ascertained the capability to induce the sensitivity of *C. albicans* to fluconazole, enhancing its activity at doses nontoxic for health. Quercetin extracted from the lichen is strongly capable of suppressing the production of virulence weapons-biofilm formation, hyphal development, phospholipase, proteinase, esterase, and haemolytic activity, all events related to QS and biofilm formation by the yeast. Its activity could be also related to an increase of farnesol production, known to coregulate hyphal development, biofilm formation, and virulence factor production (Singh et al., 2015).

5 Conclusions

Several studies have clearly provided evidence that phytochemicals from terrestrial and marine plants play a key role as bioactive ingredients and have a vital responsibility in plant as well as human health. The possibilities of designing new pharmaceuticals to support the fight against pathogenic microorganisms and all the virulence phenomena associated with them, such as cellular communication processes, adhesion, and biofilm production, are promising and warrant further exploration.

Glossary

Algae A group of organisms, not belonging to a systematic taxon, with a simple structure, autotrophic, unicellular or multicellular, which produce chemical energy by photosynthesis, generating oxygen, and which do not differ in real tissues.

Antifungine A chemical substance, natural or synthetic, capable of inhibiting the growth of fungal organisms, such as yeasts and molds.

Antimicrobial A chemical substance, natural or synthetic, capable of killing microorganisms or inhibiting their growth.

Diatoms Photosynthetic unicellular algae, ranging in size from 20 to 500 μm.

Gram negative and Gram positive bacteria Those bacteria that remain pink after Gram staining are defined as Gram negative. They are opposite to Gram positive bacteria, which instead remain colored in blue-violet at the beginning of the Gram procedure. This is due to the fact that Gram negative bacteria show a thin cell wall, consisting of no more than 5% of peptidoglycan (unlike Gram positive ones, where peptidoglycan accounts for about 50%–90% of the wall itself), which allows the dye to penetrate and color the cell, allowing the bleach to penetrate and discolor the cell.

Las One of the quorum sensing regulatory systems in *Pseudomonas aeruginosa*, responsible also for a degradation of elastin content in lung tissue human, with subsequent pulmonary hemorrhages associated with *P. aeruginosa* infections.

Lux Set of genes, called the *lux* operon, codifying for the luciferin-luciferase bacterial system in the Gram negative bioluminescent bacterium *Vibrio fischeri*.

***N*-Acyl homoserine lactones (abbreviated as AHLs or *N*-AHLs)** A class of signaling molecules involved in bacterial quorum sensing.

Polyphenols Natural antioxidants present in plants that may be useful in preventing the oxidation of lipoproteins in contrast to free radicals, and with positive effect on heart, senescence, cancer, and in contrast to microbial infections.

Quorum quenching The activity carried out by synthetic or natural compounds to block or limit the QS mechanism, enabling bacteria and fungi to restrict the expression of specific genes involved in the cell-cell mechanisms of communication and pathogenicity.

Quorum sensing A transcriptional regulation system dependent on microbial cell density that is a mechanism that many bacterial cells of the same species use to communicate with each other.

Violacein A natural pigment (MW 343.33) that gives "violet-blueberry" color to violacein-producing bacteria such as *Chromobacterium violaceum* and *Janthinobacterium lividum*. These microorganisms can produce it as an accessory molecule, probably increasing their ability to adapt to and resist adverse environmental conditions. It is used in some tests to evaluate QS activity.

Abbreviations

ABC	ATP-binding cassette
AHLs	*N*-acyl homoserine lactones
AI-1	autoinducer 1
AI-2	autoinducer 2
LAB	lactic acid bacteria
ncRNAs	not-codifying RNAs
PGUs	phloroglucinol units
QQ	quorum quenching
QS	quorum sensing
QSI	quorum sensing inhibitors
QSM	quorum sensing molecule

References

Abachi, S., Lee, S., Rupasinghe, H.P., 2016. Molecular mechanisms of inhibition of *Streptococcus* species by phytochemicals. Molecules 21 (2), 215.

Adonizio, A., Kong, K.F., Mathee, K., 2008. Inhibition of quorum sensing controlled virulence factor production in *Pseudomonas aeruginosa* by South Florida plant extracts. Antimicrob. Agents Chemother. 52, 198–203.

Al-Hussaini, R., Mahasneh, A.R., 2009. Microbial growth and quorum sensing antagonist activities of herbal plants extracts. Molecules 14, 3425–3435.

Al-Refi, M.R., 2016. Antimicrobial, Anti-Biofilm, Anti-Quorum Sensing and Synergistic Effects of Some Medicinal Plants Extracts. Master thesis, The Islamic University, Gaza, pp. 1–150.

Aura, A.M., 2008. Microbial metabolism of dietary phenolic compounds in the colon. Phytochem. Rev. 7, 407–429.

Bacha, K., Tariku, Y., Gebreyesus, F., Zerihun, S., Mohammed, A., Weiland-Bräuer, N., Mulat, M., 2016. Antimicrobial and anti-quorum sensing activities of selected medicinal plants of Ethiopia: implication for development of potent antimicrobial agents. BMC Microbiol. 16 (1), 139.

Bassler, B.L., 2002. Small talk: cell-to-cell communication in bacteria. Cell 109, 421–424.

Bäumler, A.J., Sperandio, V., 2016. Interactions between the microbiota and pathogenic bacteria in the gut. Nature 535, 85–93.

Bialonska, D., Kasimsetty, S.G., Schrader, K.K., Ferreira, D., 2009. The effect of pomegranate (*Punica granatum* L.) by-products and ellagitannins on the growth of human gut bacteria. J. Agric. Food Chem. 57, 8344–8349.

Bixler, G.D., Bhushan, B., 2012. Biofouling: lessons from nature. Philos. Trans. R. Soc. A Math. Phys. Eng. Sci. 370, 2381–2417.

Borriello, S.P., Setchell, K.D.R., Axelson, M., Lawson, A.M., 1985. Production and metabolism of lignans by the human faecal flora. J. Appl. Bacteriol. 58, 37–43.

Boustie, J., Tomasi, S., Grube, M., 2011. Bioactive lichen metabolites: alpine habitats as an untapped source. Phytochem. Rev. 10, 287–307.

Brackman, G., Hillaert, U., van Calenbergh, S., Nelis, H.J., Coenye, T., 2009. Use of quorum sensing inhibitors to interfere with biofilm formation and development in *Burkholderia multivorans* and *Burkholderia cenocepacia*. Res. Microbiol. 160, 144–151.

Capitani, C.D., Carvalho, A.C.L., Rivelli, D.P., Barros, S.B.M., Castro, I.A., 2009. Evaluation of natural and synthetic compounds according to their antioxidant activity using a multivariate approach. Eur. J. Lipid Sci. Technol. 111, 1090–1099.

Cerdá, B., Espin, J.C., Parra, S., Martinez, P., Tomas-Barberan, F.A., 2004. The potent in vitro antioxidant ellagitannins from pomegranate juice are metabolized into bioavailable but poor antioxidant hydroxy-6H-dibenzopyran-6-one derivatives by the colonic microflora in healthy humans. Eur. J. Nutr. 43, 205–220.

Chang, W., Li, Y., Zhang, L., Cheng, A., Liu, Y., Lou, H., 2012. Retigeric acid B enhances the efficacy of azoles combating the virulence and biofilm formation of *Candida albicans*. Biol. Pharm. Bull. 35, 1794–1801.

Chang, C.Y., Krishnan, T., Wang, H., Chen, Y., Yin, W.F., Chong, Y.M., Chan, K.G., 2014. Non-antibiotic quorum sensing inhibitors acting against N-acyl homoserine lactone synthase as druggable target. Sci. Rep. 4, 7245.

Chen, F., Gao, Y., Chen, X., Yu, Z., Li, X., 2013. Quorum quenching enzymes and their application in degrading signal molecules to block quorum sensing-dependent infection. Int. J. Mol. Sci. 14, 17477–17500.

Chernin, L.S., Winson, M.K., Thompson, J.M., Haran, S., Bycroft, B.W., Chet, I., Williams, P., Gordon, S., Stewart, A.B., 1998. Chitinolytic activity in *Chromobacterium violaceum*: substrate analysis and regulation by quorum sensing. J. Bacteriol. 180, 4435–4441.

Culberson, C.F., 1963. The lichen substances of the genus *Evernia*. Phytochemistry 2, 335–340.

Deryabin, D.G., Tolmacheva, A.A., 2015. Antibacterial and anti-quorum sensing molecular composition derived from *Quercus cortex* (oak bark) extract. Molecules 20, 17093–17108.

Dong, Y.H., Wang, L.H., Zhang, L.H., 2007. Quorum quenching microbial infections: mechanisms and implications. Philos. Trans. R. Soc. Lond. B 362 (1483), 1201–1211.

El-Hamid, M.I.A., 2016. A new promising target for plant extracts: inhibition of bacterial quorum sensing. J. Mol. Biol. Biotechnol. 1, 1.

Eom, S.H., Kim, Y.M., Kim, S.K., 2012. Antimicrobial effect of phlorotannins from marine brown algae. Food Chem. Toxicol. 50, 3251–3255.

Espin, J.C., Larrosa, M., Garcia-Conesa, M.T., Tomas Barberan, F., 2013. Biological significance of urolithins, the gut microbial ellagic acid-derived metabolites: the evidence so far. Evid. Based Complement. Alternat. Med. 2013, 27042718.

Feucht, W., Schwalb, P., Zinkernagel, V., 2000. Complexation of fungal structures with monomeric and prooligomeric flavanols. J. Plants Dis. Prot. 107, 106–110.

Fiamegos, Y.C., Panagiotis, L., Kastritis, X., Vassiliki, E., Han, H., Bonvin, A.M.J.J., Vervoort, J., Lewis, K., Hamblin, M.R., Tegos, G.P., 2011. Antimicrobial and efflux pump inhibitory activity of caffeoylquinic acids from *Artemisia absinthium* against gram-positive pathogenic bacteria. PLoS One, 6, e18127.

Fratianni, F., Coppola, R., Sada, A., Mendiola, J., Ibanez, E., Nazzaro, F., 2010. A novel functional probiotic product containing phenolics and anthocyanins. Int. J. Probiotics Prebiotics 5, 85–90.

Fratianni, F., Coppola, R., Nazzaro, F., 2011. Phenolic composition and antimicrobial and antiquorum sensing activity of an ethanolic extract of peels from the apple cultivar Annurca. J. Med. Food 14, 957–963.

Fratianni, F., Nazzaro, F., Marandino, A., Fusco, M.R., Coppola, R., De Feo, V., De Martino, L., 2013. Biochemical composition, antimicrobial activities, and anti-quorum-sensing activities of ethanol and ethyl acetate extracts from *Hypericum connatum* Lam. (*Guttiferae*). J. Med. Food 16, 454–459.

Gao, M., Teplitski, M., Robinson, J.B., Bauer, W.D., 2003. Production of substances by *Medicago truncatula* that affect bacterial quorum sensing. Mol. Plant-Microbe Interact. 16, 827–834.

Giménez-Bastida, J.A., Truchado, P., Larrosa, M., Espín, J.C., Tomás-Barberán, F.A., Allende, A., García-Conesa, M.T., 2012. Urolithins, ellagitannin metabolites produced by colon microbiota, inhibit quorum sensing in *Yersinia enterocolitica*: phenotypic response and associated molecular changes. Food Chem. 132, 1465–1474.

Gökalsın, B., Sesal, N.C., 2016. Lichen secondary metabolite evernic acid as potential quorum sensing inhibitor against *Pseudomonas aeruginosa*. World J. Microbiol. Biotechnol. 32, 150.

Gonthier, M.P., Remesy, C., Scalbert, A., Cheynier, V., Souquet, J.M., Poutanen, K., Aura, A.M., 2006. Microbial metabolism of caffeic acid and its esters chlorogenic and caftaric acids by human faecal microbiota *in vitro*. Biomed. Pharmacother. 60, 536–540.

Granato, D., Koot, A., Schnitzler, E., van Ruth, S.M., 2015. Authentication of geographical origin and crop system of grape juices by phenolic compounds and antioxidant activity using chemometrics. J. Food Sci. 80, C584–C593.

Gupta, S., Abu-Ghannam, N., 2011. Recent developments in the application of seaweeds or seaweed extracts as a means for enhancing the safety and quality attributes of foods. Innov. Food Sci. Emerg. Technol. 12, 600–609.

Hermund, D.B., 2018. Antioxidant properties of seaweed-derived substances. In: Qin, Y. (Ed.), Bioactive Seaweeds for Food Applications. Elsevier, Amsterdam, pp. 201–221.

Hollman, P.C., Katan, M.B., 1999. Dietary flavonoids: intake, health effects and bioavailability. Food Chem. Toxicol. 37, 937–942.

Hornby, J.M., Jensen, E.C., Lisec, A.D., Tasto, J.J., Jahnke, B., Shoemaker, R., Dussault, P., Nickerson, K.W., 2001. Quorum sensing in the dimorphic fungus *Candida albicans* is mediated by farnesol. Appl. Environ. Microbiol. 67, 2982–2992.

Hwa In Lee, S., Barancelli, G.V., Mendes de Camargo, T., Corassin, C.H., Rosim, R.E., Gomes da Cruz, A., Cappato, L.P., Fernandes de Oliveira, C.A., 2017. Biofilm-producing ability of *Listeria monocytogenes* isolates from Brazilian cheese processing plants. Food Res. Int. 91, 88–91.

Karlsson, P.C., Huss, U., Jenner, A., Halliwell, B., Bohlin, L., Rafter, J.J., 2005. Human fecal water inhibits COX-2 in colonic HT-29 cells: role of phenolic compounds. J. Nutr. 135, 2343–2349.

Keppler, K., Humpf, H.U., 2005. Metabolism of anthocyanins and theirphenolic degradation products by the intestinal microflora. Bioorg. Med. Chem. 13, 5195–5205.

Koh, C.L., Sam, C.K., Yin, W.F., Tan, L.Y., Krishnan, T., Chong, Y.M., Chan, K.G., 2013. Plant-derived natural products as sources of anti-quorum sensing compounds. Sensors 13, 6217–6228.

Korukluoglu, M., Gulgor, G., 2017. Anti quorum sensing activity of plants. In: Mendez-Vilas, A. (Ed.), Antimicrobial research: Novel Bioknowledge and Educational Programs. Formatex Research Center, Badajoz, pp. 529–535.

Kosanic, M., Rankovic, B., 2011. Antioxidant and antimicrobial properties of some lichens and their constituents. J. Med. Food 14, 1–7.

Kroon, P.A., Williams, G., 1999. Hydroxycinnamates in plants and food: current and future perspectives. J. Sci. Food Agric. 79, 355–361.

Kuhnle, G., Spencer, J.P., Schroeter, H., Shenoy, B., Debnam, E.S., Srai, S.K., Rice-Evans, C., Hahn, U., 2000. Epicatechin and catechin are O-methylated and glucuronidated in the small intestine. Biochem. Biophys. Res. Commun. 277, 507–512.

Kuipers, O.P., de Ruyter, P.G.G.A., Kleerebezem, M., de Vos, W.M., 1998. Quorum sensing-controlled gene expression in lactic acid bacteria. J. Biotechnol. 64, 15–21.

Leach, J.E., Lloyd, L.A., McGee, J.D., Hilaire, E., Wang, X., Guikema, J.A., 2007. Trafficking of plant defense response compounds. In: Keen, N.T., Mayama, S., Leach, J.E., Tsuyumu, S. (Eds.), Delivery and Perception of Pathogen Signals in Plants. The American Phytopathological Society (APS) Press, St. Paul, MN, pp. 1–268.

Lebeer, S., De Keersmaecker, S.C.J., Verhoeven, T.L.A., Fadda, A.A., Marchal, K., Vanderleyden, J., 2007. Functional analysis of luxS in the probiotic strain *Lactobacillus rhamnosus* GG reveals a central metabolic role important for growth and biofilm formation. J. Bacteriol. 189, 860–871.

Lee, H.C., Jenner, A.M., Low, C.D., Lee, Y.K., 2006. Effect of tea phenolics and their aromatic faecal bacterial metabolites on intestinal microbiota. Res. Microbiol. 157, 876–884.

Lee, K.M., Lim, J., Nam, S., Young Yoon, M., Kwon, Y.K., Jung, B.Y., Park, Y.J., Park, S., Yoon, S.S., 2011. Inhibitory effects of broccoli extract on *Escherichia coli* O157:H7 quorum sensing and *in vivo* virulence. FEMS Microbiol. Lett. 321, 67–74.

Liu, Z.Y., Zeng, H., Zeng, M.Y., 2014. Primary studies on screening of marine algae polyphenols for quorum sensing inhibitor and their activities. J. Food Saf. Qual. 5, 4097–4101.

Lopes, G., Sousa, C., Silva, L.R., Pinto, E., Andrade, P.B., Bernardo, J., Mouga, T., Valentao, P., 2012. Can phlorotannins purified extracts constitute a novel pharmacological alternative for microbial infections with associated inflammatory conditions? PLoS One 7(2), e31145.

Lopes, G., Pinto, E., Andrade, P.B., Valentão, P., 2013. Antifungal activity of phlorotannins against dermatophytes and yeasts: approaches to the mechanism of action and influence on *Candida albicans* virulence factor. PLoS One 8(8), e72203.

Mahmoudi, E., Tarzaban, S., Khodaygan, P., 2014. Dual behaviour of plants against bacterial quorum sensing: inhibition or excitation. J. Plant Pathol. 96, 295–301.

Manach, C., Hubert, J., Llorach, R., Scalbert, A., 2009. The complex links between dietary phytochemicals and human health deciphered by metabolomics. Mol. Nutr. Food Res. 53, 1303–1315.

Martínez, J.H.I., Castañeda, H.G.T., 2013. Preparation and chromatographic analysis of phlorotannins. J. Chromatogr. Sci. 51, 825–838.

Meselhy, M.R., Nakamura, N., Hattori, M., 1997. Biotransformation of (−)-epicatechin-3-O-gallate by human intestinal bacteria. Chem. Pharm. Bull. 45, 888–893.

Millot, M., Girardot, M., Dutreix, L., Mambu, L., Imbert, C., 2017. Antifungal and anti-biofilm activities of acetone lichen extracts against *Candida albicans*. Molecules 22, 651–662.

Mitrovic, T., Stamenkovic, S., Cvetkovic, V., Tošić, S., Stanković, M., Radulovic, N., Mladenovic, M., Stankovic, M., Topuzovic, M., Radojević, I., Stefanović, O., Vasic, S., Comic, L., 2014. *Platismatia glaucia* and *Pseudevernia furfuracea* lichens as sources of antioxidant, antimicrobial and antibiofilm agents. EXCLI J. 13, 938–953.

Musthafa, K.S., Ravi, A.V., Annapoorani, A., Packiavathy, I.S.V., Pandian, S.K., 2010. Evaluation of anti-quorum-sensing activity of edible plants and fruits through inhibition of the N-acyl-homoserine lactone system in *Chromobacterium violaceum* and *Pseudomonas aeruginosa*. Chemotherapy 56, 333–339.

Nagayama, K., Iwamura, Y., Shibata, T., Hirayama, I., Nakamura, T., 2002. Bactericidal activity of phlorotannins from the brown alga *Ecklonia kurome*. Antimicrob. Agents Chemother. 50, 889–893.

Nazzaro, F., Fratianni, F., Coppola, R., 2013. Quorum sensing and phytochemicals. Int. J. Mol. Sci. 14, 12607–12619.

Nazzaro, F., Fratianni, F., d'Acierno, A., Coppola, R., 2015. Gut microbiota and polyphenols: a strict connection enhancing human health. In: Ravishankar Rai, V. (Ed.), Advances in Food Biotechnology, first ed. John Wiley & Sons, Ltd., New York, pp. 335–349

Nazzaro, F., Fratianni, F., Coppola, R., De Feo, V., 2017. Essential oils and antifungal activity. Pharmaceuticals 10, 86.

Nithyanand, P., Beema Shafreen, R.M., Muthamil, S., Karutha Pandian, S., 2015. Usnic acid inhibits biofilm formation and virulent morphological traits of *Candida albicans*. Microbiol. Res. 179, 20–28.

Nomura, H., Isshiki, Y., Sakuda, K., Sakuma, K., Kondo, S., 2013. Effects of oakmoss and its components on biofilm formation of *Legionella pneumophila*. Biol. Pharm. Bull. 36, 833–837.

Noumi, E., Snoussi, M., Merghnia, A., Nazzaro, F., Quindós, G., Akdamar, G., Mastouria, M., Al-Sienid, A., Ceylan, O., 2017. Phytochemical composition, anti-biofilm and anti-quorum sensing potential of fruit, stem and leaves of *Salvadora persica* L. methanolic extracts. Microb. Pathog. 109, 169–176.

Panzella, L., Napolitano, A., 2017. Natural phenol polymers: recent advances in food and health applications. Antioxidants 6, 30.

Parkar, S.G., Stevenson, D.E., Skinner, M.A., 2008. The potential influence of fruit polyphenols on colonic microflora and human gut health. Int. J. Food Microbiol. 124, 295–298.

Pascual-Teresa, S., Sanchez-Ballesta, M.T., 2007. Anthocyanins: from plant to health. Phytochem. Rev. 7, 281–299.

Peleg, A.Y., Hogan, D.A., Mylonakis, E., 2010. Medically important bacterial-fungal interactions. Nat. Rev. Microbiol. 8, 340–349.

Plaza, M., Amigo-Benavent, M., del Castillo, M., Ibáñez, E., Herrero, M., 2010. Facts about the formation of new antioxidants in natural samples after subcritical water extraction. Food Res. Int. 43, 2341–2348.

Pompilio, A., Riviello, A., Crocetta, V., Di Giuseppe, F., Pomponio, S., Sulpizio, M., Di Ilio, C., Angelucci, S., Barone, L., Di Giulio, A., Di Bonaventura, G., 2016. Evaluation of antibacterial and antibiofilm mechanisms by usnic acid against methicillin-resistant *Staphylococcus aureus*. Future Microbiol 11, 1315–1338.

Possemiers, S., Bolca, S., Eeckhaut, E., Depypere, H., Verstraete, W., 2007. Metabolism of isoflavones, lignans and prenylflavonoids by intestinal bacteria: producer phenotyping and relation with intestinal community. FEMS Microbiol. Ecol. 6, 1372–1383.

Quirós Sauceda, A.E., Pacheco-Ordaz, R., Ayala-Zavala, J.F., Mendoza, A.H., González-Córdova, A.F., Vallejo-Galland, B., González-Aguilar, G.A., 2017. Impact of fruit dietary fibers and polyphenols on modulation of the human gut microbiota. In: Yahia, E.M. (Ed.), Fruit and Vegetable Phytochemicals: Chemistry and Human Health. vol. 1. John Wiley & Sons Ltd., pp. 405–422.

Rasamiravaka, T., Jedrzejowski, A., Kiendrebeogo, M., Rajaonson, S., Randriamampionona, D., Rabemanantsoa, S., Andriantsimahavandy, A., Rasamindrakotroka, A., El Jaziri, M., Vandeputte, O.M., 2013. Endemic Malagasy *Dalbergia* species inhibit quorum sensing in *Pseudomonas aeruginosa* PAO1. Microbiology 159 (Pt 5), 924–938.

Rasmussen, T.B., Givskov, M., 2006. Quorum sensing inhibitors: a bargain of effects. Microbiology 152, 895–904.

Reading, N.C., Sperandio, V., 2006. Quorum sensing: the many languages of bacteria. FEMS Microbiol. Lett. 254, 1–11.

Rechner, A.R., Smith, M.A., Kuhnle, G., Gibson, G.R., Debnam, E.A., Srai, S.K.S., Moore, K.P., Rice-Evans, C., 2004. A colonic metabolism of dietary polyphenol: influence of structure on microbial fermentation products. Free Radic. Biol. Med. 36, 212–225.

Ristic, S., Rankovic, B., Kosanić, M., Stamenkovic, S., Stanojković, T., Sovrlić, M., Manojlović, N., 2016. Biopharmaceutical potential of two *Ramalina* lichens and their metabolites. Curr. Pharm. Biotechnol. 17, 651–658.

Rudrappa, T., Bais, H.P., 2008. Curcumin, a known phenolic from *Curcuma longa*, attenuates the virulence of *Pseudomonas aeruginosa* PAO1 in whole plant and animal pathogenicity models. J. Agric. Food Chem. 56, 1955–1962.

Rutherford, S.T., Bassler, B.L., 2012. Bacterial quorum sensing: its role in virulence and possibilities for its control. Cold Spring Harb. Perspect. Med. 2, a012427.

Sarabhai, S., Sharma, P., Capalash, N., 2013. Ellagic acid derivatives from *Terminalia chebula* Retz. Down regulate the expression of quorum sensing genes to attenuate *Pseudomonas aeruginosa* PAO1 virulence. PLoS One 8, e53441.

Savoia, D., 2012. Plant-derived antimicrobial compounds: alternatives to antibiotics. Future Microbiol 7, 979–990.

Scalbert, A., Williamson, G., 2000. Dietary intake and bioavailability of polyphenols. J. Nutr. 130, 2073S–2085S.

Schuster, M., Sexton, D.J., Diggle, S.P., Greenberg, E.P., 2013. Acyl-homoserine lactone quorum sensing: from evolution to application. Annu. Rev. Microbiol. 67, 43–63.

Selma, M.V., Espin, J.C., Tomas-Barberan, F.A., 2009. Interaction between phenolics and gut microbiota: role in human health. J. Agric. Food Chem. 57, 6485–6501.

Setchell, K.D.R., Brown, N.M., Zimmer-Nechemias, L., Brashear, W.T., Wolfe, B., Kirscher, A.S., Heubi, J.E., 2002. Evidence for lack of absorption of soy isoflavone glycosides in humans, supporting the crucial role of intestinal metabolism for bioavailability. Am. J. Clin. Nutr. 76, 447–453.

Shannon, E., Abu-Ghannam, N., 2016. Antibacterial derivatives of marine algae: an overview of pharmacological mechanisms and applications. Mar. Drugs 14 (81), 1–23.

Shibata, T., Kawaguchi, S., Hama, Y., Inagaki, M., Yamaguchi, K., Nakamura, T., 2004. Local and chemical distribution of phlorotannins in brown algae. J. Appl. Phycol. 16, 291–296.

Shukla, V., Bhathena, Z., 2016. Broad spectrum anti-quorum sensing activity of tannin-rich crude extracts of Indian medicinal plants. Scientifica 2016, 1–8.

Shukla, V., Joshi, G.P., Rawat, M., 2010. Lichens as a potential natural source of bioactive compounds: a review. Phytochem. Rev. 9, 303–314.

Singh, B.N., Upreti, D.K., Singh, B.R., Pandey, G., Verma, S., Roy, S., Naqvi, A.H., Rawat, A.K., 2015. Quercetin sensitizes fluconazole-resistant *Candida albicans* to induce apoptotic cell death by modulating quorum sensing. Antimicrob. Agents Chemother. 59, 2153–2168.

Smith, A.H., Mackie, R.I., 2004. Effect of condensed tannins on bacterial diversity and metabolic activity in the rat gastrointestinal tract. Appl. Environ. Microbiol. 70, 1104–1115.

Sturme, M.H.J., Nakayama, J., Molenaar, D., Murakami, Y., Kunugi, R., Fujii, T., Vaughan, E.E., Kleerebezem, M., De Vos, W.M., 2005. An agr-like two-component regulatory system in *Lactobacillus plantarum* is involved in production of a novel cyclic peptide and regulation of adherence. J. Bacteriol. 187, 5224–5235.

Suga, H., Smith, K.M., 2003. Molecular mechanisms of bacterial quorum sensing as a new drug target. Curr. Opin. Chem. Biol. 7, 586–591.

Thomas, N.V., Kim, S.K., 2011. Potential pharmacological applications of polyphenolic derivatives from marine brown algae. Environ. Toxicol. Pharmacol. 32, 325–335.

Thompson, J.A., Oliveira, R.A., Djukovic, A., Ubeda, C., Xavier, K.B., 2015. Manipulation of the quorum-sensing signal AI-2 affects the antibiotic-treated gut microbiota. Cell Rep. 10, 1861–1871.

Tolmacheva, A.A., Rogozhin, E.A., Deryabin, D.G., 2014. Antibacterial and quorum sensing regulatory activities of some traditional Eastern-European medicinal plants. Acta Pharma. 64, 173–186.

Truchado, P., Larrosa, M., Castro-Ibanez, I., Allende, A., 2015. Plant food extracts and phytochemicals: their role as quorum sensing inhibitors. Trends Food Sci. Technol. 43, 189–204.

Van Duynhoven, J., Vaughan, E.E., Jacobs, D.M., Kemperman, R.A., Van Velzen, E.J., Gross, G., Roger, L.C., Possemiers, S., Smilde, A.K., Doré, J., Westerhuis, J.A., Van de Wiele, T., 2011. Metabolic fate of polyphenols in the human superorganism. Proc. Natl. Acad. Sci. U. S. A. 108, 4531–4538.

Vandeputte, O.M., Kiendrebeogo, M., Rajaonson, S., Diallo, B., Mol, A., El Jaziri, M., Braucher, M., 2010. Identification of catechin as one of the flavonoids from *Combretum albiflorum* bark extract that reduces the production of quorum-sensing-controlled virulence factors in *Pseudomonas aeruginosa* PAO1. Appl. Environ. Microbiol. 76, 243–253.

Vandeputte, O.M., Kiendrebeogo, M., Rasamiravaka, T., Stévigny, C., Duez, P., Rajaonson, S., Diallo, B., Mol, A., Baucher, M., El Jazir, M., 2011. The flavanone naringenin reduces the production of quorum sensing-controlled virulence factors in *Pseudomonas aeruginosa* PAO1. Microbiology 157, 2120–2132.

Vattem, D.A., Mihalik, K., Crixell, S.H., McClean, R., C, J., 2007. Dietary phytochemicals as quorum sensing inhibitors. Fitoterapia 78, 302–310.

Vazquez-Armenta, F.J., Bernal-Mercado, A.T., Lizardi-Mendoza, J., Silva-Espinoza, B.A., Cruz-Valenzuela, M.R., Gonzalez-Aguilar, G.A., Nazzaro, F., Fratianni, F., Ayala-Zavala, J.F., 2017. Phenolic extracts from grape stems inhibit *Listeria monocytogenes* motility and adhesion to food contact surfaces. J. Adhes. Sci. Technol. 32, 1–19.

Vijayavel, K., Martinez, J.A., 2010. *In vitro* antioxidant and antimicrobial activities of two Hawaiian marine Limu: *Ulva fasciata* (Chlorophyta) and *Gracilaria salicornia* (Rhodophyta). J. Med. Food 13, 1494–1499.

Vikram, A., Jayaprakasha, G.K., Jesudhasan, P.R., Pillai, S.D., Patil, B.S., 2010. Suppression of bacterial cell-cell signalling, biofilm formation and type III secretion system by citrus flavonoids. J. Appl. Microbiol. 10, 515–527.

Walle, T., Hsieh, F., Delegge, M.H., Oatis, J.E., Walle, U.K., 2004. High absorption but very low bioavailability of oral resveratrol in humans. Drug Metab. Dispos. 32, 1377–1382.

Whiteley, M., Diggle, S.P., Greenberg, E.P., 2017. Progress in and promise of bacterial quorum sensing research. Nature 551, 313–320.

Yang, K., Duley, M.L., Zhu, J., 2018. Metabolomics study reveals enhanced inhibition and metabolic dysregulation in *Escherichia coli* induced by *Lactobacillus acidophilus*-fermented black tea extract. J. Agric. Food Chem. 66, 1386–1393.

PART 4

Applications

CHAPTER 9

Pseudomonas aeruginosa *Quorum Sensing and Biofilm Inhibition*

Barış Gökalsın[*,a], Didem Berber[†,a], Nüzhet Cenk Sesal[†,a]
[*]*Marmara University, Department of Biology, Institute of Pure and Applied Sciences, Istanbul, Turkey,*
[†]*Marmara University, Department of Biology, Faculty of Arts and Sciences, Istanbul, Turkey*

1 Introduction

Antibiotics have been used against infections widely since their discovery. However, misuse of these drugs resulted in the adaptation of microorganisms by gaining antibiotic-resistance. There are major concerns that antibiotics will become dysfunctional within a short period of time, prevention and treatment of infections will become impossible, and classical infections will be resurrected as one of the major causes of mortality (Van Hecke et al., 2017). The World Health Organization reports that we are in a shortage of treatment options against antibiotic-resistant bacteria, and an international coordinated effort is essential to overcome this situation (WHO, 2017). Laxminarayan et al. (2016) estimates that an average of 214,000 neonatal sepsis deaths are caused by antibiotic-resistant pathogens. It is also reported that approximately 23,000 people die every year in the United States due to antibiotic-resistant infections (CDC, 2017). Hospitalized patients suffer the consequences of antibiotic-resistant bacteria directly, and among these bacteria, *Pseudomonas aeruginosa* can be difficult to treat due to multidrug resistance.

P. aeruginosa is a Gram negative opportunistic pathogen that is responsible for approximately 10% of all nosocomial infections (Diekema et al., 1999). *P. aeruginosa* can cause infections in lungs, bloodstream, urinary tract, and surgical sites. Although it mostly causes infections in immunocompromised patients, healthy people can also get infected. Multidrug resistant (MDR) *P. aeruginosa* is presented as the main cause of high mortality rates in cystic fibrosis (CF) patients. CF is a disease that is encountered frequently and affects approximately 70,000 people in the world, known to degrade lung functions by affecting the respiratory system (Rivas Caldas and Boisrame, 2015). With concentrated treatments, CF patients can have a life

---

[a] All authors contributed equally to this work.

Quorum Sensing. https://doi.org/10.1016/B978-0-12-814905-8.00009-5

expectancy of 35–50 years. Left untreated, many CF patients can die at a young age. MDR biofilm forms make the disease very hard to treat with antibiotics.

P. aeruginosa colonies are known to produce a polysaccharide matrix and attach to surfaces when they reach a certain population density, in a macroscopically seen and hard to remove biofilm form. It is reported that this biofilm form is up to 1000–3000 times more resistant to antibiotics compared with the same species in planktonic form (Olson et al., 2002). Biofilm form provides many advantages, such as amassing nutrition and protecting the microorganisms against disinfectants, antibiotics, UV light, pH, moisture, and heat fluctuations, organisms that feed on bacteria and viruses (Hall-Stoodley et al., 2004).

The evident response against *P. aeruginosa* infections is to develop new treatment methods in addition to the efforts of preventing infections and reorganizing antibiotic usages. The antivirulence approach is the leading progression toward this goal. Antivirulence does not aim to exterminate pathogens directly but to alternately inhibit virulence (Fuqua and Greenberg, 2002). Accordingly, specific biological information is needed for the targets of antivirulence methods. There are many recognized virulence factors, such as bacterial toxins, surface proteins, immunoevasion factors, and adhesins. It is now understood that most of these factors in addition to numerous behaviors are controlled by quorum sensing (QS) systems, including bioluminescence, swarming, conjugation, protease activity, and also biofilm formation.

2 Quorum Sensing and Quorum Quenching

It has been established that planktonic cells cannot exist freely in environments because they have to compete with other coexisting organisms and survive in extreme conditions. Therefore, they communicate via QS. Higher levels of bacterial population density trigger QS system for intraspecies, interspecies or interkingdom interactions. Small and diffusible chemical signaling molecules called autoinducers (AI) are secreted into bacterial local milieu. Three types of main AIs were identified. AI-1, called also AHLs (*N*-acylated L-homoserine lactones), are utilized by Gram negative bacteria. Autoinducing peptides (AIP) are used by Gram positive bacteria and autoinducer-2 (AI-2) by both Gram positive and negative bacteria for interspecies interaction (Rutherford and Bassler, 2012). The QS system consists of five elements that are responsible for QS regulation: AI molecules, signal synthase enzymes, receptors, regulators, and genes. Most pathogenic bacteria orchestrate their virulence factors, biofilm formation, and antibiotic resistance by QS systems (Li and Tian, 2012).

The term *quorum sensing* was firstly introduced in 1970 in *Vibrio fischeri* and *Vibrio harveyi*, which are known marine bacteria with a characteristic luminescent feature. Through the study of the bioluminescence of these bacteria, the details of QS systems were revealed. QS has been defined as a sensory system of bacteria to detect the environmental changes with regard to population density in their surrounding milieu. The luminescent characteristics of the aforementioned bacteria can be easily monitored after the lag phase of bacterial growth. To reach the maximum luminescence in early log-phase cultures as in the stationary phase, cell-free supernatant from the stationary phase can be added to the medium. AI molecules are small diffusible compounds that are released to the surrounding environment of the bacteria. The concentration of these signaling molecules remains at lesser levels in low cell densities, whereas it accumulates up to a threshold of concentration at higher cell densities.

The first identified AI of *V. fischeri* (VAI) was *N*-3-(oxohexanoyl)-homoserine lactone (3OC6HSL) (Eberhard et al., 1981). It has been well established that the biosynthesis of VAI is encoded by *luxI* gene in a positive feedback mechanism. The increase in bacterial population density leads to the accumulation of VAI molecules at higher concentrations in the surrounding milieu. The interaction between these signal molecules and LuxR gene leads to transcription of the *luxICDABE* operon. Then, the luciferase is encoded by the *luxCDABE* operon, and QS activation becomes visible through luminescence. On the other hand, lower concentrations of VAI due to low cell densities are insufficient to activate the *luxR* and *luxI* genes, and luciferase is not encoded. This information proves convenient for several studies focused on the QS mechanism.

It is understood that AHL molecules only function as QS signals when the bacterial cell population density increases and certain threshold levels of AHL are obtained. For the regulation of the QS system, the synthesis of AHLs via AHL synthase and the accumulation of these signals at higher concentrations due to population density are needed. Therefore, QS systems can be evaluated by AHL monitorization. AHL accumulation depends on physical and chemical factors in bacterial communities. For example, AHL molecules can mostly diffuse across the membranes.

Several pathogenic bacteria such as *P. aeruginosa* cause high mortality and morbidity rates despite high doses of antibiotics, especially in immunocompromised patients (Borges et al., 2016). These pathogenic bacteria develop various ways to avoid bactericidal and bacteriostatic effects of antibiotics, such as horizontal gene transfer and spontaneous mutations (Kalia, 2013). Also, the failure of antibiotic treatments occurs due to group transfer, redox mechanism, or enzymatic hydrolysis (Hentzer and Givskov, 2003). Therefore, recent studies have focused on developing alternative strategies to prevent bacterial resistance against antibiotics by the disruption of bacterial QS. These anti-QS approaches are called *quorum quenching* (QQ; Hentzer and Givskov, 2003). Several natural or synthetic QS inhibitor (QSI) compounds from plants, animals, fungi, bacteria, and algae have been discovered and examined for their QSI potentials. Moreover, technological advances in computational sciences and in silico methods allow a rapid screening of these compounds. Their biological and therapeutic effects have been reported by many studies, as explained in this chapter.

2.1 Quorum Sensing Systems and Biofilm in P. aeruginosa

P. aeruginosa QS systems regulate crucial functions such as virulence, motility, biofilm formation, and secondary metabolites production. Like other Gram negative bacteria,

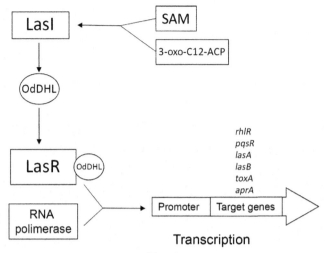

Fig. 1

A basic diagram of *las* signaling network in *P. aeruginosa*. The *las* system detects 3-oxo-C12-HSL produced by signal synthase LasI, via the transcriptional regulator LasR, and regulates virulence factors via two-component signal transduction system.

P. aeruginosa also utilizes AHLs for its main QS systems. It is known that *P. aeruginosa* has four hierarchically connected QS systems for interspecies communication: *las, rhl, pqs,* and *iqs* (Daniels et al., 2004; Lee and Zhang, 2015).

The *las* system consists of the transcriptional regulator LasR, signal synthase LasI, and the autoinducer *N*-3-oxododecanoyl-homoserine lactone (3-oxo-C12-HSL, OdDHL), as shown in Fig. 1. The *rhl* system similarly has RhlR, RhlI, and the autoinducer *N*-butanoyl-homoserine lactone (C4-HSL, BHL). The *pqs* system has PqsR as its regulator, *pqsABCDE-phnAB* operon, and PqsH for signal synthesis, and 2-alkyl-4-quinolones (AQs) as its signal molecules, including 2-heptyl-4- hydroxyquinolone (HHQ) and the 2-heptyl-3-hydroxy-4(1H)-quinolone named pseudomonas quinolone signal (PQS). Finally, the recently discovered *iqs* system utilizes 2-(2-hydroxyphenyl)-thiazole-4-carbaldehyde as its signal and is related to environmental stress.

The main QS systems *las, rhl,* and *pqs* regulate the production of many virulence factors, such as elastase, exotoxin A, rhamnolipids, pyocyanin, lipase, pyoverdine, lectins, and so on. It is also thought that the efflux pump system of MDR bacteria is related to a *pqs* system. Among these three systems, *las* governs other systems hierarchically, while *pqs* seems to mediate between *las* and *rhl* systems while regulating some virulence factors (Lee and Zhang, 2015). However, the hierarchy of QS systems in *P. aeruginosa* may change and adapt according to the environmental stress. For example, an *iqs* system can take over *las* functions in the case of severe phosphate depletion stress, or the *pqs* system can activate without the *las* system. Therefore, the virulence pathways can change according to environmental conditions and stress. Other regulators such as QscR and RsaL can inhibit signal production, maintaining a balance for this complex overall signal mechanisms.

QS is known to be involved in the biofilm formation of *P. aeruginosa*. QS mutant strains form flat and weak biofilms compared with wild type strains, although it is important to consider the culturing conditions. It is accepted that QS plays a role in biofilm formation, but some of the proposed mechanisms remain debatable.

The first step in biofilm formation is the attachment of bacteria to a surface. Flagellar motility (type IV pili) and adhesins are important factors in this stage. After irreversible attachment, microcolonies are formed, increasing QS communication. The maturation of biofilm starts accordingly. The 3D structure of mature biofilm varies according to environmental conditions, as well as the amount of factors produced by the bacteria. These factors are exopolysaccharides (EPS) like alginate, structural DNA, iron chelator pyoverdine, and surfactants like rhamnolipid, most of which are directly controlled by QS metabolism. The right amount of rhamnolipid secretion is crucial for a mature biofilm and the next step: dispersion. Overexpression of rhamnolipid causes the biofilm to disperse, thus allowing the bacteria to colonize other surfaces (Boyle et al., 2013).

The role of *P. aeruginosa* QS systems on biofilm formation during these stages seems obvious. However, there have been contradictory and diverse results and opinions. QS-biofilm relation is usually studied by using flow-cell systems. These systems have small channels through which media is circulated constantly. Bacteria strains form biofilm structures in these channels, and they are monitored by confocal laser microscopes. The strains are QS-deficient mutants that are compared with their wild-type counterparts. The varying results of these studies concerning QS relation to biofilm formation lead to the common opinion that culturing and environmental conditions have significant impacts on biofilm structures. However, it is known that QS systems have important effects on biofilm formation, as previously explained, and their inhibition seems a reasonable approach from a therapeutic point of view (Joo and Otto, 2012).

2.2 Screening for Quorum Sensing Inhibitors

QSI molecules have to be efficient, stable, and practicable compounds with low-molecular weight and high specificity for signal regulators. It is important not to cause adverse effects for bacteria and the host. Furthermore, these compounds must not be affected by hydrolytic enzymes of the host. On the other hand, some compounds bind to receptors and activate them acting as agonists and cause an upregulation in QS-related genes. QSIs are preferred to display antagonistic effects on their inhibition targets.

Quorum quenching (QQ) is a general term which is used for all processes targeting to inhibit QS system. The purpose of QQ approaches is to disrupt bacterial communication without killing bacteria or preventing their growth. There are several targets for QS inhibition. A summary of QQ approaches for *P. aeruginosa* are shown in Fig. 2.

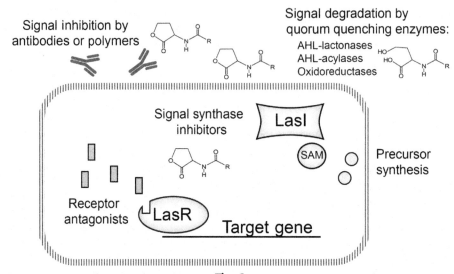

Fig. 2

QQ approaches and QSIs for *P. aeruginosa las* system. These approaches mainly focus on autoinducer signal synthesis inhibition including the synthesis of their precursors, inhibiting or degrading signal molecules and blocking the detection of signal molecules.

2.2.1 Inhibition of Quorum Sensing Signal Biosynthesis

Inhibition of QS signal biosynthesis is one of the QQ approaches used against *P. aeruginosa*. In Gram negative bacteria, enoyl-acyl-carrier-protein (ACP) reductase (ENR) and *S*-adenoysl methionine (SAM) may be targeted for *N*-Acyl homoserine lactone (AHL) synthesis (Dong et al., 2007). In AI-2 QS inhibition, the synthesis of 4,5-dihydroxy 2,3 pentanedione (DPD), which is formed from cleavage of *S*-ribosyl-L-homocysteine SHR by LuxS enzyme, can also be a QQ target (Galloway et al., 2011).

Since bacterial antibiotic-resistance is a major global health care problem, *P. aeruginosa* QS systems have been studied in detail. Most pathogen bacteria such *as P. aeruginosa* coordinate its pathogenesis via QS system. The genes responsible for AHL synthesis and accumulation have been targeted as an alternative approach, and the potential of AHL synthase as an antimicrobial target has been revealed. For this purpose, repressor genes have been used to decrease the transcription of *luxI* homologues. *dksA,* a suppressor gene for *rhlI* from *P. aeruginosa*, has been isolated by Branny et al. (2001). The *rhlI* gene is responsible for the transcription of C4-HSL synthase. On the other hand, the repressor gene *qscR* targets the *lasI* gene and modulates the synthesis of QS signals, along with the virulence factors in *P. aeruginosa*. It was reported that the mutant *qscR* gene results in premature signals and also premature transcriptions in the following steps. Other QS signal synthesis repressor genes are known, like *rsaL* in *P. aeruginosa*.

NADH-dependent ENR (FabI) is responsible for the synthesis of acyl-ACP and also the formation of acyl chains for AHLs. Despite the fact that triclosan, an antibacterial agent, can inhibit FabI enzyme production and C_4HSL, it was reported that the resistance of *P. aeruginosa* cannot be prevented by triclosan due to efflux pump system (LaSarre and Federle, 2013). The relationship between efflux pump systems and QS is a compelling topic in this regard, and further investigations may reveal intriguing prospects.

2.2.2 QS Signal Degradation and Inactivation

Degradation of QS signals may be achieved chemically, metabolically, or enzymatically. In chemical degradation, alkaline pH values cause the lactone ring to open, whereas acidic pH values lead to recyclization (Rasmussen and Givskov, 2006). However, QS signal degradation is mostly handled by enzymes, and inactivation can occur using antibodies.

2.2.3 Inhibition of Signal Detection

The inhibition of signal molecule detection may be achieved by competing antagonist molecules by binding to the receptor before the signal molecules. Inactivation of the signal receptors provides an inhibition in virulence factor expression. Most natural QSIs are known to show activity against *P. aeruginosa* through the inhibition of LasR, RhlR, and PqsR.

2.3 Natural Quorum Sensing Inhibitors

Natural QSIs are mostly compounds, extracts, enzymes, and antibodies obtained from natural sources.

2.3.1 Quorum Quenching Enzymes

AHL-lactonases, acylases, oxidoreductases, paraxonases, and 2,4-dioxygenase (Hod) are reported to be QQ enzymes.

AHL-Lactonases

AHL lactonases are involved in the group of metalloproteins, and they form acyl homoserine by hydrolyzing ester bond of the HSL ring. This group of enzymes shows significant specificity for AHL molecules due to a highly conserved HSL ring (LaSarre and Federle, 2013). The autoinducer inactivation gene (AiiA), firstly described lactonase enzyme, has been discovered in *Bacillus genus*—*B. anthracis, B. cereus, B. mycoides, B. subtilis, B. thuringiensis, Arthrobacter* spp., *Acidobacteria, Agrobacterium* spp., and *Klebsiella* spp., and demonstrated to degrade AHL (Kalia and Purohit, 2011).

It was also reported that AiiA alleles in *P. aeruginosa* and *Bacillus thailandensis* inhibited the aggregation of AHLs. AttM, found in the plant pathogen *Agrobacterium tumefaciens*, exhibits low levels of similarity to *AiiA*, despite the same conserved HXDH motif. Also, AiiB, AhlD,

AhlK, AidC, QlcA, BipB01, BipB04, BipB05, BipB07, QsdA, AiiM, AidH, and QsdH have been reported to act as lactonase enzymes. They differ in their DNA sequence and dependence for metal ions (LaSarre and Federle, 2013).

Mammalian enzymes, paraoxonases 1, 2, and 3 (PON1, PON2, PON3), exhibit lactonase activity. Hraiech et al. (2014) discovered a new variant molecule for *Sso*Pox (hyperthermostable lactonase). They demonstrated that *Sso*Pox-I (phosphotriesterase-like lactonase) had a QQ potential in a rat model of acute pneumonia, and the survival rates were increased by the intratracheal application of *Sso*Pox-I. Also, *lasB* activity, pyocianin synthesis, proteolytic activity, biofilm formation, and lung tissue damage were reduced.

Recently, Tang et al. (2015) reported that protease and pyocyanin production by *P. aeruginosa* was inhibited by MomL, newly identified AHL lactonase from *Muricauda olearia*. The researchers evaluated the effects of MomL on *P. aeruginosa* virulence in a *Caenorhabditis elegans* model, and virulence inhibitions were observed.

Acylases

The amide bond between HSL and acyl side chain are cleaved by this group of enzymes. After this cleavage, a fatty acid chain and an HSL moiety are formed. The specificity of these enzymes are reported to be the length of acyl chain and substitution on the third position of the chain (LaSarre and Federle, 2013).

Pseudomonas have an acylase activity for its own AHL (Grandclement et al., 2016). The amidase encoding genes called *pvdQ* (PA2385), *quiP* (PA1032), and *hacB* (PA0305) were revealed in *P. aeruginosa* (Huang et al., 2006; Wahjudi et al., 2011; Sio et al., 2006).

The acylase enzyme from *Aspergillus melleus* was immobilized into polyurethane coatings by Grover et al. (2016). This immobilized enzyme inhibited the biofilm formation and also pyocyanin production of *P. aeruginosa* ATCC 10145 and PAO1 strains. Sunder et al. (2017) investigated the effects of penicillin V acylases (PVAs) against *Pectobacterium atrosepticum* and *A. tumefaciens*. Researchers transferred these enzymes to *P. aeruginosa*, and reported an inhibition in virulence factors and biofilm formation, and also an increase in survival rate in an insect model of acute infection.

Paraoxonases

Paraoxonases are mammalian enzymes acting as QQ enzymes. There are three types of paraoxonases (PON1, PON2, and PON3). PONs have been described as lactonase-like enzymes regarding the disruption of QS system (Grandclement et al., 2016). Stoltz et al. (2007) reported that overexpressed PON2 caused $3OC_{12}$-HSL degradation in murine tracheal epithelial cells.

Recently, SsoPox-W263I was tested in *P. aeruginosa* strains obtained from patients with diabetic foot ulcers. An inhibition was reported in virulence factors (proteases and pyocyanin production). The researchers indicated that SsoPox-W263I was more efficient when compared with 5-fluorouracil and C-30 (Guendouze et al., 2017). Moreover, human serum paraoxonase 1 (hPON1) displayed a reduction in pyocyanin, rhamnolipid, elastase, staphylolytic LasA protease, and alkaline protease activities (Aybey and Demirkan, 2016).

Oxidoreductases

Another group of enzymes, the oxidoreductases, oxidize or reduce the acyl side chain of AHL without degradation. These enzymes function as QSIs by modifying the C_3 keto group of the fatty acid side of AHL molecules. The inactivation AHL-mediated biofilm formation of *P. aeruginosa* by BpiB09, NADP-dependent oxidoreductase from a metagenome-derived clone is also reported (Weiland-Brauer et al., 2016).

2,4-Dioxygenase (Hod)

Hod, another QQ enzyme, causes noctanoylanthranilic acid and carbon monoxide to be formed from PQS. Pustelny et al. (2009) reported that exogenous addition of heterocyclic-ring-cleaving enzyme Hod from *Arthrobacter* sp. Rue61a into *P. aeruginosa* cultures inhibits key virulence factors and tissue damage in a plant leaf infection model.

2.3.2 Antibodies

QQ antibodies mainly target HSLs in *P. aeruginosa* for signal inactivation. On the other hand, other factors playing a role in signal synthesis can also be targeted. Immunopharmacotherapeutic approaches employing monoclonal or polyclonal antibodies have been investigated to attenuate QS, controlling virulence regulation and biofilm formation, as shown in Table 1.

Table 1 Quorum quenching antibodies

Antibody	Study	Target	References
3-oxo-C12-HSL-bovine serum albumin (BSA) conjugate	3-oxo-C12-HSL-carrier protein conjugate against pulmonary tumour necrosis factor (TNF)-alpha and apoptosis in macrophages	*P. aeruginosa* C12HSL	Miyairi et al. (2006)
HSL-2 and HSL-4 monoclonal antibodies (mAbs)	MAbs	*P. aeruginosa* HSLs	Palliyil et al. (2014)
MAbs	MAbs targeting DNA-binding tips of DNABII proteins	Biofilm disruption of *P. aeruginosa*, MAbs-antibiotic therapy	Novotny et al. (2016)
RS2-1G9	Monoclonal antibody	*P. aeruginosa* C12HSL	Kaufmann et al. (2006)
XYD-11G2	Monoclonal antibody	*P. aeruginosa* C12HSL	De Lamo Marin et al. (2007)

Miyairi et al. (2006) investigated the effects of active immunization with OdDHL-carrier protein conjugate on mice with *P. aeruginosa* lung infections. Their results show that mice with specific antibodies to OdDHL in serum have a higher survival rate against *P. aeruginosa* infections. Immune serum also increased the cell viability of OdDHL-induced apoptosis in macrophages. Palliyil et al. (2014) used sheep immunization and recombinant antibody technology to generate monoclonal antibodies (MAbs) and detect *P. aeruginosa* HSL molecules. They achieved nanomolar sensitivity to detect HSLs in urine. They used nematode and mouse models to compare survival rates of infected groups treated with HSL MAbs with control groups, presenting a significant increase. These studies present that antibodies against QS signal molecules can be a viable approach for supplemental treatment methods.

2.3.3 Natural Quorum Sensing Inhibitor Compounds and Extracts

Natural QSIs are produced by several organisms, such as bacteria, algae, animals, plants, or fungi, and their QS inhibitory effects have been demonstrated by many studies. These inhibitors exhibit high diversity in their biochemical structure. Unfortunately, there is a lack of excessive information about the functions of molecular structures or chemical groups of the QSIs on QS-mediated pathways (LaSarre and Federle, 2013).

Bacteria

Some potential QSIs have been reported from the members of various phyla of bacteria such as *Actinobacteria, Firmicutes, Cyanobacteria, Bacteroidetes, and Proteobacteria*. It is not uncommon to isolate bacteria from other organisms and study their QSI activities. This approach seems plausible, considering the competitive environment. A number of bacteria show their QSI activities via enzymes as previously explained. Known AHL lactonase enzymes with QSI activity are MomL, *hqiA,* AiiA$_{AI96}$, *Sso*Pox, SsoPox-W263I, AiiM, AttM, AiiB, Ahl, AhlD, MLR6805, DlhR, Qsd, AidC, AhlS, Aii20J, QsdA, GKL, MCP, AidH, AiiO, QsdH, QlcA, BpiB01, BpiB04, BpiB05, and BpiB07. MacQ, penicillin V acylases, penicillin G acylase KcPGA, AiiD, PvdQ QuiP HacA, HacB, AibP, AhlM, AiiC, and Aac are acylases. LsrK and LsrG are inhibitors for AI-2 mediated QS. Hod and CarAB targets PQS and DSF pathways, respectively. CYP102A1 is an oxidoreductase enzyme (LaSarre and Federle, 2013; Grandclement et al., 2016).

Devaraj et al. (2017) reported the QSI potentials of 147 soil actinobacterial strains against swarming motility and pyocyanin production in *P. aeruginosa* PAO1, with positive results. They observed that three actinobacterial strains (*Micromonospora, Rhodococcus,* and *Streptomyces*) inhibited violacein production of *Chromobacterium violaceum* CV026, and also swarming and pyocyanin production of *P. aeruginosa* PAO1. Yaniv et al. (2017) tested the QSI potentials of library clones including 2500 bacterial artificial chromosomes (BAC) from the Red Sea planktonic microbiome against the indicator organism, *C. violaceum*. They found that an active compound, 14-A5, can inhibit the QS pathways, and it can also reduce biofilm formation in *P. aeruginosa*.

Chang et al. (2017) screened the marine bacterial strains from the surface waters of the North Atlantic Ocean and evaluated for their anti-QS potentials. *Rhizobium* sp. NAO1 extracts were found to have AHL-based QS analogue molecules. They observed that *Rhizobium* sp. NAO1 has inhibitory effects on QS system and also biofilm formation on *C. violaceum* and *P. aeruginosa* PAO1, respectively. They also studied the virulence factors such as siderophores and elastase activity of *P. aeruginosa* PAO1, and certain inhibitions were noted. The researchers detected an increased susceptibility to aminoglycoside antibiotics when applied with secondary metabolite products of *Rhizobium* sp. NAO1 due to the inhibition of *P. aeruginosa* biofilm formation.

An ethyl acetate (EtOAc) extract of *Rheinheimera aquimaris* QSI02 was tested for QQ activity against *C. violaceum* CV026. They utilized bioassay-guided isolation protocol and detected an active diketopiperazine factor, cyclo (Trp-Ser) from *R. aquimaris*. The diketopiperazine factor was shown to suppress production of pyocyanin, elastase activity, and biofilm formation in *P. aeruginosa* PAO1 (Sun et al., 2016).

Muller et al. (2014) isolated and identified *Rhodococcus erythropolis* as a PQS-degrading bacterium from soil. They reported that *R. erythropolis* strain BG43 had a potential to degrade HHQ and PQS, which are QS molecules of *P. aeruginosa*, into anthranilic acid and also can transform 2-heptyl-4-hydroxyquinoline-*N*-oxide to PQS. Two sets of PQS-inducible genes, responsible for encoding enzymes in pathways of HHQ hydroxylation to PQS and the degradation of PQS to anthranilate, were identified on a plasmid pRLCBG43 of strain BG43—namely, aqdA1B1C1 and aqdA2B2C2. It is assumed that these genes play an important role in the expression of dioxygenases for PQS cleavage in *P. aeruginosa*. The AqdC proteins firstly identified enzymes that cleave PQS. The potential for inhibition of pyocyanin, rhamnolipid, and pyoverdine production was demonstrated by expression of PQS dioxygenase gene aqdC1 or aqdC2 in *P. aeruginosa* PAO1 (Muller et al., 2015).

Algae

Several active compounds such as phlorotannins regarding QS inhibition can be obtained from macroalgae and microalgae. Numerous marine-derived QSIs have been reported from various marine organisms. Moreover, some macroalgae are believed to defend themselves against surface-associated bacteria (Saurav et al., 2017).

Rajamani et al. (2008) evaluated the effect of lumichrome, a natural compound from the green algae *Chlamydomonas reinhardtii* CC-2137, on *P. aeruginosa*. The lumichrome is a derivative of the vitamin riboflavin. They detected an increase in the *luxCDABE* gene expression by riboflavin and lumichrome, indicating the specificity of these compounds to the LasR receptor, and they reported that activated LasR by lumichrome could bind to LasI promoter.

Also, riboflavin and lumichrome could bind to the same binding pocket of LasR, which was confirmed by docking methods.

Delisea pulchra, red Australian macroalgae, was shown to have antimicrobial effects due to its secondary metabolites. They produce brominated and chlorinated furanones, which are similar to AHL molecules in structure, and they may bind easily to the signal receptors (Shannon and Abu-Ghannam, 2016).

Animals

Several QSIs have been identified from animals, most of which are utilized in antibiofouling. Sesterterpene metabolites manoalide, manoalide monoacetate, and secomanoalide from *Luffariella variabilis* and ethyl acetate extract of *Hyalinella punctate* are known for their QSI effects against *P. aeruginosa* (Skindersoe et al., 2008; Pejin et al., 2016).

Costantino et al. (2017) isolated a γ-lactone called plakofuranolactone, which is responsible for LasI/R system from Indonesian marine sponge *Plakortis* cf. *Lita.* extracts. This compound was tested on *Escherichia coli* pSB401 and *C. violaceum* CV026 against short acyl chain signals for QQ potential, but no inhibitory effect was detected. On the other hand, they observed an inhibition in AHL-induced bioluminescence of C6-HSL detecting pSB401 and C12-HSL detecting pSB1075 monitor strains, and a decrease in protease activity of *P. aeruginosa* PAO1.

Skindersoe et al. (2008) screened 284 marine samples from the Great Barrier Reef for their QSI activities via two QSI selector systems (QSIS1 and QSIS2). The three C25 sesterterpene metabolites (manoalide, manoalide monoacetate, and secomanoalide) from *Luffariella variabilis* were demonstrated to have A QSI effect on *lasB::gfp* [ASV] fusion.

Hyalinella punctate, a freshwater bryozoan, was tested for its antibiofilm and anti-QS activities (Pejin et al., 2016). The ethyl acetate extracts of *H. punctata* had significant antibiofilm activity for *P. aeruginosa* PAO1. These extracts also were effective in twitching motility and in the inhibition of pyocyanin production by *P. aeruginosa* PAO1.

Quintana et al. (2015) evaluated QSI activity of 39 extracts belonging to 26 sponges, 7 soft corals, 5 algae, and 1 zooanthid. QS inhibition was found to be considerably effective in the crude extracts of *Eunicea laciniata*, *Svenzea tubulosa*, *Ircinia felix*, and *Neopetrosia carbonaria*. The researchers isolated furanosesterterpenes from the crude extract from the sponge *I. felix* with moderate anti-QS potential.

Mai et al. (2015) isolated the compounds isonaamine A, isonaamidine A, isonaamine D, leucettamine D, and di-isonaamidine A from the crude extracts of *Leucetta chagosensis*. Isonaamine D and isonaamidine A inhibited three QS pathways of *Vibrio harveyi*. Isonaamidine A had the greatest inhibitory effect on the AI-2 biosensor.

QQ potentials of 78 natural products from marine organisms (sponges, algae, fungi, tunicates, cyanobacteria, terrestrial plants) were evaluated by utilizing *C. violaceum* CV017. Demethoxy encecalin, midpacamide, tenuazonic acid, hymenialdisin, microcolins A and B, and kojic acid were found to be potent and abundant compounds for QS inhibition. Midpacamide and tenuazonic acid were found to be toxic to *E. coli* pSB401 and *E. coli* pSB1075. QS-dependent luminescence in *E. coli* pSB1075 (C12-HSL monitor) was reduced by demethoxy encecalin and hymenialdisin, whereas in *E. coli*, pSB401 was reduced by hymenialdisin, demethoxy encecalin, microcolins A and B, and kojic acid (Dobretsov et al., 2011).

Díaz et al. (2015) isolated five lipid compounds from soft coral *Eunicea* sp., and three terpenoids and sterols from *Eunicea fusca*. These compounds were tested against *Ochrobactrum pseudogringnonense, Alteromonas macleodii, V. harveyi, P. aeruginosa,* and *S. aureus*. They showed effectiveness depending on the bacteria. The batyl alcohol 1 and fuscoside E peracetate 6 were found to have potent antibiofilm effect with less toxicity.

Plants

Many natural QSIs can be provided from various plants some of which are vegetables and edible fruits. These plant-derived QSI molecules can act as agonists or antagonists. The opinion that plants can control the QS system has arisen from the knowledge that plants do not have any immune mechanism similar to humans. The idea is that they might fight against other QS pathogens via anti-QS compounds, most of which are secondary metabolites (Kalia, 2013). It is possible to extract these molecules from plant tissues (root, leaf, etc.), but the production quantities differ among plants due to their development phase. The chemical nature of these molecules is sometimes similar to the QS signaling molecules. To decide the safety of these molecules, the toxicity parameter must also be taken into consideration.

Secondary metabolites obtained from plants are of great importance due to their antimicrobial, antifungal, and antitumor properties (and more). The antimicrobial activity—exhibiting plant compounds are listed as phenolics, phenolic acids, quinones, saponins, FLs, tannins, coumarins, terpenoids, and alkaloids. Also, antibiofilm activities of plant compounds are provided by naringenin, oroidin, salicylic acid, ursolic acid, cinnamaldehyde, methyl eugenol, extracts of garlic, and edible fruits (Asfour, 2018).

Quercetin is a commonly known flavonoid with antioxidant and supplementary effect. Gopu et al. (2015) evaluated the anti-QS potential of a quercetin on *P. aeruginosa*, and they reported the inhibition of several virulence factors such as alginate, pyocyanin, protease, elastase and exopolysaccharide (EPS) production, biofilm formation, and motility.

Ursolic acid is a known, nontoxic compund with various pharmacological effects. Ren et al. (2005) created a library of 13,000 compounds and screened them for their potential to inhibit biofilm formation against *V. harveyi* and *P. aeruginosa* PAO1. They observed that ursolic acid (10 μg/mL) obtained from *Diospyros dendo* inhibited biofilm formation.

The possible inhibitory effect of methanolic leaf extracts of *Acer palmatum*, *Acer pseudosieboldianum*, and *Cercis chinensis* were tested on biofilm formation, swarming motility, pyocyanin production, and *Caenorhabditis elegans* killing activity of *P. aeruginosa* PAO1. These extracts were found to be successful in the inhibition of biofilm formation, swarming motility, and AI production (Niu et al., 2017).

Jakobsen et al. (2012a) isolated ajoene (4,5,9-trithiadodeca-1,6,11-triene 9-oxide), the sulfur-containing compound, from garlic extract. The researchers developed synthetic ajoene to evaluate QSI potential in vitro and in vivo. They indicated that this synthetic ajoene was successful in the inhibition of virulence genes involved in *P. aeruginosa*. Also, ajoene was reported to have synergistic effect with tobramycin in *P. aeruginosa* biofilms.

Packiavathy et al. (2014) investigated the anti-QS activity of curcumin from *Curcuma longa* against *Escherichia coli*, *P. aeruginosa* PAO1, *Proteus mirabilis*, and *Serratia marcescens*. The biofilm formation, exopolysaccharide (EPS) and alginate production, swimming and swarming motility were inhibited, and biofilm susceptibility enhancement to conventional antibiotics were detected.

The inhibitors of ENR, triclosan, and green tea epigallocatechin gallate (EGCG) were tested for QS inhibitory potentials by Yang et al. (2010). EGCG was reported to have higher binding affinity to ENR of *P. aeruginosa*, indicating its QQ effect. Moreover, this compound inhibited the swarming motility and biofilm formation of *P. aeruginosa*.

Vandeputte et al. (2011) evaluated the inhibitory effects of commercially available flavonoids (apigenin, luteolinflavonols, kaempferol, quercetin, myricetin, naringenin, naringin, eriodictyol, taxifolin, trans-benzylideneacetophenone) on *P. aeruginosa* PAO1 and *C. violaceum* CV026. Naringenin inhibited elastase activity, biofilm formation, *lasB*, *lasI*, *lasR*, *rhlA*, *rhlI*, *rhlR* gene expressions, production of *N*-(3-oxododecanoyl)-L-homoserine lactone (3-oxo-C12-HSL) and *N*-butanoyl-L-homoserine lactone (C4-HSL) in *P. aeruginosa* PAO1. Also, taxifolin was found to inhibit pyocyanin production and elastase activity *of P. aeruginosa* PAO1 in this study.

Phenylacetic acid is a compound known for its antifungal, antioxidant, and antiinflammatory properties. Musthafa et al. (2012b) evaluated the anti-QS potential of phenylacetic acid and observed an inhibitory effect on protease and elastase activities, EPS production, and swimming motility of PAO1.

Zingerone, from ginger root, has been shown to have inhibitory effects on swimming, swarming and twitching motility, biofilm formation, production of rhamnolipid, elastase, protease, and pyocyanin in *P. aeruginosa* by Kumar et al. (2015). Moreover, anti-QS potential of zingerone was also evaluated by molecular docking, and this compound was found to block QS signal receptors by binding to them (TraR, LasR, RhlR, and PqsR).

Fungi

Numerous secondary metabolites with QSI potentials are produced by fungi. The amount of this metabolite production can vary according to environmental conditions. Fungal QSIs are considerably important for drug and food industry. Patulin and penicillic acid are reported secondary metabolites with QSI potentials from *Penicillium radicicola* and *Penicillium coprobium* by Rasmussen et al. (2005). These mycotoxins were found to modulate QS related genes in *P. aeruginosa*, indicating their QSI activities (45% and 60% inhibition by patulin and penicillic acid, respectively). Also, they detected that patulin could increase the potential of tobramycin against biofilm forms. This compound was also found to be considerably effective in the clearance of *P. aeruginosa* in a mouse model of pulmonary infection when compared with a control group.

Agaricus blazei water extracts were also evaluated on *P. aeruginosa*. PAO1 QS-regulated virulence factors and biofilm formation. Sub-MIC concentrations of the extract (without inhibiting the growth of bacteria) exhibited reduction in virulence factors of *P. aeruginosa*, such as pyocyanin production, twitching, and swimming motility. Also, the biofilm formation was inhibited in a concentration-dependent manner at sub-MIC values (Soković et al., 2014).

Farnesol, a sesquiterpene alcohol, and tyrosol, a phenylethanoid, isolated from *Candida albicans* are reported to be QQ molecules and biofilm inhibitors. It is known that phenazine compounds are secreted by several bacteria such as *Pseudomonas* spp. These compounds play important roles, especially in bacterial virulence. For example, pyocyanin from *P. aeruginosa* is responsible from colonization in patients' lungs suffering from cystic fibrosis. Cugini et al. (2010) reported that farnesol increased the levels of PQS and C4HSL in *lasR* mutant strains. As a result, the high levels of PQS led to increase in the phenazine (pyocyanin) production from *P. aeruginosa* in the co-culture of *C. albicans* and *P. aeruginosa*. This increased PQS was reported to be dependent on *rhlR*, *rhlI*, and *pqsH*. Farnesol induced the *pqsH* transcription in *lasR* mutants by RhlR activation. Also, they observed that PqsR activity was inhibited by farnesol at low cell densities.

sRNAs

It may also be possible to achieve QS inhibition by targeting small RNAs (sRNAs), which are posttranscriptional regulators. PhrS was identified as an activator of PqsR synthesis by Sonnleitner et al. (2011). Jakobsen et al. (2017) demonstrated that previously known QSI ajoene may also act as an inhibitor against sRNAs RsmY and RsmZ in *P. aeruginosa*. sRNAs are presented as a new target for antivirulence approaches.

There's an abundance of QSIs reported from natural sources against *P. aeruginosa*, many of which are listed in Table 2.

Table 2 Natural QSI compounds against *P. aeruginosa*

QSI	Inhibition	References
[6]-Gingerol, [6]-shogaol and zingerone from *Zingiber officinale*	Biofilm formation, pyocyanin production	Kumar et al. (2015) and Kim et al. (2015)
Allicin and ajoene from *Allium sativum*	Virulence, LasI, LasR, sRNAs RsmY, and RsmZ	Jakobsen et al. (2012a)
Baicalein (flavonoid) Scutellaria baicalensis Georgi.	Biofilm formation	Zeng et al. (2008)
Baicalin hydrate, cinnamaldehyde and hamamelitannin	Increase biofilm susceptibility to treatment with antibiotics	Brackman et al. (2011)
Casbane diterpene (diterpenoid) *Croton nepetaefolius* Baill.	Biofilm formation	Carneiro et al. (2010)
Cassipourol, β-sitosterol and α-amyrin, terpenoids from *Platostoma rotundifolium*	Biofilm formation, EPS production, motility	Rasamiravaka et al. (2017)
Catechin	Pyocyanin production, elastase activity, biofilm production	Vandeputte et al. (2010)
Chrysophanol, nodakenetin, shikonin and emodin from traditional Chinese herbs	Biofilm formation	Ding et al. (2011)
Coumarate ester (phenolic compound) *Dalbergia trichocarpa*	Biofilm formation, pyocyanin, LasB elastase production, proteolytic activity, motility	Rasamiravaka et al. (2013)
Curcumin from *Curcuma longa* L.	Biofilm formation, EPS and alginate production, swimming and swarming motility, biofilm susceptibility enhancement to antibiotics	Packiavathy et al. (2014)
Delftia tsuruhatensis	Biofilm formation	Singh et al. (2017)
Dihydroxybergamottin and bergamottin	Biofilm formation	Girennavar et al. (2008)
Ellagic acid derivatives	Biofilm formation	Sarabhai et al. (2013)
Epigallocatechin gallate from green tea, triclosan (5-Chloro-2-(2,4-dichlorophenoxy)phenol)	Biofilm formation and swarming motility, show synergistic activity with ciprofloxacin	Yang et al. (2010)
Eriodictyol	Pyocyanin production, elastase activity	Vandeputte et al. (2011)
Eugenol from *Syzygium aromaticum*	Biofilm formation	Zhou et al. (2013)
Evernic acid	LasB, RhlA, biofilm formation	Gokalsin and Sesal (2016)
Farnesol	Pyocyanin production, PQS	Cugini et al. (2010)
Iberin	RhlA, LasB	Jakobsen et al. (2012b)
Malabaricone C from *Myristica cinnamomea*	Pyocyanin production, elastase activity, biofilm formation	Chong et al. (2011)
Manolide, manolide monoacetate and secomanoalide from marine sponge *Luffariella variabilis*	Biofilm formation	Skindersoe et al. (2008)
Methyl eugenol from *Cuminum cyminum*	Biofilm formation, motility	Packiavathy et al. (2012)
Naringenin (flavonoid) commercial	Biofilm formation, elastase activity, *lasB, lasI, lasR, rhlA, rhlI, rhlR* gene expression, AHL production	Vandeputte et al. (2011)

Table 2 **Natural QSI compounds against *P. aeruginosa*—Cont'd**

QSI	Inhibition	References
p-Coumaroyl-hydroxy-ursolic acid (coumarate ester of triterpen) *Diospyros dendo* Welw.	Biofilm formation	Hu et al. (2006)
Patulin (furopyranone) and penicillic acid (furanone) from *Penicillium* species	LasR, RhlR	Rasmussen et al. (2005)
Phenylacetic acid	Pyocyanin production, elastase activity, biofilm production, swimming motility	Musthafa et al. (2012b)
Riboflavin and its derivative lumichrome	LasR	Rajamani et al. (2008)
Rosmarinic acid	Biofilm formation	Walker et al. (2004)
Rosmarinic acid, naringin, chlorogenic acid, morin and mangiferin	LasR, RhlR, biofilm formation, protease, elastase and haemolysin production	Annapoorani et al. (2012)
Salicylic acid	AHL production, twitching and swimming motility, protease activity, LasR, Rhl, PQS activity, rhamnolipid production pyoverdine production, biofilm formation	Yang et al. (2009) and Bandara et al. (2006)
Salicylic acid, tannic acid and trans-cinnamaldehyde	Swarming activity and pyocyanin production	Chang et al. (2014)
Saponins, ginsenosides, and polysaccharides from Panax ginseng	AHL synthesis, LasA, LasB	Song et al. (2010)
Sesquiterpenoid viridiflorol and triterpenoids, ursolic and betulinic acids, from the liverwort *Lepidozia chordulifera*	Biofilm formation and elastolytic activity	Gilabert et al. (2015)
Solenopsin A (alkaloid) *Solenopsis invicta* (insect; ant)	Biofilm formation	Park et al. (2008)
Taxifolin	Pyocyanin production, elastase activity	Vandeputte et al. (2011)
Ursolic acid (triterpenoid) *Diospyros dendo* Welw.	Biofilm formation	Ren et al. (2005)
Zeaxanthin (carotenoid)	LasB, RhlA, biofilm formation	Gokalsin et al. (2017)

2.3.4 Bacteriophage

Bacteria are lysed through phage infections and therefore developed various defense mechanisms against these viruses such as clustered regularly interspaced short palindromic repeats (CRISPRs). Although information on CRISPR-Cas regulation is limited, some studies suggest that QS systems play an active role (Patterson et al., 2016; Hoyland-Kroghsbo et al., 2017).

Hoyland-Kroghsbo et al. (2017) reported that QS modulators can activate or suppress the CRISPR-Cas system in *P. aeruginosa*. Accordingly, they refer to the possibility of QSI-phage combined multitherapies. Qin et al. (2017) also report that QS should be involved in *P. aeruginosa* defense mechanism against phage K5 infections. On the contrary, temperate bacteriophage D3112 and JBD30 were determined to prefer QS capable *P. aeruginosa* instead of QS deficient strains in a competitive population in *Galleria mellonella*. Nevertheless, Mumford and Friman (2017) revealed that lytic PT7 phage causes a decrease in the population of LasR deficient *P. aeruginosa*, whereas an increase in PAO1 strain, in the presence of competitors *Staphylococcus aureus* and *Stenotrophomonas maltophilia*.

2.4 Synthetic Quorum Sensing Modulators

Although most QSIs are obtained from natural products, another approach is the development of synthetic QSIs. Such compounds are sometimes derived from natural ligands. In addition, there are structural mimics of HSLs and structurally unrelated compounds (Table 3).

AHL molecules have a head and a tail part in their structure: a HSL moiety with a tail of *N*-acyl residue. Usually a synthesized AHL mimic has one part intact and the other part derived, hoping to create a more robust molecule with antagonistic effect. In their study, Biswas et al. (2017) synthesized and tested several AHL mimics acetoxy-glucosamides, hydroxy-glucosamides, and 3-oxo-glucosamides against the *P. aeruginosa* MH602 strain. They showed that the strongest QSI compound 9b, a hydroxy-glucosamide, can inhibit the *P. aeruginosa las* system by 79.1%. Docking studies also revealed the binding poses of these compounds. Morkunas et al. (2012) have synthesized a collection of abiotic OdDHL mimics and have shown that some are capable of inhibiting QS and pyocyanin production. Hodgkinson et al. (2012) also synthesized and evaluated OdDHL mimics and found a number of new compounds that can modulate the *las* system of *P. aeruginosa*.

AHL analogues are mostly studied for their ability to bind with the receptors active site, but prevent the detection of QS signals. The idea is to synthesize compounds that act as antagonists. However, some modifications might end up with causing agonistic effects. For example, meta-bromo-thiolactone (mBTL) acts as an agonist in the absence of AHLs but antagonizes both LasR and RhlR in *P. aeruginosa* when in the presence of natural AIs (O'loughlin et al., 2013).

PQS signaling has been demonstrated to be responsible for virulence factor production and biofilm maturation in *P. aeruginosa*. The precursor HHQ, which is produced from anthranilate and a β-keto-fatty acid, is later converted into PQS (LaSarre and Federle, 2013). Depending on ligand-based or fragment-based approaches, agonists/antagonists are employed for their QSI properties by researchers. Targeting anthranilate is presented as an efficient approach for inhibition of PQS. The methyl anthranilate and halogenated anthranilate analogues were demonstrated to inhibit PQS biosynthesis (Calfee et al., 2001). Another alternative

Table 3 Synthetic QSI compounds, including nanoparticles

Compound	Inhibition	References
(z)-5-Octylidenethiazolidine-2, 4-dione (TZD-C8)	LasI	Lidor et al. (2015)
2,5-Piperazinedione	LasR	Musthafa et al. (2012a)
2-Amino-oxadiazoles	PqsR	Zender et al. (2013)
2-Nitrophenyl derivatives	PqsD	Storz et al. (2013)
3-Nitro phenylacetanoyl HL (C14)	LasR	Geske et al. (2008a)
4-I N-phenylacetyl-L-homoserine lactones (PHL)	LasR	Geske et al. (2007)
4-Nitro-pyridine-N-oxide (NPO)	Virulence	Rasmussen et al. (2005)
5-Aryl-ureidothiophene-2-carboxylic acids	PqsD	Sahner et al. (2013)
Benzamidobenzoic acids	PqsD	Hinsberger et al. (2014) and Weidel et al. (2013)
Chloro-pyridine pharmacophore	LasR	Marsden et al. (2010)
Compound 1	PqsR	Lu et al. (2014)
Furanone C-30	Virulence factors	Hentzer et al. (2003)
Furanone F2, F3 and F4	QscR	Liu et al. (2010)
Fusaric acid analogues	*las, rhl* systems	Tung et al. (2017)
Hydroxy-glucosamides	*las, rhl* systems	Biswas et al. (2017)
Long-chain 4-aminoquinolines	PQS, biofilm formation	Aleksić et al. (2017)
Meta-bromo-thiolactone	LasR and RhlR	O'loughlin et al. (2013)
Mycofabricated silver nanoparticles	Biofilm formation, LasA protease, LasB elastase, pyocyanin, pyoverdin, pyochelin, rhamnolipid, and alginate	Singh et al. (2015)
N-Decanoyl cyclopentylamide	LasR and RhlR	Ishida et al. (2007)
OdDHL-mimics	LasR, pyocyanin production, biofilm formation	Morkunas et al. (2012)
OdDHL-mimics which incorporated an (hetero)aromatic head group	LasR	Hodgkinson et al. (2012)
Phenylpropionyl-homoserine lactones	LasR	Geske et al. (2008b)
Quinazolinone (QZN)	PqsR	Ilangovan et al. (2013)
Selenium nanoparticles with honey phytochemicals	LasR, biofilm formation, elastase	Prateeksha et al. (2017)
Silver nanoparticles	Biofilm formation	Barapatre et al. (2016)
Silver nanoparticles impregnated in dressing	Biofilm formation	Velazquez-Velazquez et al. (2015)
ZnO nanoparticles	Biofilm formation, elastase, pyocyanin	Garcia-Lara et al. (2015)

approach is to target the PqsD, a key enzyme in the biosynthesis of HHQ and PQS. Recently, *S*-phenyl-L-cysteine sulfoxide was reported to inhibit kynureninase, an enzyme that catalyzes the cleavage of kynurenine into anthranilic acid and 3-hydroxyanthranilic acid (Kasper et al., 2016).

There are also synthetic modulators that are unrelated to AHLs. For example, Lidor et al. (2015) have synthesized and studied thiazolidinedione type molecules, and observed that the compound named (z)-5-octylidenethiazolidine-2, 4-dione (TZD-C8) has strong motility and biofilm inhibition properties. Furthermore, they explored the QSI properties and in silico structural affinity, finding out that TZD-C8 has significant inhibition potential for the *las* system.

Tung et al. (2017) have synthesized 40 novel fusaric acid analogues via microwave assisted synthesis and investigated their QSI potentials against *las* and *rhl* Qs systems of *P. aeruginosa*. They found that one of the analogues is capable of inhibiting the *las* system and related virulence factors.

Moreover, several studies investigated the QSI potentials of nanoparticles with promising results. Singh et al. (2015) evaluated the QS inhibition properties of mycofabricated silver nanoparticles (Ag-NPs) employing metabolites of *Rhizopus arrhizus* BRS-07 against *P. aeruginosa*. It was shown that these Ag-NPs can inhibit QS-regulated virulence factors, including LasA protease, LasB elastase, pyocyanin, pyoverdin, pyochelin, rhamnolipid, and alginate. Barapatre et al. (2016) biosynthesized Ag-NPs via lignin-degrading fungi *Aspergillus flavus* and *Emericella nidulans*. They observed antimicrobial effects as well as strong antibiofilm effects. On the other hand, Velazquez-Velazquez et al. (2015) have impregnated Ag-NPs in dressings and tested them against *P. aeruginosa* biofilms. They suggest that Ag-NP impregnated dressings can reduce or prevent bacterial growth in wound environments.

In 2017, Prateeksha et al. studied selenium nanoparticles (Se-NPs) with honey phytochemicals against *P. aeruginosa* biofilm and QS. They utilize Se-NPs as vectors for a drug delivery system. They have shown that the nano-scaffold demonstrated a greater QS and biofilm inhibition in both in vitro and in vivo, compared to its counterparts. Molecular dockings suggested that nano-scaffold assists honey phytochemicals in binding to the LasR receptor cavity.

As expected, the number of candidate compounds can be too much to handle easily, and they can be modified in various ways. Since there are many options on how these modifications can be made, many novel synthetic QSIs are based on structural scaffolds that are determined by high-throughput and virtual screens of small molecules. Another idea to modulate QS is to inhibit AHL synthase enzymes by deriving synthetic analogues of their intermediates as previously described. SAM is an intriguing intermediate, considering it is a precursor for both AI1 and AI2 (Kalia, 2013).

3 Conclusion and Opinion

There is a growing concern that antibiotics will soon be ineffective against common infections in the near future due to the increase in bacterial antibiotic resistance. Therefore, the subsequent response is antivirulence strategy with the leading application of QQ. A multitude of QSIs against *P. aeruginosa* have been discovered to date with natural compounds in abundance, as presented in this chapter. Without a doubt, there are many more natural compounds, and their synthetic derivatives are waiting to be investigated for their QSI potentials. Modern technology allows us to perform screens swiftly by employing carefully developed monitor strains. In silico methods like building compound libraries for virtual screenings by computer-aided simulations also pick up speed. These methods save time, costs, and labor in laboratories; therefore it is essential to analyze and deal with their limitations.

Compared with planktonic bacteria, biofilm forms have a complex structure. Therefore, it should be assumed that QS signal production and concentration is not homogenous. The diffusion of signal molecules would also vary, causing some bacteria to detect more AHL molecules than others. Moreover, flat or mushroom structures of biofilms must have different signal gradients. All these factors and more should be investigated for a better understanding of the system flow and selecting the inhibitors accordingly.

Many studies nowadays focus on multitherapies with QQ approach. Multitherapies can be performed in combination with two or more QSIs, enzymes, antibodies, or antibiotics. QSIs are known to increase bacterial susceptibility to antibiotics and phage infections. Moreover, infection models and applications against multispecies and strains keep this approach more realistic. Certainly there are some disadvantages and unknown parameters regarding QQ therapies, and it even might be possible for bacteria to gain resistance against QSIs. Further research will confidently allow us to utilize QSIs effectively and sort out the drawbacks.

Glossary

Acyl-homoserine lactone A small and diffusible signaling molecule involved in quorum sensing mechanism.

Antibiotic-resistance The ability of microorganisms to resist the effects of antibiotics that could successfully treat them before.

Antivirulence therapy Alternative treatment approach to inhibit bacterial virulence factors without any extermination of pathogens.

Biofilm A *layer* of prokaryotic microorganisms that attach to surfaces and secrete exopolysaccharides.

Cystic fibrosis A disease that is encountered frequently and affects approximately 70,000 people in the world, known to degrade lung functions by affecting respiratory system.

Multidrug resistance Antimicrobial resistance exhibited by microorganisms against multiple antimicrobial drugs.

Quorum quenching Alternative strategy to disrupt bacterial communication without killing bacteria or preventing their growth.

Quorum sensing inhibitors Natural or synthetic compounds inhibiting a quorum sensing mechanism.

Quorum sensing Bacterial communication system controlling numerous behaviors, including bioluminescence, swarming, conjugation, protease activity, and biofilm formation, and so on, via small and diffusible chemical signaling molecules called autoinducers.

Virtual screening A computer-aided technique that is used in drug discovery to search chemical libraries, including compounds having particular properties.

Abbreviations

3OC6HSL	*N*-3-(oxohexanoyl)-homoserine lactone
3-oxo-C12-HSL, OdDHL	*N*-3-oxododecanoyl-homoserine lactone
Ag-NP	Silver Nanoparticles
AHL	*N*-acylated L-homoserine lactone
AI	autoinducer
AI-2	autoinducer-2
AiiA	autoinducer inactivation gene
AIP	autoinducing Peptide
AQ	2-alkyl4-quionolones
BAC	bacterial artificial chromosomes
C4-HSL, BHL	*N*-butanoyl-L-homoserine lactone
CDC	The Centers for Disease Control and Prevention
CF	cystic fibrosis
CRISPR	clustered regularly interspaced short palindromic repeat
DPD	4,5-dihydroxy 2,3 pentanedione
EGCG	epigallocatechin gallate
ENR	enoyl-acyl-carrier-protein (ACP) reductase
EPS	exopolysaccharide
FabI	NADH-dependent ENR
HHQ	2-heptyl-4-hydroxyquinolone
Hod	2,4-dioxygenase
hPON1	human serum paraoxonase 1
mBTL	meta-bromo-thiolactone
MDR	multidrug resistant
PON1	paraoxonase 1
PON2	paraoxonase 2
PON3	paraoxonase 3
PQS	pseudomonas quinolone signal
PVA	penicillin V acylase
QQ	quorum quenching
QS	quorum sensing
QSI	QS inhibitor

QSIS	QSI selector system
SAM	*S*-adenoysl methionine
Se-NP	selenium nanoparticles
sRNA	small RNA
SsoPox	hyperthermostable lactonase
TZD-C8	(*z*)-5-octylidenethiazolidine-2, 4-dione
VAI	AI of *V. fischeri*

Acknowledgment

The authors would like to thank Scientific and Technological Research Council of Turkey for their support (TÜBİTAK - 315S092).

References

Aleksić, I., Šegan, S., Andrić, F., Zlatović, M., Moric, I., Opsenica, D.M., Senerovic, L., 2017. Long-chain 4-aminoquinolines as quorum sensing inhibitors in *Serratia marcescens* and *Pseudomonas aeruginosa*. ACS Chem. Biol. 12 (5), 1425–1434.

Annapoorani, A., Umamageswaran, V., Parameswari, R., Pandian, S.K., Ravi, A.V., 2012. Computational discovery of putative quorum sensing inhibitors against LasR and RhlR receptor proteins of *Pseudomonas aeruginosa*. J. Comput. Aided Mol. Des. 26 (9), 1067–1077.

Asfour, H.Z., 2018. Antiquorum sensing natural compounds. J. Microsc. Ultrastruct. 6 (1), 1–10.

Aybey, A., Demirkan, E., 2016. Inhibition of quorum sensing-controlled virulence factors in *Pseudomonas aeruginosa* by human serum paraoxonase. J. Med. Microbiol. 65 (2), 105–113.

Bandara, M.B., Zhu, H., Sankaridurg, P.R., Willcox, M.D., 2006. Salicylic acid reduces the production of several potential virulence factors of *Pseudomonas aeruginosa* associated with microbial keratitis. Invest. Ophthalmol. Vis. Sci. 47 (10), 4453–4460.

Barapatre, A., Aadil, K.R., Jha, H., 2016. Synergistic antibacterial and antibiofilm activity of silver nanoparticles biosynthesized by lignin-degrading fungus. Bioresour. Bioprocess. 3 (1), 8.

Biswas, N.N., Yu, T.T., Kimyon, O., Nizalapur, S., Gardner, C.R., Manefield, M., Griffith, R., Black, D.S., Kumar, N., 2017. Synthesis of antimicrobial glucosamides as bacterial quorum sensing mechanism inhibitors. Bioorg. Med. Chem. 25 (3), 1183–1194.

Borges, A., Abreu, A.C., Dias, C., Saavedra, M.J., Borges, F., Simões, M., 2016. New perspectives on the use of phytochemicals as an emergent strategy to control bacterial infections including biofilms. Molecules 21 (7), 877.

Boyle, K.E., Heilmann, S., van Ditmarsch, D., Xavier, J.B., 2013. Exploiting social evolution in biofilms. Curr. Opin. Microbiol. 16 (2), 207–212.

Brackman, G., Cos, P., Maes, L., Nelis, H.J., Coenye, T., 2011. Quorum sensing inhibitors increase the susceptibility of bacterial biofilms to antibiotics *in vitro* and *in vivo*. Antimicrob. Agents Chemother. 55 (6), 2655–2661.

Branny, P., Pearson, J.P., Pesci, E.C., Kohler, T., Iglewski, B.H., Van Delden, C., 2001. Inhibition of quorum sensing by a *Pseudomonas aeruginosa dksA* homologue. J. Bacteriol. 183 (5), 1531–1539.

Calfee, M.W., Coleman, J.P., Pesci, E.C., 2001. Interference with Pseudomonas quinolone signal synthesis inhibits virulence factor expression by *Pseudomonas aeruginosa*. Proc. Natl. Acad. Sci. U. S. A. 98 (20), 11633–11637.

Carneiro, V.A., Santos, H.S., Arruda, F.V., Bandeira, P.N., Albuquerque, M.R., Pereira, M.O., Henriques, M., Cavada, B.S., Teixeira, E.H., 2010. Casbane diterpene as a promising natural antimicrobial agent against biofilm-associated infections. Molecules 16 (1), 190–201.

Centers for Disease Control and Prevention, 2017. Available from: https://www.cdc.gov/drugresistance/about.html (Accessed 2017).

Chang, C.Y., Krishnan, T., Wang, H., Chen, Y., Yin, W.F., Chong, Y.M., Tan, L.Y., Chong, T.M., Chan, K.G., 2014. Non-antibiotic quorum sensing inhibitors acting against N-acyl homoserine lactone synthase as druggable target. Sci. Rep. 4, 7245.

Chang, H., Zhou, J., Zhu, X., Yu, S., Chen, L., Jin, H., Cai, Z., 2017. Strain identification and quorum sensing inhibition characterization of marine-derived *Rhizobium* sp. NAO1. R. Soc. Open Sci. 4(3), 170025.

Chong, Y.M., Yin, W.F., Ho, C.Y., Mustafa, M.R., Hadi, A.H., Awang, K., Narrima, P., Koh, C.L., Appleton, D.R., Chan, K.G., 2011. Malabaricone C from *Myristica cinnamomea* exhibits anti-quorum sensing activity. J. Nat. Prod. 74 (10), 2261–2264.

Costantino, V., Della Sala, G., Saurav, K., Teta, R., Bar-Shalom, R., Mangoni, A., Steindler, L., 2017. Plakofuranolactone as a quorum quenching agent from the Indonesian sponge *Plakortis* cf. *lita*. Mar. Drugs. 15(3).

Cugini, C., Morales, D.K., Hogan, D.A., 2010. *Candida albicans*-produced farnesol stimulates Pseudomonas quinolone signal production in LasR-defective *Pseudomonas aeruginosa* strains. Microbiology 156 (Pt 10), 3096–3107.

Daniels, R., Vanderleyden, J., Michiels, J., 2004. Quorum sensing and swarming migration in bacteria. FEMS Microbiol. Rev. 28 (3), 261–289.

De Lamo Marin, S., Xu, Y., Meijler, M.M., Janda, K.D., 2007. Antibody catalyzed hydrolysis of a quorum sensing signal found in Gram-negative bacteria. Bioorg. Med. Chem. Lett. 17 (6), 1549–1552.

Devaraj, K., Tan, G.Y.A., Chan, K.G., 2017. Quorum quenching properties of *Actinobacteria* isolated from Malaysian tropical soils. Arch. Microbiol. 199 (6), 897–906.

Díaz, Y.M., Laverde, G.V., Gamba, L.R., Wandurraga, H.M., Arévalo-Ferro, C., Rodríguez, F.R., Beltrán, C.D., Hernández, L.C., 2015. Biofilm inhibition activity of compounds isolated from two *Eunicea* species collected at the Caribbean Sea. Rev. Bras 25, 605–611.

Diekema, D.J., Pfaller, M.A., Jones, R.N., Doern, G.V., Winokur, P.L., Gales, A.C., Sader, H.S., Kugler, K., Beach, M., 1999. Survey of bloodstream infections due to gram-negative bacilli: frequency of occurrence and antimicrobial susceptibility of isolates collected in the United States, Canada, and Latin America for the SENTRY Antimicrobial Surveillance Program, 1997. Clin. Infect. Dis. 29 (3), 595–607.

Ding, X., Yin, B., Qian, L., Zeng, Z., Yang, Z., Li, H., Lu, Y., Zhou, S., 2011. Screening for novel quorum-sensing inhibitors to interfere with the formation of *Pseudomonas aeruginosa* biofilm. J. Med. Microbiol. 60 (Pt 12), 1827–1834.

Dobretsov, S., Teplitski, M., Bayer, M., Gunasekera, S., Proksch, P., Paul, V.J., 2011. Inhibition of marine biofouling by bacterial quorum sensing inhibitors. Biofouling 27 (8), 893–905.

Dong, Y.H., Wang, L.Y., Zhang, L.H., 2007. Quorum-quenching microbial infections: mechanisms and implications. Philos. Trans. R. Soc. B Biol. Sci. 362 (1483), 1201–1211.

Eberhard, A., Burlingame, A.L., Eberhard, C., Kenyon, G.L., Nealson, K.H., Oppenheimer, N.J., 1981. Structural identification of autoinducer of *Photobacterium fischeri* luciferase. Biochemistry 20 (9), 2444–2449.

Fuqua, C., Greenberg, E.P., 2002. Listening in on bacteria: acyl-homoserine lactone signalling. Nat. Rev. Mol. Cell Biol. 3 (9), 685–695.

Galloway, W.R., Hodgkinson, J.T., Bowden, S.D., Welch, M., Spring, D.R., 2011. Quorum sensing in Gram-negative bacteria: small-molecule modulation of AHL and AI-2 quorum sensing pathways. Chem. Rev. 111 (1), 28–67.

Garcia-Lara, B., Saucedo-Mora, M.A., Roldan-Sanchez, J.A., Perez-Eretza, B., Ramasamy, M., Lee, J., Coria-Jimenez, R., Tapia, M., Varela-Guerrero, V., Garcia-Contreras, R., 2015. Inhibition of quorum-sensing-dependent virulence factors and biofilm formation of clinical and environmental *Pseudomonas aeruginosa* strains by ZnO nanoparticles. Lett. Appl. Microbiol. 61 (3), 299–305.

Geske, G.D., O'neill, J.C., Miller, D.M., Mattmann, M.E., Blackwell, H.E., 2007. Modulation of bacterial quorum sensing with synthetic ligands: systematic evaluation of N-acylated homoserine lactones in multiple species and new insights into their mechanisms of action. J. Am. Chem. Soc. 129 (44), 13613–13625.

Geske, G.D., Mattmann, M.E., Blackwell, H.E., 2008a. Evaluation of a focused library of N-aryl L-homoserine lactones reveals a new set of potent quorum sensing modulators. Bioorg. Med. Chem. Lett. 18 (22), 5978–5981.

Geske, G.D., O'neill, J.C., Miller, D.M., Wezeman, R.J., Mattmann, M.E., Lin, Q., Blackwell, H.E., 2008b. Comparative analyses of N-acylated homoserine lactones reveal unique structural features that dictate their ability to activate or inhibit quorum sensing. ChemBioChem 9 (3), 389–400.

Gilabert, M., Marcinkevicius, K., Andujar, S., Schiavone, M., Arena, M.E., Bardon, A., 2015. Sesqui- and triterpenoids from the liverwort *Lepidozia chordulifera* inhibitors of bacterial biofilm and elastase activity of human pathogenic bacteria. Phytomedicine 22 (1), 77–85.

Girennavar, B., Cepeda, M., Soni, K.A., Vikram, A., Jesudhasan, P., Jayaprakasha, G., Pillai, S., Patil, B., 2008. Grapefruit juice and its furocoumarin inhibits autoinducer signaling and biofilm formation in bacteria. Int. J. Food Microbiol. 125 (2), 204–208.

Gokalsin, B., Sesal, N.C., 2016. Lichen secondary metabolite evernic acid as potential quorum sensing inhibitor against *Pseudomonas aeruginosa*. World J. Microbiol. Biotechnol. 32 (9), 150.

Gokalsin, B., Aksoydan, B., Erman, B., Sesal, N.C., 2017. Reducing virulence and biofilm of *Pseudomonas aeruginosa* by potential quorum sensing inhibitor carotenoid: Zeaxanthin. Microb. Ecol. 74 (2), 466–473.

Gopu, V., Meena, C.K., Shetty, P.H., 2015. Quercetin influences quorum sensing in food borne bacteria: *in-vitro* and *in-silico* evidence. PLoS One 10(8), e0134684.

Grandclement, C., Tannieres, M., Morera, S., Dessaux, Y., Faure, D., 2016. Quorum quenching: role in nature and applied developments. FEMS Microbiol. Rev. 40 (1), 86–116.

Grover, N., Plaks, J.G., Summers, S.R., Chado, G.R., Schurr, M.J., Kaar, J.L., 2016. Acylase-containing polyurethane coatings with anti-biofilm activity. Biotechnol. Bioeng. 113 (12), 2535–2543.

Guendouze, A., Plener, L., Bzdrenga, J., Jacquet, P., Remy, B., Elias, M., Lavigne, J.P., Daude, D., Chabriere, E., 2017. Effect of quorum quenching lactonase in clinical isolates of *Pseudomonas aeruginosa* and comparison with quorum sensing inhibitors. Front. Microbiol. 8, 227.

Hall-Stoodley, L., Costerton, J.W., Stoodley, P., 2004. Bacterial biofilms: from the natural environment to infectious diseases. Nat. Rev. Microbiol. 2 (2), 95–108.

Hentzer, M., Givskov, M., 2003. Pharmacological inhibition of quorum sensing for the treatment of chronic bacterial infections. J. Clin. Investig. 112 (9), 1300–1307.

Hentzer, M., Wu, H., Andersen, J.B., Riedel, K., Rasmussen, T.B., Bagge, N., Kumar, N., Schembri, M.A., Song, Z., Kristoffersen, P., Manefield, M., Costerton, J.W., Molin, S., Eberl, L., Steinberg, P., Kjelleberg, S., Hoiby, N., Givskov, M., 2003. Attenuation of *Pseudomonas aeruginosa* virulence by quorum sensing inhibitors. EMBO J. 22 (15), 3803–3815.

Hinsberger, S., De Jong, J.C., Groh, M., Haupenthal, J., Hartmann, R.W., 2014. Benzamidobenzoic acids as potent PqsD inhibitors for the treatment of *Pseudomonas aeruginosa* infections. Eur. J. Med. Chem. 76, 343–351.

Hodgkinson, J.T., Galloway, W.R., Wright, M., Mati, I.K., Nicholson, R.L., Welch, M., Spring, D.R., 2012. Design, synthesis and biological evaluation of non-natural modulators of quorum sensing in *Pseudomonas aeruginosa*. Org. Biomol. Chem. 10 (30), 6032–6044.

Hoyland-Kroghsbo, N.M., Paczkowski, J., Mukherjee, S., Broniewski, J., Westra, E., Bondy-Denomy, J., Bassler, B.L., 2017. Quorum sensing controls the *Pseudomonas aeruginosa* CRISPR-Cas adaptive immune system. Proc. Natl. Acad. Sci. U. S. A. 114 (1), 131–135.

Hraiech, S., Hiblot, J., Lafleur, J., Lepidi, H., Papazian, L., Rolain, J.M., Raoult, D., Elias, M., Silby, M.W., Bzdrenga, J., Bregeon, F., Chabriere, E., 2014. Inhaled lactonase reduces *Pseudomonas aeruginosa* quorum sensing and mortality in rat pneumonia. PLoS One 9(10), e107125.

Hu, J.F., Garo, E., Goering, M.G., Pasmore, M., Yoo, H.D., Esser, T., Sestrich, J., Cremin, P.A., Hough, G.W., Perrone, P., Lee, Y.S., Le, N.T., O'neil-Johnson, M., Costerton, J.W., Eldridge, G.R., 2006. Bacterial biofilm inhibitors from *Diospyros dendo*. J. Nat. Prod. 69 (1), 118–120.

Huang, J.J., Petersen, A., Whiteley, M., Leadbetter, J.R., 2006. Identification of QuiP, the product of gene PA1032, as the second acyl-homoserine lactone acylase of *Pseudomonas aeruginosa* PAO1. Appl. Environ. Microbiol. 72 (2), 1190–1197.

Ilangovan, A., Fletcher, M., Rampioni, G., Pustelny, C., Rumbaugh, K., Heeb, S., Camara, M., Truman, A., Chhabra, S.R., Emsley, J., Williams, P., 2013. Structural basis for native agonist and synthetic inhibitor recognition by the *Pseudomonas aeruginosa* quorum sensing regulator PqsR (MvfR). PLoS Pathog. 9(7), e1003508.

Ishida, T., Ikeda, T., Takiguchi, N., Kuroda, A., Ohtake, H., Kato, J., 2007. Inhibition of quorum sensing in *Pseudomonas aeruginosa* by N-acyl cyclopentylamides. Appl. Environ. Microbiol. 73 (10), 3183–3188.

Jakobsen, T.H., Van Gennip, M., Phipps, R.K., Shanmugham, M.S., Christensen, L.D., Alhede, M., Skindersoe, M.E., Rasmussen, T.B., Friedrich, K., Uthe, F., Jensen, P.O., Moser, C., Nielsen, K.F., Eberl, L., Larsen, T.O., Tanner, D., Hoiby, N., Bjarnsholt, T., Givskov, M., 2012a. Ajoene, a sulfur-rich molecule from garlic, inhibits genes controlled by quorum sensing. Antimicrob. Agents Chemother. 56 (5), 2314–2325.

Jakobsen, T.H., Bragason, S.K., Phipps, R.K., Christensen, L.D., Van Gennip, M., Alhede, M., Skindersoe, M., Larsen, T.O., Hoiby, N., Bjarnsholt, T., Givskov, M., 2012b. Food as a source for quorum sensing inhibitors: iberin from horseradish revealed as a quorum sensing inhibitor of *Pseudomonas aeruginosa*. Appl. Environ. Microbiol. 78 (7), 2410–2421.

Jakobsen, T.H., Warming, A.N., Vejborg, R.M., Moscoso, J.A., Stegger, M., Lorenzen, F., Rybtke, M., Andersen, J.B., Petersen, R., Andersen, P.S., Nielsen, T.E., Tolker-Nielsen, T., Filloux, A., Ingmer, H., Givskov, M., 2017. A broad range quorum sensing inhibitor working through sRNA inhibition. Sci. Rep. 7 (1), 9857.

Joo, H.S., Otto, M., 2012. Molecular basis of in vivo biofilm formation by bacterial pathogens. Chem. Biol. 19 (12), 1503–1513.

Kalia, V.C., 2013. Quorum sensing inhibitors: an overview. Biotechnol. Adv. 31 (2), 224–245.

Kalia, V.C., Purohit, H.J., 2011. Quenching the quorum sensing system: potential antibacterial drug targets. Crit. Rev. Microbiol. 37 (2), 121–140.

Kasper, S.H., Bonocora, R.P., Wade, J.T., Musah, R.A., Cady, N.C., 2016. Chemical inhibition of kynureninase reduces *Pseudomonas aeruginosa* quorum sensing and virulence factor expression. ACS Chem. Biol. 11 (4), 1106–1117.

Kaufmann, G.F., Sartorio, R., Lee, S.-H., Mee, J.M., Altobell, L.J., Kujawa, D.P., Jeffries, E., Clapham, B., Meijler, M.M., Janda, K.D., 2006. Antibody interference with N-acyl homoserine lactone-mediated bacterial quorum sensing. J. Am. Chem. Soc. 128 (9), 2802–2803.

Kim, H.S., Lee, S.H., Byun, Y., Park, H.D., 2015. 6-Gingerol reduces *Pseudomonas aeruginosa* biofilm formation and virulence via quorum sensing inhibition. Sci. Rep. 5, 8656.

Kumar, L., Chhibber, S., Kumar, R., Kumar, M., Harjai, K., 2015. Zingerone silences quorum sensing and attenuates virulence of *Pseudomonas aeruginosa*. Fitoterapia 102, 84–95.

LaSarre, B., Federle, M.J., 2013. Exploiting quorum sensing to confuse bacterial pathogens. Microbiol. Mol. Biol. Rev. 77 (1), 73–111.

Laxminarayan, R., Matsoso, P., Pant, S., Brower, C., Rottingen, J.A., Klugman, K., Davies, S., 2016. Access to effective antimicrobials: a worldwide challenge. Lancet 387 (10014), 168–175.

Lee, J., Zhang, L., 2015. The hierarchy quorum sensing network in *Pseudomonas aeruginosa*. Protein Cell 6 (1), 26–41.

Li, Y.H., Tian, X., 2012. Quorum sensing and bacterial social interactions in biofilms. Sensors (Basel) 12 (3), 2519–2538.

Lidor, O., Al-Quntar, A., Pesci, E.C., Steinberg, D., 2015. Mechanistic analysis of a synthetic inhibitor of the *Pseudomonas aeruginosa* LasI quorum-sensing signal synthase. Sci. Rep. 5, 16569.

Liu, H.B., Lee, J.H., Kim, J.S., Park, S., 2010. Inhibitors of the *Pseudomonas aeruginosa* quorum-sensing regulator, QscR. Biotechnol. Bioeng. 106 (1), 119–126.

Lu, C., Maurer, C.K., Kirsch, B., Steinbach, A., Hartmann, R.W., 2014. Overcoming the unexpected functional inversion of a PqsR antagonist in *Pseudomonas aeruginosa*: an *in vivo* potent antivirulence agent targeting *pqs* quorum sensing. Angew. Chem. Int. Ed. 53 (4), 1109–1112.

Mai, T., Tintillier, F., Lucasson, A., Moriou, C., Bonno, E., Petek, S., Magre, K., Al Mourabit, A., Saulnier, D., Debitus, C., 2015. Quorum sensing inhibitors from *Leucetta chagosensis* Dendy, 1863. Lett. Appl. Microbiol. 61 (4), 311–317.

Marsden, D.M., Nicholson, R.L., Skindersoe, M.E., Galloway, W.R., Sore, H.F., Givskov, M., Salmond, G.P., Ladlow, M., Welch, M., Spring, D.R., 2010. Discovery of a quorum sensing modulator pharmacophore by 3D small-molecule microarray screening. Org. Biomol. Chem. 8 (23), 5313–5323.

Miyairi, S., Tateda, K., Fuse, E.T., Ueda, C., Saito, H., Takabatake, T., Ishii, Y., Horikawa, M., Ishiguro, M., Standiford, T.J., Yamaguchi, K., 2006. Immunization with 3-oxododecanoyl-L-homoserine lactone-protein conjugate protects mice from lethal *Pseudomonas aeruginosa* lung infection. J. Med. Microbiol. 55 (Pt 10), 1381–1387.

Morkunas, B., Galloway, W.R., Wright, M., Ibbeson, B.M., Hodgkinson, J.T., O'connell, K.M., Bartolucci, N., Della Valle, M., Welch, M., Spring, D.R., 2012. Inhibition of the production of the *Pseudomonas aeruginosa* virulence factor pyocyanin in wild-type cells by quorum sensing autoinducer-mimics. Org. Biomol. Chem. 10 (42), 8452–8464.

Muller, C., Birmes, F.S., Niewerth, H., Fetzner, S., 2014. Conversion of the *Pseudomonas aeruginosa* quinolone signal and related alkylhydroxyquinolines by *Rhodococcus sp.* strain BG43. Appl. Environ. Microbiol. 80 (23), 7266–7274.

Muller, C., Birmes, F.S., Ruckert, C., Kalinowski, J., Fetzner, S., 2015. *Rhodococcus erythropolis* BG43 genes mediating *Pseudomonas aeruginosa* quinolone signal degradation and virulence factor attenuation. Appl. Environ. Microbiol. 81 (22), 7720–7729.

Mumford, R., Friman, V.P., 2017. Bacterial competition and quorum-sensing signalling shape the eco-evolutionary outcomes of model *in vitro* phage therapy. Evol. Appl. 10 (2), 161–169.

Musthafa, K.S., Balamurugan, K., Pandian, S.K., Ravi, A.V., 2012a. 2,5-Piperazinedione inhibits quorum sensing-dependent factor production in *Pseudomonas aeruginosa* PAO1. J. Basic Microbiol. 52 (6), 679–686.

Musthafa, K.S., Sivamaruthi, B.S., Pandian, S.K., Ravi, A.V., 2012b. Quorum sensing inhibition in *Pseudomonas aeruginosa* PAO1 by antagonistic compound phenylacetic acid. Curr. Microbiol. 65 (5), 475–480.

Niu, K., Kuk, M., Jung, H., Chan, K., Kim, S., 2017. Leaf extracts of selected gardening trees can attenuate quorum sensing and pathogenicity of *Pseudomonas aeruginosa* PAO1. Indian J. Microbiol. 1–10.

Novotny, L.A., Jurcisek, J.A., Goodman, S.D., Bakaletz, L.O., 2016. Monoclonal antibodies against DNA-binding tips of DNABII proteins disrupt biofilms in vitro and induce bacterial clearance in vivo. EBioMedicine 10, 33–44.

O'loughlin, C.T., Miller, L.C., Siryaporn, A., Drescher, K., Semmelhack, M.F., Bassler, B.L., 2013. A quorum-sensing inhibitor blocks *Pseudomonas aeruginosa* virulence and biofilm formation. Proc. Natl. Acad. Sci. U. S. A. 110 (44), 17981–17986.

Olson, M.E., Ceri, H., Morck, D.W., Buret, A.G., Read, R.R., 2002. Biofilm bacteria: formation and comparative susceptibility to antibiotics. Can. J. Vet. Res. 66 (2), 86–92.

Packiavathy, I.A., Agilandeswari, P., Musthafa, K.S., Pandian, S.K., Ravi, A.V., 2012. Antibiofilm and quorum sensing inhibitory potential of *Cuminum cyminum* and its secondary metabolite methyl eugenol against Gram negative bacterial pathogens. Food Res. Int. 45 (1), 85–92.

Packiavathy, I.A., Priya, S., Pandian, S.K., Ravi, A.V., 2014. Inhibition of biofilm development of uropathogens by curcumin—an anti-quorum sensing agent from *Curcuma longa*. Food Chem. 148, 453–460.

Palliyil, S., Downham, C., Broadbent, I., Charlton, K., Porter, A.J., 2014. High-sensitivity monoclonal antibodies specific for homoserine lactones protect mice from lethal *Pseudomonas aeruginosa* infections. Appl. Environ. Microbiol. 80 (2), 462–469.

Park, J., Kaufmann, G.F., Bowen, J.P., Arbiser, J.L., Janda, K.D., 2008. Solenopsin A, a venom alkaloid from the fire ant *Solenopsis invicta*, inhibits quorum-sensing signaling in *Pseudomonas aeruginosa*. J. Infect. Dis. 198 (8), 1198–1201.

Patterson, A.G., Jackson, S.A., Taylor, C., Evans, G.B., Salmond, G.P.C., Przybilski, R., Staals, R.H.J., Fineran, P.C., 2016. Quorum sensing controls adaptive immunity through the regulation of multiple CRISPR-Cas systems. Mol. Cell 64 (6), 1102–1108.

Pejin, B., Ciric, A., Karaman, I., Horvatovic, M., Glamoclija, J., Nikolic, M., Sokovic, M., 2016. *In vitro* antibiofilm activity of the freshwater bryozoan *Hyalinella punctata*: a case study of *Pseudomonas aeruginosa* PAO1. Nat. Prod. Res. 30 (16), 1847–1850.

Prateeksha, Singh, B.R., Shoeb, M., Sharma, S., Naqvi, A.H., Gupta, V.K., Singh, B.N., 2017. Scaffold of selenium nanovectors and honey phytochemicals for inhibition of *Pseudomonas aeruginosa* quorum sensing and biofilm formation. Front. Cell. Infect. Microbiol. 7, 93.

Pustelny, C., Albers, A., Buldt-Karentzopoulos, K., Parschat, K., Chhabra, S.R., Camara, M., Williams, P., Fetzner, S., 2009. Dioxygenase-mediated quenching of quinolone-dependent quorum sensing in *Pseudomonas aeruginosa*. Chem. Biol. 16 (12), 1259–1267.

Qin, X., Sun, Q., Yang, B., Pan, X., He, Y., Yang, H., 2017. Quorum sensing influences phage infection efficiency via affecting cell population and physiological state. J. Basic Microbiol. 57 (2), 162–170.

Quintana, J., Brango-Vanegas, J., Costa, G.M., Castellanos, L., Arévalo, C., Duque, C., 2015. Marine organisms as source of extracts to disrupt bacterial communication: bioguided isolation and identification of quorum sensing inhibitors from *Ircinia felix*. Rev. Bras 25, 199–207.

Rajamani, S., Bauer, W.D., Robinson, J.B., Farrow 3rd, J.M., Pesci, E.C., Teplitski, M., Gao, M., Sayre, R.T., Phillips, D.A., 2008. The vitamin riboflavin and its derivative lumichrome activate the LasR bacterial quorum-sensing receptor. Mol. Plant-Microbe Interact. 21 (9), 1184–1192.

Rasamiravaka, T., Jedrzejowski, A., Kiendrebeogo, M., Rajaonson, S., Randriamampionona, D., Rabemanantsoa, C., Andriantsimahavandy, A., Rasamindrakotroka, A., Duez, P., El Jaziri, M., Vandeputte, O.M., 2013. Endemic malagasy *Dalbergia* species inhibit quorum sensing in *Pseudomonas aeruginosa* PAO1. Microbiology 159 (Pt 5), 924–938.

Rasamiravaka, T., Ngezahayo, J., Pottier, L., Ribeiro, S.O., Souard, F., Hari, L., Stevigny, C., Jaziri, M.E., Duez, P., 2017. Terpenoids from *Platostoma rotundifolium* (Briq.) A. J. Paton alter the expression of quorum sensing-related virulence factors and the formation of biofilm in *Pseudomonas aeruginosa* PAO1. Int. J. Mol. Sci. 18(6).

Rasmussen, T.B., Givskov, M., 2006. Quorum sensing inhibitors: a bargain of effects. Microbiology 152 (Pt 4), 895–904.

Rasmussen, T.B., Skindersoe, M.E., Bjarnsholt, T., Phipps, R.K., Christensen, K.B., Jensen, P.O., Andersen, J.B., Koch, B., Larsen, T.O., Hentzer, M., Eberl, L., Hoiby, N., Givskov, M., 2005. Identity and effects of quorum-sensing inhibitors produced by *Penicillium* species. Microbiology 151 (Pt 5), 1325–1340.

Ren, D., Zuo, R., Gonzalez Barrios, A.F., Bedzyk, L.A., Eldridge, G.R., Pasmore, M.E., Wood, T.K., 2005. Differential gene expression for investigation of *Escherichia coli* biofilm inhibition by plant extract ursolic acid. Appl. Environ. Microbiol. 71 (7), 4022–4034.

Rivas Caldas, R., Boisrame, S., 2015. Upper aero-digestive contamination by *Pseudomonas aeruginosa* and implications in cystic fibrosis. J. Cyst. Fibros. 14 (1), 6–15.

Rutherford, S.T., Bassler, B.L., 2012. Bacterial quorum sensing: its role in virulence and possibilities for its control. Cold Spring Harb. Perspect. Med. 2 (11).

Sahner, J.H., Brengel, C., Storz, M.P., Groh, M., Plaza, A., Muller, R., Hartmann, R.W., 2013. Combining *in silico* and biophysical methods for the development of *Pseudomonas aeruginosa* quorum sensing inhibitors: an alternative approach for structure-based drug design. J. Med. Chem. 56 (21), 8656–8664.

Sarabhai, S., Sharma, P., Capalash, N., 2013. Ellagic acid derivatives from *Terminalia chebula* Retz. downregulate the expression of quorum sensing genes to attenuate *Pseudomonas aeruginosa* PAO1 virulence. PLoS One 8 (1), e53441.

Saurav, K., Costantino, V., Venturi, V., Steindler, L., 2017. Quorum sensing inhibitors from the sea discovered using bacterial N-acyl-homoserine lactone-based biosensors. Mar. Drugs 15 (3), 53.

Shannon, E., Abu-Ghannam, N., 2016. Antibacterial derivatives of marine algae: an overview of pharmacological mechanisms and applications. Mar. Drugs 14 (4).

Singh, B.R., Singh, B.N., Singh, A., Khan, W., Naqvi, A.H., Singh, H.B., 2015. Mycofabricated biosilver nanoparticles interrupt *Pseudomonas aeruginosa* quorum sensing systems. Sci. Rep. 5, 13719.

Singh, V.K., Mishra, A., Jha, B., 2017. Anti-quorum sensing and anti-biofilm activity of *Delftia tsuruhatensis* extract by attenuating the quorum sensing-controlled virulence factor production in *Pseudomonas aeruginosa*. Front. Cell. Infect. Microbiol. 7, 337.

Sio, C.F., Otten, L.G., Cool, R.H., Diggle, S.P., Braun, P.G., Bos, R., Daykin, M., Camara, M., Williams, P., Quax, W.J., 2006. Quorum quenching by an N-acyl-homoserine lactone acylase from *Pseudomonas aeruginosa* PAO1. Infect. Immun. 74 (3), 1673–1682.

Skindersoe, M.E., Ettinger-Epstein, P., Rasmussen, T.B., Bjarnsholt, T., De Nys, R., Givskov, M., 2008. Quorum sensing antagonism from marine organisms. Mar. Biotechnol. (NY) 10 (1), 56–63.

Soković, M., Ćirić, A., Glamočlija, J., Nikolić, M., Van Griensven, L.J., 2014. *Agaricus blazei* hot water extract shows anti quorum sensing activity in the nosocomial human pathogen *Pseudomonas aeruginosa*. Molecules 19 (4), 4189–4199.

Song, Z., Kong, K.F., Wu, H., Maricic, N., Ramalingam, B., Priestap, H., Quirke, J.M.E., Høiby, N., Mathee, K., 2010. Panax ginseng has anti-infective activity against opportunistic pathogen *Pseudomonas aeruginosa* by inhibiting quorum sensing, a bacterial communication process critical for establishing infection. Phytomedicine 17 (13), 1040–1046.

Sonnleitner, E., Gonzalez, N., Sorger-Domenigg, T., Heeb, S., Richter, A.S., Backofen, R., Williams, P., Huttenhofer, A., Haas, D., Blasi, U., 2011. The small RNA PhrS stimulates synthesis of the *Pseudomonas aeruginosa* quinolone signal. Mol. Microbiol. 80 (4), 868–885.

Stoltz, D.A., Ozer, E.A., Ng, C.J., Yu, J.M., Reddy, S.T., Lusis, A.J., Bourquard, N., Parsek, M.R., Zabner, J., Shih, D.M., 2007. Paraoxonase-2 deficiency enhances *Pseudomonas aeruginosa* quorum sensing in murine tracheal epithelia. Am. J. Phys. Lung Cell. Mol. Phys. 292 (4), L852–L860.

Storz, M.P., Brengel, C., Weidel, E., Hoffmann, M., Hollemeyer, K., Steinbach, A., Müller, R., Empting, M., Hartmann, R.W., 2013. Biochemical and biophysical analysis of a chiral PqsD inhibitor revealing tight-binding behavior and enantiomers with contrary thermodynamic signatures. ACS Chem. Biol. 8 (12), 2794–2801.

Sun, S., Dai, X., Sun, J., Bu, X., Weng, C., Li, H., Zhu, H., 2016. A diketopiperazine factor from *Rheinheimera aquimaris* QSI02 exhibits anti-quorum sensing activity. Sci. Rep. 6, 39637.

Sunder, A.V., Utari, P.D., Ramasamy, S., Van Merkerk, R., Quax, W., Pundle, A., 2017. Penicillin V acylases from gram-negative bacteria degrade N-acylhomoserine lactones and attenuate virulence in *Pseudomonas aeruginosa*. Appl. Microbiol. Biotechnol. 101 (6), 2383–2395.

Tang, K., Su, Y., Brackman, G., Cui, F., Zhang, Y., Shi, X., Coenye, T., Zhang, X.H., 2015. MomL, a novel marine-derived N-acyl homoserine lactonase from *Muricauda olearia*. Appl. Environ. Microbiol. 81 (2), 774–782.

Tung, T.T., Jakobsen, T.H., Dao, T.T., Fuglsang, A.T., Givskov, M., Christensen, S.B., Nielsen, J., 2017. Fusaric acid and analogues as Gram-negative bacterial quorum sensing inhibitors. Eur. J. Med. Chem. 126, 1011–1020.

Van Hecke, O., Wang, K., Lee, J.J., Roberts, N.W., Butler, C.C., 2017. Implications of antibiotic-resistance for patients' recovery from common infections in the community: a systematic review and meta-analysis. Clin. Infect. Dis. 65 (3), 371–382.

Vandeputte, O.M., Kiendrebeogo, M., Rajaonson, S., Diallo, B., Mol, A., El Jaziri, M., Baucher, M., 2010. Identification of catechin as one of the flavonoids from *Combretum albiflorum* bark extract that reduces the production of quorum-sensing-controlled virulence factors in *Pseudomonas aeruginosa* PAO1. Appl. Environ. Microbiol. 76 (1), 243–253.

Vandeputte, O.M., Kiendrebeogo, M., Rasamiravaka, T., Stevigny, C., Duez, P., Rajaonson, S., Diallo, B., Mol, A., Baucher, M., El Jaziri, M., 2011. The flavanone naringenin reduces the production of quorum sensing-controlled virulence factors in *Pseudomonas aeruginosa* PAO1. Microbiology 157 (Pt 7), 2120–2132.

Velazquez-Velazquez, J.L., Santos-Flores, A., Araujo-Melendez, J., Sanchez-Sanchez, R., Velasquillo, C., Gonzalez, C., Martinez-Castanon, G., Martinez-Gutierrez, F., 2015. Anti-biofilm and cytotoxicity activity of impregnated dressings with silver nanoparticles. Mater. Sci. Eng. C Mater. Biol. Appl. 49, 604–611.

Wahjudi, M., Papaioannou, E., Hendrawati, O., Van Assen, A.H., Van Merkerk, R., Cool, R.H., Poelarends, G.J., Quax, W.J., 2011. PA0305 of *Pseudomonas aeruginosa* is a quorum quenching acylhomoserine lactone acylase belonging to the Ntn hydrolase superfamily. Microbiology 157 (Pt 7), 2042–2055.

Walker, T.S., Bais, H.P., Deziel, E., Schweizer, H.P., Rahme, L.G., Fall, R., Vivanco, J.M., 2004. *Pseudomonas aeruginosa*-plant root interactions. Pathogenicity, biofilm formation, and root exudation. Plant Physiol. 134 (1), 320–331.

Weidel, E., De Jong, J.C., Brengel, C., Storz, M.P., Braunshausen, A., Negri, M., Plaza, A., Steinbach, A., Muller, R., Hartmann, R.W., 2013. Structure optimization of 2-benzamidobenzoic acids as PqsD inhibitors for *Pseudomonas aeruginosa* infections and elucidation of binding mode by SPR, STD NMR, and molecular docking. J. Med. Chem. 56 (15), 6146–6155.

Weiland-Brauer, N., Kisch, M.J., Pinnow, N., Liese, A., Schmitz, R.A., 2016. Highly effective inhibition of biofilm formation by the first metagenome-derived AI-2 quenching enzyme. Front. Microbiol. 7, 1098.

World Health Organization, 2017. Antibacterial agents in clinical development. World Health Organization, Geneva.

Yang, L., Rybtke, M.T., Jakobsen, T.H., Hentzer, M., Bjarnsholt, T., Givskov, M., Tolker-Nielsen, T., 2009. Computer-aided identification of recognized drugs as *Pseudomonas aeruginosa* quorum-sensing inhibitors. Antimicrob. Agents Chemother. 53 (6), 2432–2443.

Yang, L., Liu, Y., Sternberg, C., Molin, S., 2010. Evaluation of enoyl-acyl carrier protein reductase inhibitors as *Pseudomonas aeruginosa* quorum-quenching reagents. Molecules 15 (2), 780–792.

Yaniv, K., Golberg, K., Kramarsky-Winter, E., Marks, R., Pushkarev, A., Beja, O., Kushmaro, A., 2017. Functional marine metagenomic screening for anti-quorum sensing and anti-biofilm activity. Biofouling 33 (1), 1–13.

Zender, M., Klein, T., Henn, C., Kirsch, B., Maurer, C.K., Kail, D., Ritter, C., Dolezal, O., Steinbach, A., Hartmann, R.W., 2013. Discovery and biophysical characterization of 2-amino-oxadiazoles as novel antagonists of PqsR, an important regulator of *Pseudomonas aeruginosa* virulence. J. Med. Chem. 56 (17), 6761–6774.

Zeng, Z., Qian, L., Cao, L., Tan, H., Huang, Y., Xue, X., Shen, Y., Zhou, S., 2008. Virtual screening for novel quorum sensing inhibitors to eradicate biofilm formation of *Pseudomonas aeruginosa*. Appl. Microbiol. Biotechnol. 79 (1), 119–126.

Zhou, L., Zheng, H., Tang, Y., Yu, W., Gong, Q., 2013. Eugenol inhibits quorum sensing at sub-inhibitory concentrations. Biotechnol. Lett. 35 (4), 631–637.

CHAPTER 10

Cyclic Peptides in Neurological Disorders: The Case of Cyclo(His-Pro)

Ilaria Bellezza, Matthew J. Peirce, Alba Minelli

Department of Experimental Medicine, University of Perugia, Perugia, Italy

1 Introduction

Bacterial quorum sensing (QS) is a cell-to-cell communication system based on the production, secretion, and detection of signals, named autoinducers. These signals are generated in response to variations in population level variables such as the density of bacterial population and changes in the makeup of its constituent species, and simultaneously regulate gene expression and coordinate collective behaviors. In this way, diverse bacterial communities acquire the capacity to act as a group and switch on virulence factor production, biofilm formation, and resistance development. Indeed, many microbial processes would be ineffective if performed by a single bacterial cell acting alone (Ng and Bassler, 2009; Bassler and Vogel, 2013; Cornforth et al., 2014; Schluter et al., 2016). QS processes have been found in Gram negative and Gram positive bacteria species, but interestingly, numerous examples of "interkingdom" signals (i.e., signals that cross taxonomic borders, for example between prokaryotic and eukaryotic cells) have also been identified. From the beginning, prokaryotes and eukaryotes survive and coexist by sensing and responding to signals produced and released by the other (Rosier et al., 2016). Once synthesized, the signal, whose concentration is proportional to population density, is secreted. When the levels of signal exceed a threshold value, the signal is detected by a QS receptor that engages a downstream transduction cascade to launch a high cell density gene expression program.

In Gram negative bacteria, AIs molecules are drawn from several chemical classes, such as acyl homoserine lactones (AHLs), alkylquinolones, α-hydroxyketones, and diffusible signal factors (fatty acid-like compounds), synthesized from common metabolites via a single synthase or a series of enzymatic reactions (Hawver et al., 2016; Ryan et al., 2015). AI can freely cross cell membranes, and when their extracellular concentration is sufficiently high, they can diffuse back to act in an autocrine manner on the producing cell by binding cytosolic receptors. The latter act as transcription factors and, upon AI binding, regulate the expression of QS genes.

Quorum Sensing. https://doi.org/10.1016/B978-0-12-814905-8.00010-1

Typically, Gram negative bacteria use QS systems homologous to the luminescence (*Lux*) controlling operon (LuxI/LuxR system) originally identified in the luminescent marine bacterium *Vibrio fischeri*. The LuxI homolog acts as an AI synthase while the LuxR homolog is a cytosolic transcription factor AI receptor which, upon AI binding, acquires transactivating capacity and mediates the transcription of target genes (Hawver et al., 2016; Ryan et al., 2015).

The homologous system in Gram positive bacteria is slightly more complex and is mediated by short oligopeptides. Autoinducer peptides (AIPs) are produced by the AIP synthase, and after proteolytic processing are actively secreted via a transporter, usually an ATP binding cassette transporter. When their extracellular concentration reaches a threshold level, they can be transported back into the cytoplasm, where they are detected by a QS transcription factor. At high cell density, the AIPs, once released into the extracellular space via a transporter, undergo posttranslational modifications and bind to transmembrane receptors, thus triggering a phosphorylation cascade that controls the downstream QS response (Cook and Federle, 2014; Hawver et al., 2016; Monnet et al., 2016). The receptors for Gram positive QS comprise a family with four members—Rap, NprR, PlcR, and PrgX (Rocha-Estrada et al., 2010)—collectively termed *RNPP*. Rap is an aspartyl phosphate phosphatase and transcriptional activator protein, whereas NprR, PlcR, and PrgX are DNA-binding transcription factors. In addition, NprR is a neutral protease regulator, PlcR is a phospholipase C regulator, and PrgX regulates the plasmid conjugative transfer (Rocha-Estrada et al., 2010; Zouhir et al., 2013; Cook and Federle 2014). A third type of molecule, a furanosyl borate diester, also called autoinducer-2 (AI-2), represents a universal signal for interspecies communication (Hense and Schuster, 2015).

Thus while specific details vary, generically, QS systems connect the release of a diffusible messenger molecule to the targeted regulation of a specific and context-appropriate program of gene expression. QS enables cooperative behavior distinct from traditional paracrine signaling, where sender and receiver cells are different; in QS the sender and receiver are frequently one and the same. Moreover, these QS behaviors have characteristic ecological functionalities and evolutionary properties (Diggle et al., 2007). QS systems integrate and process information from the physical environment to optimize their activities at a community level, enabling the cooperative and coordinated behavior of a diverse cell population. Thus QS systems are distributed over a wide taxonomic range, such as fungi, plants (algae), animals, and possibly even viruses, where host cell lysis may depend on the virus concentration (Hallmann, 2011; Hogan, 2006; Moussa et al., 2012; Sprague and Winans, 2006; Weitz et al., 2008).

2 Cyclic Peptides

Many biologically active peptides, consisting of either the canonical 20 amino acids or nonproteogenic amino acids, have been discovered in bacteria (Clardy and Walsh, 2004). These linear or cyclic peptides can be synthesized in the ribosomes and later processed or produced independently of ribosomes by peptide synthetases (Clardy et al., 2006). Peptides, as well as proteins, are bioactive molecules constituting attractive initial leads for a rational drug design (Tapeinou et al., 2015). Linear peptides, because of their susceptibility to proteolytic degradation, have not historically been molecules of choice for the pharmaceutical industry. However, the introduction of modifications can render these molecules more attractive. Indeed, cyclization leads to increased stability, *N*-methylation can increase membrane permeability and stability, the incorporation of unnatural amino acids increases specificity and stability, PEGylation (i.e., the attachment or amalgamation of polyethylene glycol (PEG) polymer chains) is capable of reducing clearance, and assorted structural constraints (e.g., disulfide bonds), as well as recent progress in "stapled" peptides, improved potency and specificity. These modifications are currently employed as promising new modalities for future therapeutics (Tapeinou et al., 2015; Verdine and Hilinski, 2012). In particular, peptides, once cyclized, have superior binding affinities and reduced entropic cost associated with receptor binding. Indeed, confining a peptide into a cyclic structure reduces the conformational freedom of its parent linear structure and enhances its metabolic stability, bioavailability, and specificity. Naturally occurring cyclic peptides have a wide variety of unusual and potent biological activities: many of these compounds control intra- and interspecies bacterial virulence and bacterial-host interactions, penetrate cells by passive diffusion, and some, like the clinically important drug cyclosporine A, are orally bioavailable (Bockus et al., 2013). Cyclic dipeptides (CDPs), such as enniatins, can be produced by plant pathogens and act as mycotoxins by disturbing the physiological behavior of the host cells. Moreover, they display several other biological effects such as insecticidal, antifungal, and antibacterial activities. Indeed, by exerting a potent cytotoxic effect on human and animal cell lines, they are considered potential anticancer drugs (Prosperini et al., 2017). Hence, cyclic peptides are promising lead compounds for drug development in a wide range of clinical settings and represent attractive biosynthetic targets.

The QS peptides described in the literature are stored in the curated Quorumpeps database (http://quorumpeps.ugent.be), giving details of the peptides and QS processes (receptor, activity, species) in which they are implicated (Wynendaele et al., 2015). The analysis of the 231 QS peptides, listed in Quorumpeps, shows that cyclic peptides occupy a distinct portion of the QS peptide space according to their size distribution, compactness, lipophilicity/hydrophobicity, presence of aromatic amino acids, and sulfur atoms. On the other hand, results of the species distribution indicate that most of the Gram positive bacteria synthesize chemically similar peptides. Notably, cyclic peptides, named θ-defensins, have also been found in leukocytes of rhesus macaques and baboons (Lehrer et al., 2012), where they are involved in defense from a range of viral (HIV-1, HSV, and influenza A viruses, coronavirus severe acute respiratory syndrome) and bacterial pathogens (*Bacillus anthracis* spores), therefore representing interesting potential therapeutic agents.

2.1 Cyclic Dipeptides

CDPs, also known as 2,5-diketopiperazines, are a family of small and biologically active molecules, mostly acting as QS effectors, that contain a family-defining CDP core/scaffold structure (Fig. 1), and are produced by proteobacterial species as well as by humans (Bellezza et al., 2014a,b; Borthwick, 2012; Cornacchia et al., 2012; Minelli et al., 2008; Mishra et al., 2017; Prasad, 1995). CDPs are a class of cyclic organic compounds in which the two nitrogen atoms of a piperazine 6-membered ring form amide linkages. The nomenclature of CDPs is indicated by the three-letter code for each of the two amino acids, plus a prefix to designate the absolute configuration (e.g., cyclo(L-Xaa-L-Yaa)). CDPs can be configured as both cis and trans-isoforms, but cis configurations are predominant (Eguchi and Kakuta, 1974). Various amino acid modifications confer diversified chemical and biological functions. CDPs exhibit better biological activity than their linear counterparts due to their higher stability, protease resistance, and conformational rigidity, all factors that increase their ability to specifically interact with biological targets (Liskamp et al., 2011; Menegatti et al., 2013). They constitute a large class of secondary metabolites produced by bacteria, fungi, plants, and animals (Borthwick, 2012; Giessen and Marahiel, 2014; Huang et al., 2010; Mishra et al., 2017; Prasad, 1995). Indeed, approximately 90% of CDP producers are bacterial (Giessen and Marahiel, 2014). The CDP scaffold can be synthesized either by purely chemically means using different solid phases or under reflux conditions in solution (Borthwick, 2012; Gonzalez et al., 2012) or, more naturally, by biosynthetic enzymes called nonribosomal peptide synthetases (NRPSs) and CDP synthases (CDPSs; Belin et al., 2012; Giessen and Marahiel, 2014). Common chemical synthesis of CDPs includes the condensation of individual amino acids at high temperature. Dipeptides substituted with an amine at one terminus and an ester at the other can also spontaneously cyclize to form a CDP. However, conditions must be optimized in order to force a cyclization reaction and to limit racemization. This is the procedure most commonly used for the chemical synthesis of CDP. Cyclization of amino dipeptide esters can also be carried out under thermal conditions, normally by refluxing them in high boiling solvents such as toluene or xylene for 24 h (Borthwick, 2012). In addition, CDPs are often products of unwanted side reactions or degradation products of oligo- and polypeptides in processed food and beverages (Borthwick and Da Costa, 2017; Prasad, 1995). They are frequently formed during the chemical degradation of products in roasted coffee, stewed

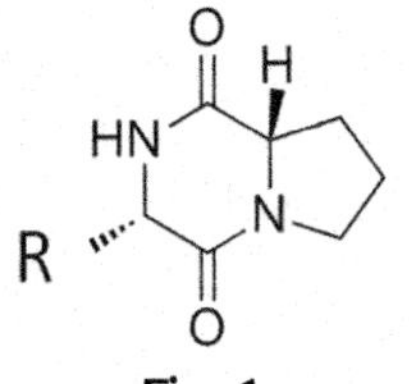

Fig. 1
Family-defining core/scaffold structure of histidine containing cyclic dipeptides.

beef, and beer (Chen et al., 2009; Gautschi et al., 1997; Ginz and Engelhardt, 2000). Nonenzymatic processes can also lead to the formation of functional CDPs in various organisms, as described for cyclo(L-His-L-Pro) (Bellezza et al., 2014a,b; Minelli et al., 2008). In mammals, cyclo(His-Pro) (CHP) is obtained from the action of pyroglutamate aminopeptidase on the thyrotropin-releasing hormone (TRH, pGlu-His-Pro). The resulting dipeptide is then nonenzymatically cyclized to CHP. The proline induces constraints that promote the cis-conformation of the peptide bond between the histidine and the proline, thereby facilitating cyclization, which generates the CDP scaffold.

As for the enzymatic pathways of CDP formation, two unrelated biosynthetic enzyme families catalyze the formation of CDPs: NRPSs and CDPSs. It has been shown that CDP scaffolds can be synthesized by one or more specialized NRPSs, either via specific biosynthetic pathways or via the premature release of dipeptidyl intermediates during the chain elongation process. The NRPS genes for certain peptides are usually organized in one operon in prokaryotes, and in a gene cluster in eukaryotes (Schwarzer et al., 2003). NRPSs are large modular enzymes, which simultaneously act as a template and as biosynthetic machinery. Each module is responsible for the incorporation of one amino acid into the final peptide, and can be further subdivided into the catalytic domains responsible for specific synthetic steps during peptide synthesis (Felnagle et al., 2008). In each module, NRPSs consist of three necessary domains: an adenylation (A) domain; a thiolation (T) domain, posttranslationally modified with a $4'$-phosphopantetheinyl ($4'$-Ppant) arm, also termed the peptidyl carrier protein (PCP) domain; and a condensation (C) domain, separated by short spacer regions of approximately 15 amino acids. The A domain selects, activates, and loads the monomer onto the PCP domain. Here, the thiol group of the $4'$-Ppantarm of the T domain mediates the nucleophilic attack of the adenylated amino acid. Subsequent peptide bond formation between two adjacent T-bound aminoacyl intermediates is catalyzed by the C domains (Belin et al., 2012). Another essential NRPS catalytic unit is the thioesterase (TE) domain, which is located in the C-terminus and catalyzes peptide release by either hydrolysis or macrocyclization. In addition, modification domains can be integrated into NRPS modules at different locations to modify the incorporated amino acids. Epimerization and *N*-methyltransferase domains catalyze the generation of D- and methylated amino acids, respectively (Koglin and Walsh, 2009; Strieker et al., 2010). NRPSs rely not only on the 20 canonical amino acids, but also use several different building blocks, including nonproteinogenic amino acids, and this contributes to the structural diversity of nonribosomal peptides and their differential biological activities (Koglin and Walsh, 2009). CDPs, once synthesized by NRPSs, can be further modified by tailoring enzymes, usually encoded by genes clustered with the NRPS genes. The majority of known NRPS-derived CDPs are produced by fungi, whereas few bacteria are recognized as NRPS-derived CDP producers. Many CDPs can be formed by dedicated NRPS pathways, such as brevianamide F, erythrochelin, ergotamine, roquefortine C, acetylaszonalenin, thaxtomin A, gliotoxin, and sirodesmin PL (Balibar and Walsh, 2006; Correia et al., 2003; García-Estrada et al., 2011; Gardiner et al., 2004; Healy et al., 2002; Maiya et al., 2006; Lazos et al., 2010; Yin et al., 2009). In a few cases, CDPs can be formed by NRPSs during the synthesis of longer peptides, as truncated side products, as in the biosynthesis of cyclo(D-Phe-L-Pro) and cyclomarazine A (Gruenewald et al., 2004; Schultz et al., 2008). Biosynthesis of CDPs can also be CDPS-mediated: CDPS are a family of tRNA-dependent peptide bond-forming enzymes that do not require amino acid charging. CDPSs share a common architecture reminiscent of the catalytic domain of class-Ic amino acid tRNA synthetases (aaRSs), such as TyrRS and TrpRS (Sauguet et al., 2011). Both CDPSs and class-Ic aaRSs comprise well conserved Rossmann-fold domains, structural features associated with binding of nucleotides such as flavin adenine dinucleotide, nicotinamide adenine dinucleotide (NAD^+), and nicotinamide adenine dinucleotide phosphate ($NADP^+$), along with a helical connective polypeptide 1 (CP1) subdomain. However, class-IcaaRSs possess signature motifs involved in ATP binding (HIGH and KMSKS sequences) that are not present in CDPSs. In addition, CDPSs do not possess a distinct tRNA-binding domain, but rather contain a large patch of positively charged residues located in helix α4, which are important for the binding of aminoacyl-tRNA substrates. All these observed differences between CDPSs and their ancestral aaRSs result in unique enzymes for CDP biosynthesis. CDPSs use amino acid tRNAs as substrates to catalyze the formation of CDP peptide bonds (Belin et al., 2012; Giessen and Marahiel, 2012; Giessen et al., 2013; Gondry et al., 2009), diverting two aminoacyl-tRNAs from their essential role in ribosomal protein synthesis for use as substrates and catalyzing the formation of the two peptide bonds required for CDP formation (Lahoud and Hou, 2010). The synthesis process is initiated by the binding of the first aminoacyl substrate, likely involving ionic interactions between the negatively-charged ribose-phosphate tRNA backbone and the positive charges in helix α4 (Bonnefond et al., 2011; Sauguet et al., 2011). Hence, by using aminoacyl-tRNAs as substrates, CDPSs represent a direct link between primary and secondary metabolism. The catalytic mechanism of CDPSs can be described using a ping-pong model. All CDPSs possess two surface-accessible pockets that contain active site residues important for substrate selection and catalysis. The different aminoacyl binding sites for the two aa-tRNA substrates are termed pocket 1 (P1) and pocket 2 (P2). Upon specific recognition of the first substrate, the first aminoacyl group is transferred to the conserved serine residue of P1. Here, interaction between the tRNA moiety and basic residues in the α4 helix generates an aminoacyl-enzyme intermediate (Moutiez et al., 2014). At the same time, the aminoacyl moiety of the second aa-tRNA interacts with P2 through the α6–α7 loop. Finally, the aminoacyl-enzyme intermediate reacts with the second aa-tRNA to generate a dipeptidyl-enzyme intermediate, which undergoes intramolecular cyclization through the involvement of a conserved tyrosine, leading to the CDP scaffold as the final product. These CDPs can be modified by closely associated tailoring enzymes. There are approximately 163 putative CDPS genes identified so far, and of these, 150 are reported in bacteria, distributed among six phyla (Actinobacteria, Bacteroidetes, Chlamydiae, Cyanobacteria, Firmicutes, and Proteobacteria). Most known CDPSs are found in Actinobacteria, with 77 CDPSs reported to date. Twelve CDPSs are distributed among four eukaryotic phyla (Ascomycota, Annelida, Ciliophora, and Cnidaria), and one archaeon (*Haloterrigena hispanica*) CDPS has also been reported (Belin et al., 2012; Giessen and Marahiel, 2014; Tommonaro et al., 2012). Some bacterial CDPSs have been fully characterized, such as albonoursin in *Streptomyces noursei*, pulcherrimin in *Bacillus subtilis*, and mycocyclosin in *Mycobacterium tuberculosis* (Belin et al., 2012; Giessen et al., 2013).

The biosynthetic enzymes are usually physically and, as alluded to previously, transcriptionally associated with tailoring enzymes that specifically modify CDP-containing natural products. Putative tailoring enzymes that modify the initially assembled CDP scaffold can be found in almost all NRPS and CDPS gene clusters, and are responsible for introducing functional groups crucial for the biological activities of CDPs. In CDPS-dependent pathways, a large variety of different modification enzymes are found in close association with the respective CDPS genes (Belin et al., 2012; Giessen and Marahiel, 2014), including different types of oxidoreductases, hydrolases, transferases, and ligases. The most prevalent putative tailoring enzymes in CDPS clusters are cyclic dipeptide oxidases (CDOs). CDOs are composed of two distinct small subunits that assemble into an apparent megadalton protein complex. Depending on the substrate, the CDO can sequentially perform one or two dehydrogenation reactions. The precise reaction mechanism for this has not been elucidated, although three different scenarios have been proposed: direct dehydrogenation, α-hydroxylation followed by loss of water, and imine formation with subsequent rearrangement of the enamine (Gondry et al., 2001). Known CDOs include at least seven distinct P450 enzymes, five different types of α-ketoglutarate/FeII-dependent oxygenases, and three distinct flavin-containing mono-oxygenases. In addition to oxidoreductases, a large number of different C-, N-, and *O*-methyltransferases, α/β-hydrolases, peptide ligases, and acyl-CoA transferases have been found in CDPS gene clusters in which different transcription factors belonging to the LuxR and MarR families, among others, are observed. They are usually involved in regulating various processes in response to environmental stimuli like toxic chemicals and antibiotics, which may hint at the biological functions of CDPS-dependent modified CDPs (Ellison and Miller, 2006). Regarding NRPS-dependent pathways, a similar variety of modification enzymes has been reported, and again, enzymes that modulate the oxidation of the CDP scaffold and side chains are the most numerous (Belin et al., 2012). One distinguishing feature of fungal NRPS gene clusters is the prevalence of different prenyltransferases, which perform prenylations and reverse prenylations at various positions of tryptophan-containing CDP scaffolds (Yu et al., 2012). Judging by the diverse set of putative modification enzymes found within NRPS and CDPS gene clusters, it is assumed that highly modified CDPs represent a diverse family of microbial natural products with varied functions.

Moreover, it is worth noting that the CDP core, besides rendering these molecules resistant to proteolysis, also enables the crossing of the intestinal barrier and blood-brain barrier (BBB; Beck et al., 2012; Teixidó et al., 2009). Thus the combination of flexibility and stability provides the CDP molecules with biological properties and a wide array of therapeutic possibilities (Bellezza et al., 2014a,b). The ability to inhibit plasminogen activator inhibitor-1

(PAI-1), enabling intervention in cardiovascular disease and blood clotting functions (Einholm et al., 2003), was the first discovered biological action of CDPs, later followed by the discovery of antibacterial (Rhee, 2004), antitumor (Nicholson et al., 2006), antifungal, and antiviral activities (Kwak et al., 2013; Kwak et al., 2014; Mishra et al., 2017). Korean fermented vegetable kimchi is a rich source of Pro-based CDPs that have activities against multidrug resistant bacteria (Liu et al., 2017), and cyclo(L-Val-L-Pro) and cyclo(L-Phe-L-Pro), produced by vegetables fermented with *Lactobacillus plantarum* LBP-K10, can inhibit the growth of *Candida albicans* (Kwak et al., 2014). Cyclo(D-Tyr-D-Phe), extracted from fermented modified nutrient broth of *Bacillus* sp. N strain associated with the rhabditid entomopathogenic nematode, shows significant antitumor activity against A549 cells without cytotoxicity for normal fibroblast cells (Kumar et al., 2013). The pleiotropic actions of CDPs are reflected in their ability to bind an array of targets: by binding with high affinity to oxytocin receptors, thus acting as antagonists, CDPs can inhibit ejaculation (Borthwick et al., 2012; Clément et al., 2013). CDPs released by the coldwater marine sponge *Geodia barretti* synergistically exert chemical defense (Sjögren et al., 2011). Moreover, CHP, a catabolic product of TRH (thyrotropin releasing hormone), can control blood glucose levels (Choi et al., 2012; Jung et al., 2011, 2016; Koo et al., 2011; Lee et al., 2015, 2013; Park et al., 2012) and, associated to zinc, has already been patented in the United States as an antidiabetic drug with no side effects in humans (Uyemura et al., 2010).

2.2 CDPs in QS Systems

CDPs have been isolated from marine sponges (Schmitz et al., 1983), Gram negative (Jayatilake et al., 1996) and Gram positive marine bacteria (Stierle et al., 1988). Therefore, marine bacteria are a promising source for this class of bioactive compounds. It is known that marine sponges are an abundant source of novel microorganisms that produce compounds with potential pharmacologic activity (Hentschel et al., 2001). Marine sponges with their surfaces and internal spaces provide a highly specialized environmental niche containing high numbers of bacteria. They exceed the bacteria contained in seawater by two or three orders of magnitude (Engel et al., 2002; Friedrich et al., 2001). Sponge-bacteria associations are widely distributed and evolutionarily ancient, with a direct relationship between the size and density of sponges and the content of bacterial associates. Several data suggest an advantageous coexistence of microorganisms and sponges (Proksch et al., 2002). Autotrophy of cyanobacteria can provide host sponges with additional carbon sources and fixed nitrogen, specific associations with heterotrophic bacteria facilitate the metabolism of a wide range of organic compounds, and associated bacterial and cyanobacterial communities produce secondary metabolites that enhance the chemical defense of the host (Abbamondi et al., 2014; De Rosa et al., 2003). There is experimental evidence that sponge-associated microflora are species-specific (Friedrich et al., 2001; Schmidt et al., 2000; Webster and Hill, 2001; Webster et al., 2001) and represent a stable population (Friedrich et al., 2001;

Webster and Hill, 2001; Webster et al., 2001) capable of communicating with the sponge itself. Marine sponge-associated bacteria secrete QS signals such as *N*-acyl homoserine lactones (AHLs; Taylor et al., 2004) and CDPs (Tommonaro et al., 2012). Holden et al. (1999) reported that several Gram negative bacteria produced and secreted CDPs which, in turn, can activate and/or antagonize other LuxR-based QS systems. *Pseudomonas putida* WCS358 can produce and secrete four CDPs, some capable of interacting with the QS LuxI and LuxR homologues (Degrassi et al., 2002). A set of CDPs was isolated from a range of Gram negative bacteria and reported to modulate LuxR, TraR, or LasR activity in sensitive AHL biosensor strains previously considered specific for AHLs (Degrassi et al., 2002; Holden et al., 1999; Park et al., 2006). Therefore, the capacity of QS signals generated in one organism has been demonstrated to regulate the behavior of a second organism. Thus CDPs, isolated either individually or as mixtures from culture supernatants of *Pseudomonas aeruginosa*, *Pseudomonas fluorescens*, *P. putida*, *Pseudomonas alcaligenes*, *Proteus mirabilis*, *Enterobacter agglomerans*, *Vibrio vulnificus*, and *Citrobacter freundii*, represent interspecies and cross-kingdom signals (Campbell et al., 2009). Cyclo(D-Ala-L-Val) and cyclo(L-Pro-L-Tyr) inhibit the activity of regulatory LuxR-type proteins that are involved in AHL-dependent QS regulation (Campbell et al., 2009; Galloway et al., 2011). In addition, the lasI-dependent QS system represses the CDP biosynthesis, and this is a determinant factor in the underlying interaction with *Arabidopsis thaliana* plants, since cyclo(L-Pro-L-Tyr), cyclo(L-Pro-L-Val), and cyclo(L-Pro-L-Phe) appear to mimic the biological role of auxin, a natural phytohormone (Ortiz-Castro et al., 2011).

In the case of specific sponge-bacteria associations, associated bacteria may prosper or at least survive with sponge material. A ribosomal RNA study of axenic cell cultures of *Suberites domuncula* showed a 16S rRNA band specific for bacteria (Thakur et al., 2003). The presence of D-amino acids and unusual amino acids in sponge peptides supports the microbial origin of sponge peptides (Fusetani and Matsunaga, 1993). CDPs attributed to the sponge *Tedania ignis* (Schmitz et al., 1983) were produced by the associated bacterium *Micrococcus* sp. (Stierle et al., 1988), and theopalauamide, a cyclic glycopeptide isolated from the sponge *Theonella swinhoei*, is included in uncultivable symbiont "*Candidatus Entotheonella palauensis*" (Schmidt et al., 2000). In addition, several CDPs that regulate bacterial-sponge interactions have been isolated from a proteobacterium of the genus *Ruegeria* associated with the marine sponge *S. domuncula*, from strains of the genera *Staphylococcus* and *Bacillus* associated with *Ircinia variabilis*, and from the marine bacteria *Vibrio* sp. associated with the marine sponge *Dysidea avara* (De Rosa et al., 2003; Mitova et al., 2004).

It is known that bacteria establish pathogenic or symbiotic relationships with their eukaryotic hosts, as in the case of *P. aeruginosa*, a well-known human and plant pathogen that proliferates in the rhizosphere, a narrow zone of soil influenced by root exudates. To overcome host defenses, *P. aeruginosa* produces toxins, adhesins, pyocyanin, and other virulence factors (Battle et al., 2008; de Abreu et al., 2014) by a QS mechanism in which CDPs play a pivotal role (González et al., 2017). *P. aeruginosa* QS is rather complex: the las and rhl systems are dependent on N-(3-oxododecanoyl)-L-homoserine lactone and N-butanoyl-L-homoserine lactone, respectively. These compounds are synthesized by the acyl-L-homoserine lactone synthases, encoded by the *lasI* and *rhlI* genes (de Kievit and Iglewski, 2000; Fuqua and Greenberg, 2002; Lee and Zhang, 2015). A third QS system involves the 2-heptyl-3-hydroxy-4-(1H)-quinolone and 2-heptyl-4-hydroxyquinolone, encoded by the *pqs* gene cluster (Gallagher et al., 2002; Lee and Zhang, 2015). All these systems connect signal transduction to transcription factors of the LysR-type, namely LasR, RhlR, and PqsR, which specifically respond to the cognate signal molecules and drive expression of hundreds of genes (Lee and Zhang, 2015). Signaling hierarchy, upstream of the pqs and rhl systems, is further defined by the signaling molecule, 2-(2-hydroxyphenyl)-thiazole-4-carbaldehyde (IQS, also called aeruginaldehyde), which is synthesized by the *ambBCDE* gene cluster and plays an important role in pathogenesis via the production of pyochelin siderophores (Dandekar and Greenberg, 2013; Lee et al., 2013; Lee and Zhang, 2015). The *ambBCDE* cluster encodes enzymes for L-2-amino-4-methoxy-trans-3-butenoic acid (AMB) biosynthesis, which occurs via a nonribosomal peptide synthase (NRPS) pathway and shows toxic effects to prokaryotes and eukaryotes (Rojas Murcia et al., 2015). By employing bioinformatics and functional approaches, González and co-workers (2017) recently identified NRPS from *P. aeruginosa* PAO1 wild-type (WT) strain and studied the role of CDPs in bacterial physiology and their interaction with plants. The authors showed that in mutants defective in putative MM-NRPS, the production of CDPs was altered and that these changes, although ineffective on virulence, interfered, at very high concentrations, with the QS systems by interacting with the binding site of the cognate AHL. By using a bacteria-plant interaction system (i.e., *P. aeruginosa-A. thaliana* co-cultivation), they observed that either the repression of root growth or the promotion of root branching exerted by selected WT and NRPS mutants was related to AHL-dependent QS status and was modified by CDP levels in vivo. CDPs are also responsible for food spoilage since QS systems govern bacterial behavior in food spoilage ecosystems (Gu et al., 2013; Skandamis and Nychas, 2012). Large yellow croaker (*Pseudosciaena crocea*), one of the most commercially important marine fish species in China, is highly susceptible to spoilage as a result of digestive enzymes and microbial activity within a short period of time postmortem even under refrigerated conditions. Microbial growth and its metabolism byproducts leads to the production of trimethylamines, organic acids, alcohols, sulfides, biogenic amines, aldehydes, and ketones with unpleasant and unacceptable off-flavors (Gram and Dalgaard, 2002). The microbial spoilage of chilled fish is chiefly connected with the presence of Gram negative proteolytic psychrotrophic bacteria, mainly *Shewanella* spp., *Pseudomonas* spp., and genera of the *Enterobacteriaceae* family (Gram and Dalgaard, 2002; Skandamis and Nychas, 2012), each species regulating the cell-cell communication by producing, secreting, and responding to small diffusible molecules to activate or repress a specific target gene expression. Various signaling compounds, including AHLs and AI-2, have been reported in spoiled milk, meat, vegetables, and aquatic product (Blana and Nychas, 2014; Liu et al., 2006; Rash et al., 2005). Several authors (Gu et al., 2013; Zhu et al., 2016) studied the specific spoilage organism (SSO) of *P. crocea* by investigating the role of QS system of SSO isolated from spoiled fish. They found that Shewanella, mainly *Shewanella baltica* and *Shewanella putrefaciens*, was the predominant genera at the end of shelf-life of *P. crocea*. In cell-free *S. baltica* culture, AI-2 and two CDPs, cyclo-(L-Pro-L-Leu) and cyclo-(L-Pro-L-Phe), were detected. The production of biofilm, trimethylamines, and putrescine in these spoilers significantly increased in the presence of cyclo-(L-Pro-L-Leu), rather than cyclo-(L-Pro-L-Phe) and 4,5-dihydroxy-2,3-pentanedione (the AI-2 precursor). Exposure to exogenous cyclo-(L-Pro-L-Leu) upregulated the transcription levels of luxR, torA, and ODC. In the fish homogenate, under refrigerated storage, exogenous cyclo-(L-Pro-L-Leu) enhanced the growth rate of the dominant bacteria, H_2S-producing bacteria, while the exogenous AI-2 precursor retarded the growth of competing bacteria, such as *Enterobacteriaceae*. Cyclo-(L-Pro-L-Leu) stimulated the accumulation of metabolites on the spoilage process of homogenate, thus confirming that the spoilage potential of *S. baltica* in *P. crocea* is regulated by QS mediated by CDPs. Finally, because of the growing importance of bacterial microbiome/bacteriome, fungal microbiome/mycobiome, the role of CDPs as mediators between oral bacteria and fungal genus has been also investigated (Brown et al., 2015). Oral candidiasis is a major complication of HIV infection (Shiboski et al., 2001; Shiboski, 2002; Thompson et al., 2010). A study by Brown et al. (2015) focused on the interaction of *Candida* with other taxa in the oral metabiome. Oral metabolites are products of the host, the oral bacterial microbiome (bacteriome), and the oral fungal microbiome (mycobiome). Functional shifts in the bacteriome and mycobiome contribute to the difference in a healthy oral environment versus oral candidiasis, and a significant shift in correlations between disease and control samples indicates an underlying metabolic change in the ecosystem. By profiling the entire oral metabiome and by using correlation difference probability network analysis, the authors proved the significant role of cyclic mono and dipeptides as QS mediators between oral bacteria and fungal genus, and hypothesized a possible contribution of CDPs to the etiology of oral candidiasis. Marchesan et al. (2015), by analyzing microbial community composition in periodontitis affected subjects, discovered the presence of several periodontal pathogens of the phylum *Synergistetes*. The authors demonstrated that *Synergistetes* phylum was strongly associated with two novel metabolites—cyclo(Leu-Pro) and cyclo(Phe-Pro)—which, by acting as QS molecules, can cause periodontal dysbiosis and periodontal disease.

Endogenous or probiotic strains have been shown to attenuate the production of virulence factor by bacterial pathogens. Indeed, Li et al. (2011) demonstrated that vaginal resident *Lactobacillus reuteri* RC-14 produces the CDPs cyclo(L-Phe-L-Pro) and cyclo(L-Tyr-L-Pro) that interfere with the staphylococcal QS system agr, a key regulator of virulence genes. This leads to the repression of the expression of toxic shock syndrome toxin-1 by the prototypical menstrual *Staphylococcus aureus* strain responsible for the menstruation-associated toxic shock syndrome (Li et al., 2011).

3 CHP in Neurological Disorders

3.1 Background

Thyrotropin-releasing hormone is a tripeptide formed by pGlu-His-Pro-NH$_2$, which is generated in the hypothalamus following the action of the pyroglutamyl-peptidase enzyme before being transformed into a linear dipeptide (His-Pro-NH$_2$), then cyclized by a nonenzymatic process at 37°C to produce CHP, also known as histidylproline diketopiperazine (Minelli et al., 2008; Prasad and Peterkofsky, 1976). In the 1970s, distribution studies showed that CHP is ubiquitous in the central nervous system—a finding that prompted significant research effort to define the biological roles of the CDP. The administration of exogenous CHP to animals is followed by a variety of biological activities, such as attenuation of ketamine anesthesia, extension of pentobarbital-induced sleep, and alleviation of some pharmacological effects of alcohol (Prasad, 2001). Moreover, it plays a significant role in modulating food intake and body core temperature, pain awareness, and by acting as an endocrine effector, inhibits prolactin secretion (Morley et al., 1981; Prasad, 1995, 2001). All these effects seem to share common dopaminergic mechanisms. Faden et al. (1981) reported that TRH and TRH-like compounds improve neurological recovery after spinal trauma and enhance cognitive function, although presenting potent endocrine, analeptic, and autonomic actions that hinder the therapeutic use of TRH. However, strikingly, the same authors also showed that CHP, the metabolic product of TRH, retains all pharmacological activities without known side effects (Faden et al., 2004, 2005). Further support for the involvement of CHP in brain function and potential implications for neurological diseases arose in 2007 when Taubert et al. (2007) showed that the CDP is a specific substrate for Organic Cation Transporter 2, a sodium-dependent transporter highly expressed in the dopaminergic brain structures classically targeted in Parkinson disease, particularly the *substantia nigra pars compacta* (Taubert et al., 2007). Not only was CHP found to co-localize in these regions; it was further shown to protect neurons from cytotoxicity induced by salsolinol, a metabolite of L-DOPA linked to Parkinson.

3.2 Current Understanding

For more than a decade, Minelli and coworkers have been actively involved in defining the effects of CHP in the brain. Indeed, the first clues of the potential application of this molecule in treating neurological disorders came in late 2006, when they discovered that CHP protects dopaminergic PC12 cells from apoptosis only in the presence of experimental conditions that cause cellular stress. Moreover, it has been shown that the treatment with CDPs activates two heat-shock proteins (Hsp), hsp27 and alpha-B-crystallin (Minelli et al., 2006), proteins implicated in the correct protein folding. Moreover, Hsps by mitigating apoptosis induced by protein misfolding are deeply involved in neurodegenerative diseases. This effect had been practically unnoticed at the time, while today it looks likely to acquire considerable significance by linking the cell-protective antiapoptotic effect of the compound to enhanced capacity to manage metabolic stresses such as the protein misfolding response. CHP attenuates the production of reactive oxygen species (ROS), and prevents glutathione (GSH) depletion caused by stressors such as glutamate, rotenone, paraquat, and beta-amyloid treatment, by triggering the nuclear accumulation of NF-E2-related factor-2 (Nrf2), a transcription factor that upregulates antioxidant-/electrophile-responsive element (ARE-EpRE)–related genes (see later for details). Based on these findings, it was reasoned that CHP, acting as a selective activator of the brain modulable Nrf2 pathway, may be a promising candidate as a neuroprotective agent acting through the induction of phase II genes (Minelli et al. 2009a,b). Oxidative stress is a condition in which the production of ROS exceeds the cellular buffering capacity. ROS are extremely reactive species that can cause irreparable damage to macromolecules, such as proteins, nucleic acids, and lipids, thus leading to cell death/genetic mutations. Neurons are terminally differentiated cells and thus extremely susceptible to oxidative stress. Indeed, they largely depend on surrounding glial cells for GSH availability (Hsu et al., 2005; Reynolds et al., 2007). GSH is an unconventional tripeptide that undergoing redox reactions can buffer increased ROS levels and repair oxidized cellular macromolecules. Several enzymes are involved in GSH action, and the majority are under the transcriptional control of Nrf2 (see later; Brigelius-Flohé and Flohé, 2011; Minelli et al., 2009a,b). Moreover, glial cells, by acting as the immune system of the central nervous system, respond to neuroinflammatory stimuli by increasing the production of reactive nitrogen species (RNS) such as nitric oxide (NO), a very diffusible molecule that can react with ROS, in particular with superoxide anion, to produce the highly reactive and toxic peroxynitrite. This condition has been recognized as nitrosative stress. The brain is therefore very sensitive to changes in redox status, and maintaining redox homeostasis is critical for preventing oxidative damage. When glial cells are overpowered by very high levels of ROS, brain cells experience oxidative stress and nitrosative stress, which act synergistically to disrupt normal neuronal processes. In fact, markers of oxidative stress and nitrosative stress are a defining feature of all neurodegenerative diseases and strongly corroborate a causal link between ROS/RNS and neurodegeneration (Gupta et al., 2014; Leszek et al., 2016; Tsang and Chung, 2009; Valko et al., 2007). Under conditions of oxidative stress, mitochondria and the process of energy generation by oxidative phosphorylation become dysfunctional, thus generating greater levels of ROS and decreasing ATP synthesis. It worth pointing out that mitochondrial dysfunction is strongly associated with neurodegenerative diseases. Indeed, in the presence of failing mitochondria, NADPH oxidase produces superoxide anions, which combined with NO, produced mainly by inducible nitric oxide synthase, generate the highly RNS peroxynitrite (Contestabile et al., 2003; Dasuri et al., 2013; Grottelli et al., 2016; Valko et al., 2007). Since mismanaged oxidative stress signals lead to apoptosis and apoptotic cells are themselves known to release ROS, one can imagine a self-perpetuating cycle of ROS-induced apoptosis driving the apoptotic release of further ROS, leading to additional apoptosis.

In mammals, oxidative stress damage is controlled mainly by the NF-E2-related factor 2 (Nrf2) Kelch-like ECH-associated protein 1 (Keap1) system, inherited from ancestors as an antistress response, aimed at preserving cellular homeostasis. Under basal conditions, Nrf2 is sequestered by cytoplasmic Keap1 and targeted for proteasomal degradation (Bellezza et al., 2010; Brigelius-Flohé and Flohé, 2011; Itoh et al., 1999). Under conditions of oxidative stress, the Nrf2-Keap1 interaction is dissolved in a dose-dependent manner, allowing Nrf2 to translocate to the nucleus where it heterodimerizes with one of the small Maf proteins. The heterodimers recognize the antioxidant response elements (AREs) that are enhancer sequences present in the regulatory regions of Nrf2 target genes, essential for the recruitment of key factors for transcription (Suzuki et al., 2013; Suzuki and Yamamoto, 2015). Nrf2 affects the expression of nearly 500 genes that encode proteins acting as redox balancing factors, detoxifying enzymes, stress response proteins, and metabolic enzymes (Fuse and Kobayashi, 2017; Hahn et al., 2015; Yang et al., 2016), thus Nrf2 can be regarded as master regulators of the oxidative stress response. It follows that CHP, with its ability to activate the Nrf2 system, can conceivably be regarded as an antioxidant compound. It was reasoned that this capacity to induce a protective antioxidation response might make the CDP a valuable treatment for neurological disease based on oxidative damage. However, since neurological disorders are multifactorial pathologies in which crucial roles are also played by endoplasmic reticulum (ER) stress, calcium loading, excitotoxicity, and inflammation, it can be hypothesized that the beneficial effects of the dipeptide could not be solely ascribed to the activation of Nrf2.

Indeed, stressful conditions lead to the activation of several pathways, including the unfolded protein response (UPR) that is induced by misfolded proteins accumulating in the lumen of the ER, a condition recognized as ER stress. ER stress leads to the activation of three stress sensor proteins located in the ER membrane PERK (protein kinase R (PKR)—like endoplasmic reticulum kinase), ATF6 (activating transcription factor 6), and IRE1 (inositol-requiring enzyme 1), via the dissociation of the molecular chaperone GRP78/Bip (binding immunoglobulin protein/78 kDa glucose-regulated protein). This results in the general inhibition of protein translation, through PERK-mediated eif2α phosphorylation, in order to alleviate ER protein load. Furthermore, through ATF6 and IRE1α branches molecular chaperones expression is upregulated to increase the folding capacity of the cell. When the stressful stimuli overcome cellular mending capacity, homeostatic conditions cannot be restored and the cell undergoes apoptosis. Indeed, a persistent stress condition causes the induction of the transcription factor CHOP (C/EBP homologous protein), which induces the cellular machinery to initiate the apoptotic program.

A role for CHP in the regulation of UPR Has been demonstrated by finding that the CDP counteracts ER stress induced by tunicamycin in microglial cells (Bellezza et al., 2014a,b). Indeed, CHP induces a protective UPR by activating eif2α and GRP78/Bip, and protects cells from apoptosis by reducing the expression of the proapoptotic protein CHOP. These molecular events significantly reduce the tunicamycin-induced decrease in cell viability (Bellezza et al.,

2014a,b). It is noteworthy that the PERK arm of UPR activates Nrf2 that, in turn, by reducing oxidative stress, can then lessen the amount of oxidized and thus misfolded proteins.

Various studies have proposed that Nrf2 plays a critical role in counteracting the NF-κB—driven inflammatory response in a variety of experimental models (Bellezza et al., 2010, 2014a, b; Brigelius-Flohé and Flohé, 2011; Sandberg et al., 2014). The term NF-κB (nuclear factor kappa-light-chain-enhancer of activated B cells) refers to a family of transcription factors that controls inflammatory responses. The most studied NF-κB family member is the p50-p65 heterodimer which, upon inflammatory stimuli, induces the expression of proinflammatory mediators. NF-κB is maintained in an inactive state in the cytoplasm through the binding to its inhibitor, inhibitor of κB (IκBα). Canonical NF-κB activation pathway relies on the activation of IKK (IκB kinase) protein kinases that phosphorylate IκBα which, in turn, is degraded by the proteasome. This event leads to NF-κB activation and nuclear translocation with the consequent upregulation of NF-κB target genes (Bellezza et al., 2010).

At the transcriptional level, NF-κB competes with transcription coactivator CREB binding protein, thus repressing Nrf2 signaling. In addition, by recruiting histone deacetylase 3 (HDAC3) and causing a local hypoacetylation, NF-κB reduces Nrf2 signaling (Wang et al., 2012). In the presence of concurrent nuclear increases in these two transcription factors, NF-κB antagonizes Nrf2-induced gene transcription, whereas all the compounds that reduce the inflammatory response by suppressing NF-κB signaling activate the Nrf2 pathway (Grottelli et al., 2016; Kim et al., 2013; Li et al., 2008; Minelli et al., 2012). This link, first suggested by studies showing that Nrf2-deficient mice exhibit a neurodegenerative phenotype (Burton et al., 2006), was substantiated by the fact that the lack of Nrf2 is associated with an increase in cytokine production (Pan et al., 2012). In the Nrf2 proximal promoter, there are several κB sites (i.e., genomic sequences recognized and bound by NF-κB); therefore in the presence of a proinflammatory stimulus such as tumor necrosis factor α (TNFα), some cells respond by upregulating Nrf2, leading to feedback suppression of cytokine gene expression (Rushworth et al., 2012). In addition, NF-κB activation can modulate Nrf2 activity as a protective anti-inflammatory mechanism via the small GTPase RAC1 (Ras-related C3 botulinum toxin substrate 1). Once activated by LPS (lipopolysaccharide), RAC1, through Nrf2 activation, upregulates HO-1 (heme-oxygenase 1) expression, which shifts the cells to a more reducing environment, essential for terminating the NF-κB activation (Cuadrado et al., 2014). The molecular mechanism of CHP action on the NF-κB system was investigated by using a mouse ear inflammation model. It was observed that CHP reduces 12-otetradecanoylphorbol-13-acetate-induced edema. Moreover, CHP interferes with the crosstalk between the antioxidant Nrf2/HO-1 and the proinflammatory NF-κB pathways in murine immortalized microglial BV2 cells challenged with the proinflammatory molecule LPS. Indeed, cyclooxygenase-2 and matrix metalloproteinase 3, two gene products governed by NF-κB, were downregulated by CHP and upregulated in heme oxygenase-1 (HO-1) knock-down cells. On the basis of these data, showing that CHP suppresses the proinflammatory NF-κB signaling via Nrf2-mediated HO-1 activation, the use of CHP as an in vivo antiinflammatory compound has been proposed (Minelli et al., 2012). It is becoming increasingly clear that neuroinflammation is one of the features shared by all neurodegenerative diseases (Bellezza et al., 2014a,b; Dinkova-Kostova et al., 2018; González-Reyes et al., 2017). Usually triggered by peripheral inflammation, the term describes a wide range of immune responses by the central nervous system cells, such as microglia, astrocytes, and blood brain barrier, each linked by a dynamic crosstalk. In the presence of prolonged and sustained inflammation, the neuroinflammatory response results in synaptic impairment, neuronal death, and eventually neurodegeneration (Boulamery and Desplat-Jégo, 2017; Lyman et al., 2014; Rustenhoven et al., 2017). As described previously, the effects of CHP could counteract this pathogenic state in two ways: by activating Nrf2 and inducing HO-1 activity, the compound might simultaneously drive a protective antioxidant response, mitigating oxidative stress damage, while inhibiting NF-κB signaling, reducing damage associated with inflammation (Minelli et al., 2012). Based on these considerations, the inhibition of glial inflammation by the CDP was hypothesized. Systemic administration of CHP exerts antiinflammatory effects in the central nervous system by downregulating systemic (hepatic) and local (cerebral) TNFα expression, thereby counteracting LPS-induced gliosis (Bellezza et al., 2014a,b). These effects are known to decrease the detrimental effect of inflammatory neurotoxins on neurons (Catorce and Gevorkian, 2016). These data suggested a beneficial effect of CHP in a neuroinflammatory setting and its potential therapeutic utility in neuroinflammatory diseases, and we suggested that the CDP might be used also to treat other neuropathological conditions. To test this possibility more directly, Minelli and coworkers tested the effects of CHP in the microglial cells of hSOD1G93A mice. These transgenic mice, expressing the human gene encoding for (superoxide dismutase 1) SOD1 mutated in Gly93-Ala (SOD1G93A), recapitulate several aspects of amyotrophic lateral sclerosis (ALS) and provide a powerful model system to identify pathophysiological mechanisms of the disease and to screen potential therapeutic compounds (Grottelli et al., 2015, 2016). In this setting, CHP acts as an antioxidant agent even in a SOD1G93A environment, and more importantly, its effects even offered the prospect of going beyond protection toward neuronal regrowth by strongly upregulating mRNA levels of the neuronal growth factor brain-derived neurotrophic factor, a molecule linked to the preservation of existing neuronal function but also the growth and differentiation of new neurons. Thus CHP may both inhibit the neuronal damage associated with oxidative stress and microglial inflammatory responses caused by SOD1 mutations, and act directly on neurons themselves to preserve and perhaps restore their function, suggesting its possible utility as a therapeutic agent to prevent or delay disease progression in ALS (Fig. 2).

3.3 Future Perspectives

Neurodegenerative diseases are multifactorial pathologies, although each disease is characterized by distinct etiopathogenetic causes. However, common pathogenic mechanisms, such as neuroinflammation, oxidative stress, and ER stress, underpin neurodegeneration. It has

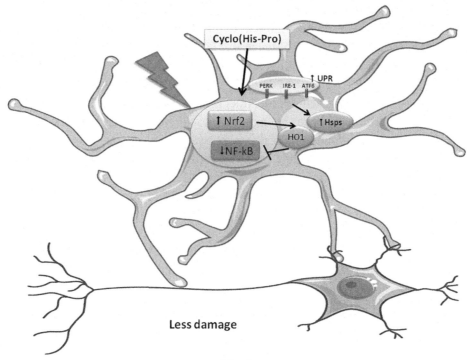

Fig. 2

Scheme of cyclo(His-Pro) action. Cyclo(His-Pro), by increasing protective unfolded protein response (UPR), by activating the antioxidant response through Nrf2 induction, and by downregulating proinflammatory response through NF-κB inhibition in microglial cells, protects neuronal cells from neurotoxic damage.

been shown that CHP can counteract each of these pathogenic pathways in several neurotoxin-exposed cellular models. So far, only the mutant SOD1 cells, the golden model for ALS, have been employed. It remains to be tested whether CHP is effective on other neurodegenerative disease-specific cellular and animal models.

Quite recently, it has been suggested that any unbalance of the gut microbiome leads to pathological signaling to the brain that might result in proinflammatory reactions, oxidative stress, and a general increase in cellular degeneration, thus contributing to multiple neurodegenerative diseases (Noble et al., 2017). The human gastrointestinal tract harbors a number of bacterial cells that outnumber by a factor of 10 the host's cells and encodes a number of genes that outnumber by a factor of 100 the host's genes. These human digestive-tract-associated microbes are now known as the *gut microbiome/microbiota*. Assessments of the number of bacterial species present in the human gut vary widely among studies, but it is generally accepted that individuals harbor more than 1000 microbial, species-level phylotypes (Lozupone et al., 2012) that can communicate via a QS mechanism (Bivar Xavier, 2018).

The role of the human gut microbiome in health and disease has been the topic of broad research, and a role for the bacterial commensals in various neurological conditions is well accepted (Byrd et al., 2018; Caballero-Villarraso et al., 2017; Cox and Weiner, 2018; Friedland and Chapman, 2017; Ho et al., 2018; Kitai and Tang, 2018; Marietta et al., 2018; Perez-Pardo et al., 2017; Roszyk and Puszczewicz, 2017; Sherwin et al., 2018; Thion et al., 2018; Yang and Duan, 2018). Indeed, the gastrointestinal tract is deeply connected with the central nervous system through the gut-brain axis, an interconnected and bidirectional network of neuroendocrine signals and immunological factors. It has been demonstrated that the gut microbiota is capable of communicating information derived from the ingested foods to the central nervous system to obtain a systemic response (Noble et al., 2017). In 1995, the presence of CHP in the gastrointestinal tract was connected to a role as a gut peptide of the entero-insular axis (Prasad, 1995). Currently, because of the highlighted role of CHP as a QS signal, capable of controlling behavior and functions of bacterial population-level responses, we propose a novel role of CHP as a modulator of the gut microbiota. Therefore, CHP, by acting directly on central nervous system cells and potentially on gut microbiota, can be considered a potential new drug for neurodegenerative diseases.

No information is currently available on the latter proposed role, since they might be validated only by preclinical trials. In this context, we hope to stimulate interest in the scientific community to test this hypothesis to produce novel therapeutic modalities for the plethora of diseases linked to gut-microbiome dysfunction.

4 Conclusion

It has been demonstrated that CHP is an endogenous CDP that can reduce oxidative and ER stress as well as inflammation, the main culprits of several neurological disorders. Thus we propose that this CDP, even orally administered, can cross the blood brain barrier and exert its beneficial effects on glial cells, whose uncontrolled response is currently recognized as one of the several causes of neuronal death. Moreover, because CHP can act as QS signal, it is plausible to suggest that this dipeptide can modulate the gut microbiome for clinical benefit in the diverse pathologies in which microbiome dysregulation is implicated.

Thus by acting directly to cease several causes of neurodegeneration and by acting indirectly on the gut microbiome linked to neurological diseases, we can potentially relieve many of the diverse contributors to neurodegenerative disease pathogenesis.

Glossary

Amyotrophic lateral sclerosis (ALS) A neurodegenerative disease characterized by muscle spasticity, rapidly progressive weakness due to muscle atrophy, and difficulty in speaking (dysarthria), swallowing (dysphagia), and breathing (dyspnea) due to degeneration of the upper and lower motor neurons. Individuals affected by the disorder may ultimately lose the ability to control all voluntary movement, although bladder and bowel function and the muscles responsible for eye movement are usually spared until the final stages of the disease. Cognitive function is generally spared for most patients.

Biofilms A structured community of bacterial cells enclosed in a self-produced protective polymeric matrix and adherent to an inert or living surface.

Blood-brain barrier (BBB) A highly selective permeability barrier that separates the circulating blood from the brain extracellular fluid (BECF) in the CNS. Formed by endothelial cells that are connected by tight junctions, it allows the passage of molecules crucial to neural function and prevents the entry of potential neurotoxins.

Cyclic dipeptides (CDPs), or 2,5-diketopiperazines Relatively simple compounds resulting from nonenzymatic cyclization of dipeptides and their amides. They are the most common peptide derivatives found in nature and are synthesized by proteobacterial species as well as by humans. CDPs are characterized by stability to proteolysis and promotion of interactions with biological targets.

Cyclic scaffold A six-membered ring that, due to its stable structural characteristics, represents a significant pharmacophore in medicinal chemistry.

Human microbiome The human body comprises around 10 trillion cells but harbors 100 trillion bacteria, for example, on the skin and in the gut. This is the human "microbiome" and has a huge impact on human health. Nevertheless, humans, in turn, can affect their microbiome by influencing the species of bacteria that take up residence in and on their bodies.

Inflammation A response of the innate immune system to harmful stimuli such as pathogens, damaged cells, or irritants. It is a protective attempt by the organism to remove the injurious stimuli and to initiate the healing process. Classical signs are pain, heat, redness, swelling, and loss of function.

Lipopolysaccharide (LPS) A glycolipid of Gram negative bacteria of the outer membrane. It is recognized by toll-like receptor 4 (TLR4) in immune cells, where it induces the activation of a proinflammatory response.

Microglia A type of nonneural cell that constitutes the resident macrophages of the brain and spinal cord and acts as the first and main form of active immune defense in the CNS.

NF-κB (nuclear factor kappa-light-chain-enhancer of activated B cells) A family of transcription factors that controls inflammatory responses. The most studied NF-κB family member is the p50-p65 heterodimer, which upon inflammatory stimuli induces the expression of proinflammatory mediators.

Quorum sensing (QS) A mechanism of cell-cell communication via secreted signaling molecules. Secreted autoinducers regulate the expression of a particular set of genes once the cell population density is sufficient to produce a threshold accumulation of the secreted autoinducer.

Reactive oxygen species (ROS) A number of reactive molecules and free radicals derived from molecular oxygen, such as singlet oxygen, superoxides, peroxides, the hydroxyl radical, and hypochlorous acid.

Thyrotropin-releasing hormone (TRH) A tripeptide hormone produced by the hypothalamus that stimulates the release of the thyroid-stimulating hormone and prolactin from the anterior pituitary.

Unfolded-protein response (UPR) An evolutionarily conserved response related to the ER stress response. The initial intent of the UPR is to adapt to the changing environment and reestablish normal ER function. When adaptation fails, ER-initiated pathways signal alarm by inducing the expression of genes encoding mediators of host defense. Excessive and prolonged ER stress triggers cell suicide, usually in the form of apoptosis, representing a last resort of multicellular organisms to dispense with dysfunctional cells.

Virulence factors Molecules expressed and secreted by pathogens (bacteria, viruses, fungi, and protozoa) that enable them to replicate and disseminate within a host in part by subverting or eluding host defenses.

Abbreviations

aaRS	amino acid tRNA synthetase
AHSL	acyl homoserine lactone
AIP	autoinducer peptides
ALS	amyotrophic lateral sclerosis
ARE	antioxidant response elements
ATF6	activating transcription factor 6
CDO	cyclic dipeptide oxidases
CDP	cyclic dipeptides
CDPS	CDP synthase
CHOP	C/EBP homologous protein
CHP	cyclo(His-Pro)
ER	endoplasmic reticulum
GRP78/Bip	binding immunoglobulin protein/78 kDa glucose-regulated protein
GSH	glutathione
HO-1	heme oxygenase-1
Hsp	heat-shock protein
IKK	IκB kinase
IRE1	inositol-requiring enzyme 1
IκBα	inhibitor of κB
NF-κB	nuclear factor kappa-light-chain-enhancer of activated B cells
Nrf2	NF-E2-related factor-2
NRPS	nonribosomal peptide synthetases
PCP	peptidyl carrier protein
PEG	polyethylene glycol
PERK	protein kinase R (PKR)-like endoplasmic reticulum kinase

QS	quorum sensing
RAC1	Ras-related C3 botulinum toxin substrate 1
RNS	reactive nitrogen species
ROS	reactive oxygen species
SOD1	superoxide dismutase 1
TNFα	tumor necrosis factor α
TRH	thyrotropin releasing hormone
UPR	unfolded protein response
WT	wild-type

References

Abbamondi, G.R., De Rosa, S., Iodice, C., Tommonaro, G., 2014. Cyclic dipeptides produced by marine sponge-associated bacteria as quorum sensing signals. Nat. Prod. Commun. 9, 229–232.

Balibar, C.J., Walsh, C.T., 2006. GliP, a multimodular nonribosomal peptide synthetase in Aspergillus fumigatus, makes the diketopiperazine scaffold of gliotoxin. Biochemistry 45, 15029–15038.

Bassler, B., Vogel, J., 2013. Bacterial regulatory mechanisms: the gene and beyond. Curr. Opin. Microbiol. 16, 109–111.

Battle, S.E., Meyer, F., Rello, J., Kung, V.L., Hauser, A.R., 2008. Hybrid pathogenicity island PAGI-5 contributes to the highly virulent phenotype of a *Pseudomonas aeruginosa* isolate in mammals. J. Bacteriol. 190, 7130–7140.

Beck, J.G., Chatterjee, J., Laufer, B., Kiran, M.U., Frank, A.O., Neubauer, S., Ovadia, O., Greenberg, S., Gilon, C., Hoffman, A., Kessler, H., 2012. Intestinal permeability of cyclic peptides: common key backbone motifs identified. J. Am. Chem. Soc. 134, 12125–12133.

Belin, P., Moutiez, M., Lautru, S., Seguin, J., Pernodet, J.-L., Gondry, M., 2012. The nonribosomal synthesis of diketopiperazines in tRNA-dependent cyclodipeptide synthase pathways. Nat. Prod. Rep. 29, 961.

Bellezza, I., Mierla, A.L., Minelli, A., 2010. Nrf2 and NF-κB and their concerted modulation in cancer pathogenesis and progression. Cancers (Basel) 2, 483–497.

Bellezza, I., Grottelli, S., Mierla, A.L., Cacciatore, I., Fornasari, E., Roscini, L., Cardinali, G., Minelli, A., 2014a. Neuroinflammation and endoplasmic reticulum stress are coregulated by cyclo(His-Pro) to prevent LPS neurotoxicity. Int. J. Biochem. Cell Biol. 51, 159–169.

Bellezza, I., Peirce, M.J., Minelli, A., 2014b. Cyclic dipeptides: from bugs to brain. Trends Mol. Med. 20, 551–558.

Bivar Xavier, K., 2018. Bacterial interspecies quorum sensing in the mammalian gut microbiota. C. R. Biol. 341 (5), 300. pii: S1631-0691(18), 30051-9.

Blana, V.A., Nychas, G.J., 2014. Presence of quorum sensing signal molecules in minced beef stored under various temperature and packaging conditions. Int. J. Food Microbiol. 173, 1–8.

Bockus, A.T., McEwen, C.M., Lokey, R.S., 2013. Form and function in cyclic peptide natural products: a pharmacokinetic perspective. Curr. Top. Med. Chem. 13, 821–836.

Bonnefond, L., Arai, T., Sakaguchi, Y., Suzuki, T., Ishitani, R., Nureki, O., 2011. Structural basis for nonribosomal peptide synthesis by an aminoacyl-tRNAsynthetase paralog. Proc. Natl. Acad. Sci. U. S. A. 108, 3912–3917.

Borthwick, A.D., 2012. 2,5-Diketopiperazines: synthesis, reactions, medicinal chemistry, and bioactive natural products. Chem. Rev. 112, 3641–3716.

Borthwick, A.D., Da Costa, N.C., 2017. 2,5-Diketopiperazines in food and beverages: taste and bioactivity. Crit. Rev. Food Sci. Nutr. 57, 718–742.

Borthwick, A.D., Liddle, J., Davies, D.E., Exall, A.M., Hamlett, C., Hickey, D.M., Mason, A.M., Smith, I.E., Nerozzi, F., Peace, S., Pollard, D., Sollis, S.L., Allen, M.J., Woollard, P.M., Pullen, M.A., Westfall, T.D., Stanislaus, D.J., 2012. Pyridyl-2,5-diketopiperazines as potent, selective, and orally bioavailable oxytocin antagonists: synthesis, pharmacokinetics, and in vivo potency. J. Med. Chem. 55, 783–796.

Boulamery, A., Desplat-Jégo, S., 2017. Regulation of neuroinflammation: what role for the tumor necrosis factor-like weak inducer of apoptosis/Fn14 pathway? Front. Immunol. 8, 1534.

Brigelius-Flohé, R., Flohé, L., 2011. Basic principles and emerging concepts in the redox control of transcription factors. Antioxid. Redox Signal. 15, 2335–2381.

Brown, R.E., Ghannoum, M.A., Mukherjee, P.K., Gillevet, P.M., Sikaroodi, M., 2015. Quorum-sensing dysbiotic shifts in the HIV-infected oral metabiome. PLoS One 10, e0123880.

Burton, N.C., Kensler, T.W., Guilarte, T.R., 2006. In vivo modulation of the Parkinsonian phenotype by Nrf2. Neurotoxicology 27, 1094–1100.

Byrd, A.L., Belkaid, Y., Segre, J.A., 2018. The human skin microbiome. Nat. Rev. Microbiol. 16, 143–155.

Caballero-Villarraso, J., Galvan, A., Escribano, B.M., Túnez, I., 2017. Interrelationships between gut microbiota and the host: paradigms, role in neurodegenerative diseases and future prospects. CNS Neurol. Disord. Drug Targets 16, 945–964.

Campbell, J., Lin, Q., Geske, G.D., Blackwell, H.E., 2009. New and unexpected insights into the modulation of Lux R-type quorum sensing by cyclic dipeptides. ACS Chem. Biol. 4, 1051–1059.

Catorce, M.N., Gevorkian, G., 2016. LPS-induced murine neuroinflammation model: main features and suitability for pre-clinical assessment of nutraceuticals. Curr. Neuropharmacol. 14, 155–164.

Chen, M.Z., Dewis, M.L., Kraut, K., Merritt, D., Reiber, L., Trinnaman, L., Da Costa, N.C., 2009. 2,5 Diketopiperazines(cyclic dipeptides) in beef: identification, synthesis, and sensory evaluation. J. Food Sci. 74, C100–C105.

Choi, S.A., Suh, H.J., Yun, J.W., Choi, J.W., 2012. Differential gene expression in pancreatic tissues of streptozocin-induced diabetic rats and genetically-diabetic mice in response to hypoglycemic dipeptide cyclo(His-Pro) treatment. Mol. Biol. Rep. 39, 8821–8835.

Clardy, J., Walsh, C., 2004. Lessons from natural molecules. Nature 432, 829–837.

Clardy, J., Fischbach, M.A., Walsh, C.T., 2006. New antibiotics from bacterial natural products. Nat. Biotechnol. 24, 1541–1550.

Clément, P., Bernabé, J., Compagnie, S., Alexandre, L., McCallum, S., Giuliano, F., 2013. Inhibition of ejaculation by the non-peptide oxytocin receptor antagonist GSK557296: a multi-level site of action. Br. J. Pharmacol. 169, 1477–1485.

Contestabile, A., Monti, B., Contestabile, A., Ciani, E., 2003. Brain nitric oxide and its dual role in neurodegeneration/neuroprotection: understanding molecular mechanisms to devise drug approaches. Curr. Med. Chem. 10, 2147–2174.

Cook, L.C., Federle, M.J., 2014. Peptide pheromone signaling in Streptococcus and Enterococcus. FEMS Microbiol. Rev. 38, 473–492.

Cornacchia, C., Cacciatore, I., Baldassarre, L., Mollica, A., Feliciani, F., Pinnen, F., 2012. 2,5-Diketopiperazines as neuroprotective agents. Mini Rev. Med. Chem. 12, 2–12.

Cornforth, D.M., Popat, R., McNally, L., Gurney, J., Scott-Phillips, T.C., Ivens, A., Diggle, S.P., Brown, S.P., 2014. Combinatorial quorum sensing allows bacteria to resolve their social and physical environment. Proc. Natl. Acad. Sci. U. S. A. 111, 4280–4284.

Correia, T., Grammel, N., Ortel, I., Keller, U., Tudzynski, P., 2003. Molecular cloning and analysis of the ergopeptine assembly system in the ergot fungus *Claviceps purpurea*. Chem. Biol. 10, 1281–1292.

Cox, L.M., Weiner, H.L., 2018. Microbiota signaling pathways that influence neurologic disease. Neurotherapeutics 15, 135–145.

Cuadrado, A., Martin-Moldes, Z., Ye, J., Lastres-Becker, I., 2014. Transcription factors NRF2 and NF-kappaB are coordinated effectors of the Rho family, GTP-binding protein RAC1 during inflammation. J. Biol. Chem. 289, 15244–15258.

Dandekar, A.A., Greenberg, E.P., 2013. Microbiology: plan B for quorum sensing. Nat. Chem. Biol. 9, 292–293.

Dasuri, K., Zhang, L., Keller, J.N., 2013. Oxidative stress, neurodegeneration, and the balance of protein degradation and protein synthesis. Free Radic. Biol. Med. 62, 170–185.

de Abreu, P.M., Farias, P.G., Paiva, G.S., Almeida, A.M., Morais, P.V., 2014. Persistence of microbial communities including *Pseudomonas aeruginosa* in a hospital environment: a potential health hazard. BMC Microbiol. 14, 118.

de Kievit, T.R., Iglewski, B.H., 2000. Bacterial quorum sensing in pathogenic relationships. Infect. Immun. 68, 4839–4849.

De Rosa, S., Mitova, M., Tommonaro, G., 2003. Marine bacteria associated with sponge as source of cyclic peptides. Biomol. Eng. 20, 311–316.

Degrassi, G., Aguilar, C., Bosco, M., Zahariev, S., Pongor, S., Venturi, V., 2002. Plant growth-promoting *Pseudomonas putida* WCS358 produces and secretes four cyclic dipeptides: cross-talk with quorum sensing bacterial sensors. Curr. Microbiol. 45, 250–254.

Diggle, S.P., Gardner, A., West, S., Griffin, A.S., 2007. Evolutionary theory about bacterial quorum sensing: when is a signal not a signal? Philos. Trans. R. Soc. Lond. Ser. B Biol. Sci. 362, 1241–1249.

Dinkova-Kostova, A.T., Kostov, R.V., Kazantsev, A.G., 2018. The role of Nrf2 signaling in counteracting neurodegenerative diseases. FEBS J. https://doi.org/10.1111/febs.14379.

Eguchi, C., Kakuta, A., 1974. Cyclic dipeptides, I. Thermodynamics of the cis-trans isomerization of the side chains in cyclic dipeptides. J. Am. Chem. Soc. 96, 3985–3989.

Einholm, A.P., Pedersen, K.E., Wind, T., Kulig, P., Overgaard, M.T., Jensen, J.K., Bødker, J.S., Christensen, A., Charlton, P., Andreasen, P.A., 2003. Biochemical mechanism of action of a diketopiperazine inactivator of plasminogen activator inhibitor-1. Biochem. J. 373, 723–732.

Ellison, D.W., Miller, V.L., 2006. Regulation of virulence by members of the MarR/SlyA family. Curr. Opin. Microbiol. 9, 153–159.

Engel, S., Jensen, P.R., Fenical, W., 2002. Chemical ecology of marine microbial defense. J. Chem. Ecol. 28, 1971–1985.

Faden, A.I., Jacobs, T.P., Holaday, J.W., 1981. Thyrotropin-releasing hormone improves neurologic recovery after spinal trauma in cats. N. Engl. J. Med. 305, 1063–1067.

Faden, A.I., Knoblach, S.M., Movsesyan, V.A., Cernak, I., 2004. Novel small peptides with neuroprotective and nootropic properties. J. Alzheimers Dis. 6, S93–S97.

Faden, A.I., Movsesyan, V.A., Knoblach, S.M., Ahmed, F., Cernak, I., 2005. Neuroprotective effects of novel small peptides in vitro and after brain injury. Neuropharmacology 49, 410–424.

Felnagle, E.A., Jackson, E.E., Chan, Y.A., Podevels, A.M., Berti, A.D., McMahon, M.D., Thomas, M.G., 2008. Nonribosomal peptide synthetases involved in the production of medically relevant natural products. Mol. Pharm. 5, 191–211.

Friedland, R.P., Chapman, M.R., 2017. The role of microbial amyloid in neurodegeneration. PLoS Pathog. 13, e1006654.

Friedrich, A.B., Fischer, I., Proksch, P., Hacker, J., Hentschel, U., 2001. Temporal variation of the microbial community associated with the mediterranean sponge *Aplysina aerophoba*. FEMS Microbiol. Ecol. 38, 105–113.

Fuqua, C., Greenberg, E.P., 2002. Listening in on bacteria: acylhomoserine lactone signalling. Nat. Rev. Mol. Cell Biol. 3, 685–695.

Fuse, Y., Kobayashi, M., 2017. Conservation of the Keap1-Nrf2 system: an evolutionary journey through stressful space and time. Molecules. 22, pii: E436.

Fusetani, N., Matsunaga, S., 1993. Bioactive sponge peptides. Chem. Rev. 93, 1793–1806.

Gallagher, L.A., McKnight, S.L., Kuznetsova, M.S., Pesci, E.C., Manoil, C., 2002. Functions required for extracellular quinolone signaling by *Pseudomonas aeruginosa*. J. Bacteriol. 184, 6472–6480.

Galloway, W.R.J.D., Hodgkinson, J.T., Bowden, S.D., Welch, M., Spring, D.R., 2011. Quorum sensing in gram-negative bacteria: small molecule modulation of AHL and AI-2 quorum sensing pathways. Chem. Rev. 111, 28–67.

García-Estrada, C., Ullán, R.V., Albillos, S.M., Fernández-Bodega, M.Á., Durek, P., vonDöhren, H., Martín, J.F., 2011. A single cluster of coregulated genes encodes the biosynthesis of the mycotoxins roquefortine C and meleagrin in *Penicillium chrysogenum*. Chem. Biol. 18, 1499–1512.

Gardiner, D.M., Cozijnsen, A.J., Wilson, L.M., Pedras, M.S.C., Howlett, B.J., 2004. The sirodesmin biosynthetic gene cluster of the plant pathogenic fungus *Leptosphaeria maculans*. Mol. Microbiol. 53, 1307–1318.

Gautschi, M., Schmid, J.P., Peppard, T.L., Ryan, T.P., Tuorto, R.M., Yang, X., 1997. Chemical characterization of diketopiperazines in beer. J. Agric. Food Chem. 45, 3183–3189.

Giessen, T.W., Marahiel, M.A., 2012. Ribosome-independent biosynthesis of biologically active peptides: application of synthetic biology to generate structural diversity. FEBS Lett. 586, 2065–2075.

Giessen, T., Marahiel, M., 2014. The tRNA-dependent biosynthesis of modified cyclic dipeptides. Int. J. Mol. Sci. 15, 14610–14631.

Giessen, T.W., Von Tesmar, A.M., Marahiel, M.A., 2013. Insights into the generation of structural diversity in a tRNA-dependent pathway for highly modified bioactive cyclic dipeptides. Chem. Biol. 20, 828–838.

Ginz, M., Engelhardt, U.H., 2000. Identification of proline-based diketopiperazines in roasted coffee. J. Agric. Food Chem. 48, 3528–3532.

Gondry, M., Lautru, S., Fusai, G., Meunier, G., Ménez, A., Genet, R., 2001. Cyclic dipeptide oxidase from *Streptomyces noursei*. Eur. J. Biochem. 268, 1712–1721.

Gondry, M., Sauguet, L., Belin, P., Thai, R., Amouroux, R., Tellier, C., Tuphile, K., Jacquet, M., Braud, S., Courcon, M., Masson, C., Dubois, S., Lautru, S., Lecoq, A., Hashimoto, S., Genet, R., Pernodet, J.L., 2009. Cyclodipeptide synthases are a family of tRNA-dependent peptide bond-forming enzymes. Nat. Chem. Biol. 5, 414–420.

Gonzalez, J.F., Ortin, I., de la Cuesta, E., Menendez, J.C., 2012. Privileged scaffolds in synthesis: 2,5-piperazinediones as templates for the preparation of structurally diverse heterocycles. Chem. Soc. Rev. 41, 6902–6915.

González, O., Ortíz-Castro, R., Díaz-Pérez, C., Díaz-Pérez, A.L., Magaña-Dueñas, V., López-Bucio, J., Campos-García, J., 2017. Non-ribosomal peptide synthases from *Pseudomonas aeruginosa* play a role in cyclodipeptide biosynthesis, quorum-sensing regulation, and root development in a plant host. Microb. Ecol. 73, 616–629.

González-Reyes, R.E., Nava-Mesa, M.O., Vargas-Sánchez, K., Ariza-Salamanca, D., Mora-Muñoz, L., 2017. Involvement of astrocytes in Alzheimer's disease from a neuroinflammatory and oxidative stress perspective. Front. Mol. Neurosci. 10, 427.

Gram, L., Dalgaard, P., 2002. Fish spoilage bacteria-problems and solutions. Curr. Opin. Biotechnol. 13, 262–266.

Grottelli, S., Bellezza, I., Morozzi, G., Peirce, M.J., Marchetti, C., Cacciatore, I., Costanzi, E., Minelli, A., 2015. Cyclo(His-Pro) protects SOD1G93A microglial cells from Paraquat-induced toxicity. J. Clin. Cell. Immunol. 6, 287.

Grottelli, S., Ferrari, I., Pietrini, G., Peirce, M.J., Minelli, A., Bellezza, I., 2016. The role of cyclo(His-Pro) in neurodegeneration. Int. J. Mol. Sci. 17, pii: E1332.

Gruenewald, S., Mootz, H.D., Stehmeier, P., Stachelhaus, T., 2004. In vivo production of artificial nonribosomal peptide products in the heterologous host *Escherichia coli*. Appl. Environ. Microbiol. 70, 3282–3291.

Gu, Q., Fu, L., Wang, Y., Lin, J., 2013. Identification and characterization of extracellular cyclic dipeptides as quorum-sensing signal molecules from *Shewanella baltica*, the specific spoilage organism of *Pseudosciaena crocea* during 4°C storage. J. Agric. Food Chem. 61, 11645–11652.

Gupta, S.P., Yadav, S., Singhal, N.K., Tiwari, M.N., Mishra, S.K., Singh, M.P., 2014. Does restraining nitric oxide biosynthesis rescue from toxins-induced parkinsonism and sporadic Parkinson's disease? Mol. Neurobiol. 49, 262–275.

Hahn, M.E., Timme-Laragy, A.R., Karchner, S.I., Stegeman, J.J., 2015. Nrf2 and Nrf2-related proteins in development and developmental toxicity: insights from studies in zebrafish (*Danio rerio*). Free Radic. Biol. Med. 88, 275–289.

Hallmann, A., 2011. Evolution of reproductive development in the volvocine algae. Sex. Plant Reprod. 24, 97–112.

Hawver, L.A., Jung, S.A., Ng, W.L., 2016. Specificity and complexity in bacterial quorum-sensing systems. FEMS Microbiol. Rev. 40, 738–752.

Healy, F.G., Krasnoff, S.B., Wach, M., Gibson, D.M., Loria, R., 2002. Involvement of a cytochrome P450 monooxygenase in thaxtomin A biosynthesis by *Streptomyces acidiscabies*. J. Bacteriol. 184, 2019–2029.

Hense, B.A., Schuster, M., 2015. Core principles of bacterial autoinducer systems. Microbiol. Mol. Biol. Rev. 79 (1), 153–169.

Hentschel, U., Schmid, M., Wagner, M., Fieseler, L., Gernert, C., Hacker, J., 2001. Isolation and phylogenetic analysis of bacteria with antimicrobial activities from the Mediterranean sponges *Aplysina aerophoba* and *Aplysina cavernicola*. FEMS Microbiol. Ecol. 35, 305–312.

Ho, L., Ono, K., Tsuji, M., Mazzola, P., Singh, R., Pasinetti, G.M., 2018. Protective roles of intestinal microbiota derived short chain fatty acids in Alzheimer's disease-type beta-amyloid neuropathological mechanisms. Expert. Rev. Neurother. 18, 83–90.

Hogan, D.A., 2006. Talking to themselves: autoregulation and quorum sensing in fungi. Eukaryot. Cell 5, 613–619.

Holden, M.T., Ram Chhabra, S., de Nys, R., Stead, P., Bainton, N.J., Hill, P.J., Manefield, M., Kumar, N., Labatte, M., England, D., Rice, S., Givskov, M., Salmond, G.P., Stewart, G.S., Bycroft, B.W., Kjelleberg, S., Williams, P., 1999. Quorum-sensing cross talk: isolation and chemical characterization of cyclic dipeptides from *Pseudomonas aeruginosa* and other gram-negative bacteria. Mol. Microbiol. 33, 1254–1266.

Hsu, M., Srinivas, B., Kumar, J., Subramanian, R., Andersen, J., 2005. Glutathione depletion resulting in selective mitochondrial complex I inhibition in dopaminergic cells is via an NO-mediated pathway not involving peroxynitrite: implications for Parkinson's disease. J. Neurochem. 92, 1091–1103.

Huang, R., Zhou, X., Xu, T., Yang, X., Liu, Y., 2010. Diketopiperazines from marine organisms. Chem. Biodivers. 7, 2809–2829.

Itoh, K., Ishii, T., Wakabayashi, N., Yamamoto, M., 1999. Regulatory mechanisms of cellular response to oxidative stress. Free Radic. Res. 31, 319–324.

Jayatilake, G.S., Thornton, M.P., Leonard, A.C., Grimwade, J.E., Baker, B.J., 1996. Metabolites from an Antarctic sponge-associated bacterium, *Pseudomonas aeruginosa*. J. Nat. Prod. 59, 293–296.

Jung, E.Y., Lee, H.S., Choi, J.W., Ra, K.S., Kim, M.R., Suh, H.J., 2011. Glucose tolerance and antioxidant activity of spent brewer's yeast hydrolysate with a high content of cyclo-His-Pro (CHP). J. Food Sci. 76, C272–C278.

Jung, E.Y., Hong, Y.H., Park, C., Suh, H.J., 2016. Effects of cyclo-His-Pro-enriched yeast hydrolysate on blood glucose levels and lipid metabolism in obese diabetic ob/ob mice. Nutr. Res. Pract. 10, 154–160.

Kim, S.W., Lee, H.K., Shin, J.H., Lee, J.K., 2013. Up-down regulation of HO-1 and iNOS gene expressions by ethyl pyruvate via recruiting p300 to Nrf2 and depriving it from p65. Free Radic. Biol. Med. 65, 468–476.

Kitai, T., Tang, W.H.W., 2018. Gut microbiota in cardiovascular disease and heart failure. Clin. Sci. (Lond.) 132, 85–91.

Koglin, A., Walsh, C.T., 2009. Structural insights into nonribosomal peptide enzymatic assembly lines. Nat. Prod. Rep. 26, 987–1000.

Koo, K.B., Suh, H.J., Ra, K.S., Choi, J.W., 2011. Protective effect of cyclo(His-Pro) on streptozotocin-induced cytotoxicity and apoptosis in vitro. J. Microbiol. Biotechnol. 21, 218–227.

Kumar, N., Gorantla, J.N., Mohandas, C., Nambisan, B., Lankalapalli, R.S., 2013. Isolation and antifungal properties of cyclo(d-Tyr-l-Leu) diketopiperazine isolated from *Bacillus* sp. associated with rhabditid entomopathogenic nematode. Nat. Prod. Res. 27, 2168–2172.

Kwak, M.K., Liu, R., Kwon, J.O., Kim, M.K., Kim, A.H., Kang, S.O., 2013. Cyclic dipeptides from lactic acid bacteria inhibit proliferation of the influenza A virus. J. Microbiol. 51, 836–843.

Kwak, M.K., Liu, R., Kim, M.K., Moon, D., Kim, A.H., Song, S.H., Kang, S.O., 2014. Cyclic dipeptides from lactic acid bacteria inhibit the proliferation of pathogenic fungi. J. Microbiol. 52, 64–70.

Lahoud, G., Hou, Y.-M., 2010. Biosynthesis: a new (old) way of hijacking tRNA. Nat. Chem. Biol. 6, 795–796.

Lazos, O., Tosin, M., Slusarczyk, A.L., Boakes, S., Cortés, J., Sidebottom, P.J., Leadlay, P.F., 2010. Biosynthesis of the putative siderophore erythrochelin requires unprecedented crosstalk between separate nonribosomal peptide gene clusters. Chem. Biol. 17, 160–173.

Lee, J., Zhang, L., 2015. The hierarchy quorum sensing network in *Pseudomonas aeruginosa*. Protein Cell 6, 26–41.

Lee, J., Wu, J., Deng, Y., Wang, J., Wang, C., Wang, J., Chang, C., Dong, Y., Williams, P., Zhang, L.-H., 2013. A cell-cell communication signal integrates quorum sensing and stress response. Nat. Chem. Biol. 9, 339–343.

Lee, H.J., Son, H.S., Park, C., Suh, H.J., 2015. Preparation of yeast hydrolysate enriched in cyclo-His-Pro (CHP) by enzymatic hydrolysis and evaluation of its functionality. Prev. Nutr. Food Sci. 20, 284–291.

Lehrer, R.I., Cole, A.M., Selsted, M.E., 2012. θ-Defensins: cyclic peptides with endless potential. J. Biol. Chem. 287, 27014–27019.

Leszek, J., Barreto, G.E., Gąsiorowski, K., Koutsouraki, E., Ávila-Rodrigues, M., Aliev, G., 2016. Inflammatory mechanisms and oxidative stress as key factors responsible for progression of neurodegeneration: role of brain innate immune system. CNS Neurol. Disord. Drug Targets 15, 1–8.

Li, W., Khor, T.O., Xu, C., Shen, G., Jeong, W.S., Yu, S., Kong, A.N., 2008. Activation of Nrf2-antioxidant signaling attenuates NFkappaB-inflammatory response and elicits apoptosis. Biochem. Pharmacol. 76, 1485–1489.

Li, J., Wang, W., Xu, S.X., Magarvey, N.A., McCormick, J.K., 2011. *Lactobacillus reuteri*-produced cyclic dipeptides quench agr-mediated expression of toxic shock syndrome toxin-1 in staphylococci. Proc. Natl. Acad. Sci. U. S. A. 108 (8), 3360–3365.

Liskamp, R.M.J., Rijkers, D.T.S., Kruijtzer, J.A.W., Kemmink, J., 2011. Peptides and proteins as a continuing exciting source of inspiration for peptidomimetics. ChemBioChem 12, 1626–1653.

Liu, M., Gray, J.M., Griffiths, M.W., 2006. Occurrence of proteolytic activity an N-acylhomoserine lactone signals in the spoilage of aerobically chill-stored proteinaceous raw foods. J. Food Prot. 69, 2729–2737.

Liu, R., Kim, A.H., Kwak, M.K., Kang, S.O., 2017. Proline-based cyclic dipeptides from Korean fermented vegetable kimchi and from *Leuconostoc mesenteroides* LBP-K06 have activities against multidrug-resistant bacteria. Front. Microbiol. 8, 761.

Lozupone, C.A., Stombaugh, J.I., Gordon, J.I., Jansson, J.K., Knight, R., 2012. Diversity, stability and resilience of the human gut microbiota. Nature 489, 220–230.

Lyman, M., Lloyd, D.G., Ji, X., Vizcaychipi, M.P., Ma, D., 2014. Neuroinflammation: the role and consequences. Neurosci. Res. 79, 1–12.

Maiya, S., Grundmann, A., Li, S.-M., Turner, G., 2006. The fumitremorgin gene cluster of *Aspergillus fumigatus*: identification of a gene encoding brevianamide F synthetase. ChemBioChem 7, 1062–1069.

Marchesan, J.T., Morelli, T., Moss, K., Barros, S.P., Ward, M., Jenkins, W., Aspiras, M.B., Offenbacher, S., 2015. Association of synergistetes and cyclodipeptides with periodontitis. J. Dent. Res. 94 (10), 1425–1431.

Marietta, E., Horwath, I., Taneja, V., 2018. Microbiome, immunomodulation, and the neuronal system. Neurotherapeutics 15, 23–30.

Menegatti, S., Hussain, M., Naik, A.D., Carbonell, R.G., Rao, B.M., 2013. mRNA display selection and solid-phase synthesis of Fc-binding cyclic peptide affinity ligands. Biotechnol. Bioeng. 110, 857–870.

Minelli, A., Bellezza, I., Grottelli, S., Pinnen, F., Brunetti, L., Vacca, M., 2006. Phosphoproteomic analysis of the effect of cyclo-[His-Pro] dipeptide on PC12 cells. Peptides 27, 105–113.

Minelli, A., Bellezza, I., Grottelli, S., Galli, F., 2008. Focus on cyclo(His-Pro): history and perspectives as antioxidant peptide. Amino Acids 35, 283–289.

Minelli, A., Conte, C., Grottelli, S., Bellezza, I., Cacciatore, I., Bolaños, J.P., 2009a. Cyclo(His-Pro) promotes cytoprotection by activating Nrf2-mediated up-regulation of antioxidant defence. J. Cell. Mol. Med. 13, 1149–1161.

Minelli, A., Conte, C., Grottelli, S., Bellezza, I., Emiliani, C., Bolaños, J.P., 2009b. Cyclo(His-Pro) up-regulates heme oxygenase 1 via activation of Nrf2-ARE signalling. J. Neurochem. 111, 956–966.

Minelli, A., Grottelli, S., Mierla, A., Pinnen, F., Cacciatore, I., Bellezza, I., 2012. Cyclo(His-Pro) exerts anti-inflammatory effects by modulating NF-κB and Nrf2 signalling. Int. J. Biochem. Cell Biol. 44, 525–535.

Mishra, A.K., Choi, J., Choi, S.J., Baek, K.H., 2017. Cyclodipeptides: an overview of their biosynthesis and biological activity. Molecules 22 (10), 1796.

Mitova, M., Tommonaro, G., Hentschel, U., Müller, W.E.G., De Rosa, S., 2004. Exocellular cyclic dipeptides from a *Ruegeria* strain associated with cell cultures of *Suberites domuncula*. Mar. Biotechnol. 6, 95–103.

Monnet, V., Juillard, V., Gardan, R., 2016. Peptide conversations in gram-positive bacteria. Crit. Rev. Microbiol. 42, 339–351.

Morley, J.E., Levine, A.S., Prasad, C., 1981. Histidyl-proline diketopiperazine decreases food intake in rats. Brain Res. 210, 475–478.

Moussa, S.H., Kuznetsov, V., Tran, T.A., Sacchettini, J.C., Young, R., 2012. Protein determinants of phage T4 lysis inhibition. Protein Sci. 21, 571–582.

Moutiez, M., Schmitt, E., Seguin, J., Thai, R., Favry, E., Belin, P., Mechulam, Y., Gondry, M., 2014. Unravelling the mechanism of non-ribosomal peptide synthesis by cyclodipeptide synthases. Nat. Commun. 5, 5141.

Ng, W.L., Bassler, B.L., 2009. Bacterial quorum-sensing network architectures. Annu. Rev. Genet. 43, 197–222.

Nicholson, B., Lloyd, G.K., Miller, B.R., Palladino, M.A., Kiso, Y., Hayashi, Y., et al., 2006. NPI-2358 is a tubulin-depolymerizing agent: in vitro evidence for activity as a tumor vascular-disrupting agent. Anti-Cancer Drugs 17, 25–31.

Noble, E.E., Hsu, T.M., Kanoski, S.E., 2017. Gut to brain dysbiosis: mechanisms linking western diet consumption, the microbiome, and cognitive impairment. Front. Behav. Neurosci. 11, 9.

Ortiz-Castro, R., Díaz-Pérez, C., Martínez-Trujillo, M., del Río, R.E., Campos-García, J., López-Bucio, J., 2011. Transkingdom signaling based on bacterial cyclodipeptides with auxin activity in plants. Proc. Natl. Acad. Sci. U. S. A. 108, 7253–7258.

Pan, H., Wang, H., Wang, X., Zhu, L., Mao, L., 2012. The absence of Nrf2 enhances NF-kappaB-dependent inflammation following scratch injury in mouse primary cultured astrocytes. Mediat. Inflamm. 2012217580.

Park, D.K., Lee, K.E., Baek, C.H., Kim, I.H., Kwon, J.H., Lee, W.K., Lee, K.H., Kim, B.S., Choi, S.H., Kim, K.S., 2006. Cyclo(Phe-Pro) modulates the expression of ompU in *Vibrio* spp. J. Bacteriol. 188, 2214–2221.

Park, S.W., Choi, S.A., Yun, J.W., Choi, J.W., 2012. Alterations in pancreatic protein expression in STZ-induced diabetic rats and genetically diabetic mice in response to treatment with hypoglycemic dipeptide cyclo(His-Pro). Cell. Physiol. Biochem. 29, 603–616.

Perez-Pardo, P., Hartog, M., Garssen, J., Kraneveld, A.D., 2017. Microbes tickling your tummy: the importance of the gut-brain axis in Parkinson's disease. Curr. Behav. Neurosci. Rep. 4, 361–368.

Prasad, C., 1995. Bioactive cyclic dipeptides. Peptides 16, 1511–1564.

Prasad, C., 2001. Role of endogenous cyclo(His-Pro) in voluntary alcohol consumption by alcohol-preferring C57Bl mice. Peptides 22, 2113–2118.

Prasad, C., Peterkofsky, A., 1976. Demonstration of pyroglutamyl peptidase and amidase activities toward thyrotropin-releasing hormone in hamster hypothalamic extracts. J. Biol. Chem. 251, 3229–3234.

Proksch, P., Edrada, R.A., Ebel, R., 2002. Drugs from the seas—current status and microbiological implications. Appl. Microbiol. Biotechnol. 59 (2–3), 125–134.

Prosperini, A., Berrada, H., Ruiz, M.J., Caloni, F., Coccini, T., Spicer, L.J., Perego, M.C., Lafranconi, A., 2017. A review of the mycotoxin enniatin B. Front. Public Health 5, 304.

Rash, M., Andersen, J.B., Nielsen, K.F., Flodgaard, L.R., Christensen, H., Givskov, M., Gram, L., 2005. Involvement of bacterial quorum-sensing signals in spoilage of bean sprouts. Appl. Environ. Microbiol. 71, 3321–3330.

Reynolds, A., Laurie, C., Mosley, R.L., Gendelman, H.E., 2007. Oxidative stress and the pathogenesis of neurodegenerative disorders. Int. Rev. Neurobiol. 82, 297–325.

Rhee, K.H., 2004. Cyclic dipeptides exhibit synergistic, broad spectrum antimicrobial effects and have anti-mutagenic properties. Int. J. Antimicrob. Agents 24, 423–427.

Rocha-Estrada, J., Aceves-Diez, A.E., Guarneros, G., de la Torre, M., 2010. The RNPP family of quorum-sensing proteins in Gram-positive bacteria. Appl. Microbiol. Biotechnol. 87, 913–923.

Rojas Murcia, N., Lee, X., Waridel, P., Maspoli, A., Imker, H.J., Chai, T., Walsh, C.T., Reimmann, C., 2015. The *Pseudomonas aeruginosa* antimetabolite L-2-amino-4-methoxy-trans-3-butenoic acid (AMB) is made from glutamate and two alanine residues via a thiotemplate-linked tripeptide precursor. Front. Microbiol. 6, 170.

Rosier, A., Bishnoi, U., Lakshmanan, V., Sherrier, D.J., Bais, H.P., 2016. A perspective on inter-kingdom signaling in plant-beneficial microbe interactions. Plant Mol. Biol. 90, 537–548.

Roszyk, E., Puszczewicz, M., 2017. Role of human microbiome and selected bacterial infections in the pathogenesis of rheumatoid arthritis. Reumatologia 55, 242–250.

Rushworth, S.A., Zaitseva, L., Murray, M.Y., Shah, N.M., Bowles, K.M., Mac Ewan, D.J., 2012. The high Nrf2 expression in human acute myeloid leukemia is driven by NF-kappaB and underlies its chemo-resistance. Blood 120, 5188–5198.

Rustenhoven, J., Jansson, D., Smyth, L.C., Dragunow, M., 2017. Brain pericytes as mediators of neuroinflammation. Trends Pharmacol. Sci. 38, 291–304.

Ryan, R.P., An, S.Q., Allan, J.H., McCarthy, Y., Dow, J.M., 2015. The DSF family of cell-cell signals: an expanding class of bacterial virulence regulators. PLoS Pathog. 11, e1004986.

Sandberg, M., Patil, J., D'Angelo, B., Weber, S.G., Mallard, C., 2014. NRF2-regulation in brain health and disease: implication of cerebral inflammation. Neuropharmacology 79, 298–306.

Sauguet, L., Moutiez, M., Li, Y., Belin, P., Seguin, J., Le Du, M.H., Thai, R., Masson, C., Fonvielle, M., Pernodet, J.L., Charbonnier, J.B., Gondry, M., 2011. Cyclodipeptide synthases, a family of class-I aminoacyl-tRNA synthetase-like enzymes involved in non-ribosomal peptide synthesis. Nucleic Acids Res. 39, 4475–4489.

Schluter, J., Schoech, A.P., Foster, K.R., Mitri, S., 2016. The evolution of quorum sensing as a mechanism to infer kinship. PLoS Comput. Biol. 12, e1004848.

Schmidt, E.W., Obraztsova, A.Y., Davidson, S.K., Faulkner, D.J., Haygood, M.G., 2000. Identification of the antifungal peptide-containing symbiont of the marine sponge *Theonella swinhoei* as a novel δ-proteobacterium, "Candidatus Entotheonella palauensis" Mar. Biol. 136, 969–977.

Schmitz, F.J., Vanderah, D.J., Hollenbeak, K.H., Enwall, C.E.L., Gopichand, Y., SenGupta, P.K., Hossain, M.B., van der Helm, D., 1983. Metabolites from the marine sponge *Tedania ignis*. A new atisanediol and several known diketopiperazines. J. Org. Chem. 48, 3941–3945.

Schultz, A.W., Oh, D.-C., Carney, J.R., Williamson, R.T., Udwary, D.W., Jensen, P.R., Gould, S.J., Fenical, W., Moore, B.S., 2008. Biosynthesis and structures of cyclomarins and cyclomarazines, prenylated cyclic peptides of marine actinobacterial origin. J. Am. Chem. Soc. 130, 4507–4516.

Schwarzer, D., Finking, R., Marahiel, M.A., 2003. Nonribosomal peptides: from genes to products. Nat. Prod. Rep. 20, 275–287.

Sherwin, E., Dinan, T.G., Cryan, J.F., 2018. Recent developments in understanding the role of the gut microbiota in brain health and disease. Ann. N. Y. Acad. Sci. 1420 (1), 5–25.

Shiboski, C.H., 2002. HIV-related oral disease epidemiology among women: year 2000 update. Oral Dis. 8 (Suppl 2), 44–48.

Shiboski, C.H., Wilson, C.M., Greenspan, D., Hilton, J., Greenspan, J.S., Moscicki, A.B., Adolescent Medicine HIV/AIDS Research Network, 2001. HIV-related oral manifestations among adolescents in a multicenter cohort study. J. Adolesc. Health 29, 109–114.

Sjögren, M., Jonsson, P.R., Dahlström, M., Lundälv, T., Burman, R., Göransson, U., Bohlin, L., 2011. Two brominated cyclic dipeptides released by the coldwater marine sponge *Geodia barretti* act in synergy as chemical defense. J. Nat. Prod. 74, 449–454.

Skandamis, P.N., Nychas, G.J., 2012. Quorum sensing in the context of food microbiology. Appl. Environ. Microbiol. 78, 5473–5482.

Sprague Jr., G.F., Winans, S.C., 2006. Eukaryotes learn to count: quorum sensing by yeast. Genes Dev. 20, 1045–1049.

Stierle, A.C., Cardellina IInd, J.H., Singleton, F.L., 1988. A marine *Micrococcus* produces metabolites ascribed to the sponge *Tedania ignis*. Experientia 44, 1021.

Strieker, M., Tanović, A., Marahiel, M.A., 2010. Nonribosomal peptide synthetases: structures and dynamics. Curr. Opin. Struct. Biol. 20, 234–240.

Suzuki, T., Yamamoto, M., 2015. Molecular basis of the Keap1-Nrf2 system. Free Radic. Biol. Med. 88, 93–100.

Suzuki, T., Motohashi, H., Yamamoto, M., 2013. Toward clinical application of the Keap1-Nrf2 pathway. Trends Pharmacol. Sci. 34, 340–346.

Tapeinou, A., Matsoukas, M.T., Simal, C., Tselios, T., 2015. Cyclic peptides on a merry-go-round; towards drug design. Biopolymers 104, 453–461.

Taubert, D., Grimberg, G., Stenzel, W., Schömig, E., 2007. Identification of the endogenous key substrates of the human organic cation transporter OCT2 and their implication in function of dopaminergic neurons. PLoS One 2, e385.

Taylor, M.W., Schupp, P.J., Baillie, H.J., Charlton, T.S., de Nys, R., Kjelleberg, S., Steinberg, P.D., 2004. Evidence for acyl homoserine lactone signal production in bacteria associated with marine sponges. Appl. Environ. Microbiol. 70, 4387–4389.

Teixidó, M., Zurita, E., Prades, R., Tarrago, T., Giralt, E., 2009. A novel family of diketopiperazines as a tool for the study of transport across the blood-brain barrier (BBB) and their potential use as BBB-shuttles. Adv. Exp. Med. Biol. 611, 227–228.

Thakur, N.L., Hentschel, U., Krasko, A., Pabel, C.T., Anil, A.C., Muller, W.E.G., 2003. Antibacterial activity of the sponge *Suberites domuncula* and its primmorphs: potential basis for epibacterial chemical defense. Aquat. Microb. Ecol. 31, 77–83.

Thion, M.S., Low, D., Silvin, A., Chen, J., Grisel, P., Schulte-Schrepping, J., Blecher, R., Ulas, T., Squarzoni, P., Hoeffel, G., Coulpier, F., Siopi, E., David, F.S., Scholz, C., Shihui, F., Lum, J., Amoyo, A.A., Larbi, A., Poidinger, M., Buttgerei, A., Lledo, P.M., Greter, M., Chan, J.K.Y., Amit, I., Beyer, M., Schultze, J.L.,

Schlitzer, A., Pettersson, S., Ginhoux, F., Garel, S., 2018. Microbiome influences prenatal and adult microglia in a sex-specific manner. Cell 172, 500–516e16.

Thompson IIIrd, G.R., Patel, P.K., Kirkpatrick, W.R., Westbrook, S.D., Berg, D., Erlandsen, J., Redding, S.W., Patterson, T.F., 2010. Oropharyngeal candidiasis in the era of antiretroviral therapy. Oral Surg. Oral Med. Oral Pathol. Oral Radiol. Endod. 109, 488–495.

Tommonaro, G., Abbamondi, G.R., Iodice, C., Tait, K., De Rosa, S., 2012. Diketopiperazines produced by the halophilic archaeon, *Haloterrigena hispanica*, activate AHL bioreporters. Microb. Ecol. 63, 490–495.

Tsang, A.H., Chung, K.K., 2009. Oxidative and nitrosative stress in Parkinson's disease. Biochim. Biophys. Acta 1792, 643–650.

Uyemura, K., Dhanani, S., Yamaguchi, D.T., Song, M.K., 2010. Metabolism and toxicity of high doses of cyclo(His-Pro) plus zinc in healthy human subjects. J. Drug Metab. Toxicol. 1, 105.

Valko, M., Leibfritz, D., Moncol, J., Cronin, M.T., Mazur, M., Telser, J., 2007. Free radicals and antioxidants in normal physiological functions and human disease. Int. J. Biochem. Cell Biol. 39, 44–84.

Verdine, G.L., Hilinski, G.J., 2012. Stapled peptides for intracellular drug targets. Methods Enzymol. 503, 3–33.

Wang, B., Zhu, X.L., Kim, Y., Li, J., Huang, S.Y., Saleem, S., et al., 2012. Histone deacetylase inhibition activates transcription factor Nrf2 and protects against cerebral ischemic damage. Free Radic. Biol. Med. 52, 928–936.

Webster, N.S., Hill, R.T., 2001. The culturable microbial community of the great barrier reef sponge *Rhopaloeides odorabile* is dominated by an α-proteobacterium. Mar. Biol. 138, 843–851.

Webster, N.S., Wilson, K.J., Blackall, L.L., Hill, R.T., 2001. Phylogenetic diversity of bacteria associated with the marine sponge *Rhopaloeides odorabile*. Appl. Environ. Microbiol. 67, 434–444.

Weitz, J.S., Mileyko, Y., Joh, R.I., Voit, E.O., 2008. Collective decision making in bacterial viruses. Biophys. J. 95, 2673–2680.

Wynendaele, E., Gevaert, B., Stalmans, S., Verbeke, F., De Spiegeleer, B., 2015. Exploring the chemical space of quorum sensing peptides. Biopolymers 104, 544–551.

Yang, H., Duan, Z., 2018. The local defender and functional mediator: Gut microbiome. Digestion 97, 137–145.

Yang, L., Palliyaguru, D.L., Kensler, T.W., 2016. Frugal chemoprevention: targeting Nrf2 with foods rich in sulforaphane. Semin. Oncol. 43, 146–153.

Yin, W.-B., Grundmann, A., Cheng, J., Li, S.-M., 2009. Acetylaszonalenin biosynthesis in *Neosartorya fischeri*: identification of the biosynthetic gene cluster by genomic mining and functional proof of the genes by biochemical investigation. J. Biol. Chem. 284, 100–109.

Yu, X., Liu, Y., Xie, X., Zheng, X.-D., Li, S.-M., 2012. Biochemical characterization of indole prenyltransferases: filling the last gap of prenylation positions by a 5-dimethylallyltryptophan synthase from *Aspergillus clavatus*. J. Biol. Chem. 287, 1371–1380.

Zhu, J., Zhao, A., Feng, L., Gao, H., 2016. Quorum sensing signals affect spoilage of refrigerated large yellow croaker (*Pseudosciaena crocea*) by *Shewanella baltica*. Int. J. Food Microbiol. 217, 146–155.

Zouhir, S., Perchat, S., Nicaise, M., Perez, J., Guimaraes, B., Lereclus, D., Nessler, S., 2013. Peptide-binding dependent conformational changes regulate the transcriptional activity of the quorum-sensor NprR. Nucleic Acids Res. 41, 7920–7933.

Further Reading

Chen, J.-H., Lan, X.-P., Liu, Y., Jia, A.-Q., 2012. The effects of diketopiperazines from *Callyspongia* sp. on release of cytokines and chemokines in cultured J774A.1 macrophages. Bioorg. Med. Chem. Lett. 22, 3177–3180.

Choi, S.A., Yun, J.W., Park, H.S., Choi, J.W., 2013. Hypoglycemic dipeptide cyclo(His-Pro) significantly altered plasma proteome in streptozotocin-induced diabetic rats and genetically-diabetic (ob/ob) mice. Mol. Biol. Rep. 40, 1753–1765.

Giessen, T.W., von Tesmar, A.M., Marahiel, M.A., 2010. A tRNA-dependent two-enzyme pathway for the generation of singly and doubly methylated ditryptophan 2,5-diketopiperazines. Biochemistry 52, 4274–4283.

Mander, P., Brown, G.C., 2005. Activation of microglial NADPH oxidase is synergistic with glial iNOS expression in inducing neuronal death: a dual-key mechanism of inflammatory neurodegeneration. J. Neuroinflammation 2, 20.

Miller, B.R., Gulick, A.M., 2016. Structural biology of nonribosomal peptide synthetases. Methods Mol. Biol. 1401, 3–29.

Reimer, J.M., Aloise, M.N., Harrison, P.M., Schmeing, T.M., 2016. Synthetic cycle of the initiation module of a formylating nonribosomal peptide synthetase. Nature 529, 239–242.

Schmitz, F.J., Schulz, M.M., Siripitayananon, J., Hossain, M.B., van der Helm, D., 1993. New diterpenes from the gorgonian Solenopodium excavatum. J. Nat. Prod. 56, 1339–1349.

Seguin, J., Moutiez, M., Li, Y., Belin, P., Lecoq, A., Fonvielle, M., Charbonnier, J.B., Pernodet, J.L., Gondry, M., 2011. Nonribosomal peptide synthesis in animals: the cyclodipeptide synthase of Nematostella. Chem. Biol. 18, 1362–1368.

Index

Note: Page numbers followed by *f* indicate figures and *t* indicate tables.

Printed in the United States
By Bookmasters